GLACIAL GEOLOGY

An Introduction for Engineers and Earth Scientists

Edited by

N. EYLES

University of Toronto, Canada

PERGAMON PRESS

OXFORD · NEW YORK · TORONTO · SYDNEY · PARIS · FRANKFURT

U.K.	Pergamon Press Ltd., Headington Hill Hall, Oxford OX3 0BW, England
U.S.A.	Pergamon Press Inc., Maxwell House, Fairview Park, Elmsford, New York 10523, U.S.A.
CANADA	Pergamon Press Canada Ltd., Suite 104, 150 Consumers Road, Willowdale. Ontario M2J 1P9, Canada
AUSTRALIA	Pergamon Press (Aust.) Pty. Ltd., P.O. Box 544, Potts Point, N.S.W. 2011, Australia
FRANCE	Pergamon Press SARL, 24 rue des Ecoles, 75240 Paris, Cedex 05, France
FEDERAL REPUBLIC OF GERMANY	Pergamon Press GmbH, Hammerweg 6, D-6242 Kronberg-Taunus, Federal Republic of Germany

First edition 1983

Library of Congress Cataloging in Publication Data
Main entry under title:
Glacial geology.
(Pergamon international library of science, technology, engineering, and social studies)
Includes bibliographical references and index.
1. Glacial epoch. 2. Glacial landforms. I. Eyles, N.
II. Series.
QE697.G56   1984   551.3'1   83-17418

British Library Cataloguing in Publication Data
Eyles, N
Glacial geology
1. Glaciers
I. Title
551.3'1   QE576
ISBN 0-08-030264-5 (Hardcover)
ISBN 0-08-030263-7 (Flexicover)

In order to make this volume available as economically and as rapidly as possible the author's typescript has been reproduced in its original form. This method unfortunately has its typographical limitations but it is hoped that they in no way distract the reader.

*Printed in Great Britain by
Redwood Burn Ltd., Trowbridge, Wiltshire.*

Rationale

There are increasing demands being placed, often with inadequate understanding of the problems involved, on the subsurface geology in formerly
glaciated terrains in connection with energy, waste disposal, engineering,
groundwater and mineral exploration projects. This has created a considerable
demand for an up-to-date summary of glacial deposition for engineers and earth
scientists working in areas of former glaciation where an understanding of
glacial geology is desirable.

This book is designed as a basic introduction to the geology of glaciated
terrains and mid to senior level undergraduates, college students and industry
and government employees engaged in engineering and earth science courses and
projects associated with glacial sediments and stratigraphies in the mid-latitudes.
Contributions have been sought from active specialists in university and industry
involved with many different aspects of glaciated terrain in North America
and Britain. The final result is a distillation of a very large and diverse
literature that crosses traditional discipline boundaries and areas of individual
expertise. Whilst a book of this nature cannot be exhaustive or hope to cover
all areas, sufficient references, cross-references and illustrations are provided
to enable an interested student to follow up various topics and to more easily
visualize certain esoteric aspects. As in any text that encompasses several
disciplines, nomenclatural problems and confusion surround many terms and
concepts. These are discussed in the text in as much detail as space permits.

The theme of what follows is that a spring board for many applied projects
in glaciated terrains is the comprehension of a small number of models that
protray glacial deposition. These models describe recurring associations
of landforms and sedimentary sequences (landsystems) that can be represented
by block diagrams and can be identified at the margins of modern glaciers and
also in mid-latitude zones affected by former Pleistocene ice sheets. Use of
these models affords a valuable means of rationalising glacial deposition
and provides an interpretive tool for subsequent engineering, sedimentological,
geotechnical, hydrogeological and stratigraphic studies.

Acknowledgements

This book was completed in the Geology Department of the University of Toronto at the Scarborough Campus. I am particularly grateful to John Andrews, Bryan Clarke, Terry Day, Ken Howard, Brian Kaye, John Westgate and other anonymous reviewers in North America and Britain for critical reading of the manuscript and helpful comments. Charles Dyer, Dave Harford, Lyn McGregor and Tony Westbrook lent me their considerable word processing, photographic and illustrative skills, as did Christine Cochrane at the University of Newcastle-Upon-Tyne, England. Kathy Willard and Elizabeth Seres assisted with library work. Pat Woodcock and June Yamazaki helped more than they will know by their typing and proof-reading skills and Ilze Buivids compiled the reference list. I cannot express sufficient gratitude to my wife Carolyn, herself a busy geologist for unwavering support, encouragement and advice.

The final vote of thanks must go to my collaborators in this project for their patience with what at times may have seemed like a long time to complete this book. We hope people will find the end product useful.

N. Eyles
Toronto, Canada
June, 1983

Contents

List of Contributors

W.F. Anderson, Lecturer

Department of Civil and Structural Engineering, University of Sheffield, England

J.E. Cocksedge, Geotechnical Engineer

Scott, Wilson, Kirkpatrick & Partners, Basingstoke, England

W.R. Dearman, Professor and Chairman

Department of Geotechnical Engineering, University of Newcastle-Upon-Tyne, Newcastle-Upon-Tyne, England

T.D. Douglas, Senior Lecturer

Department of Geography, Newcastle-Upon-Tyne Polytechnic, Newcastle-Upon-Tyne, England

N. Eyles, Assistant Professor

Department of Geology, University of Toronto, Toronto, Ontario, Canada

J.W. Lloyd, Senior Lecturer

Department of Geological Sciences, University of Birmingham, Birmingham, England

J. Menzies, Assistant Professor and Chairman

Department of Geography, Brock University, St. Catharines, Ontario, Canada

A.D. Miall, Associate Professor

Department of Geology, University of Toronto, Toronto, Ontario, Canada

M.S. Money, Lecturer

Department of Geotechnical Engineering, University of Newcastle-Upon-Tyne, Newcastle-Upon-Tyne, England

M.A. Paul, Lecturer

Department of Civil Engineering, Heriot-Watt University, Riccarton, Scotland

R.M. Quigley, Professor and Chairman

Faculty of Engineering Science, The University of Western Ontario, London, Ontario, Canada

J.A. Sladen, Geotechnical Engineer

E.B.A. Consultants Ltd., Calgary, Alberta, Canada

S. Somerville, Geotechnical Engineer   G. Wimpey & Company,
Wimpey Central Laboratories,
Hayes, Middlesex, England

A. Strachan, Research Associate   Department of Geotechnical Engineering,
University of Newcastle-Upon-Tyne,
Newcastle-Upon-Tyne, England

W. Wrigley, Geotechnical Engineer   E.B.A. Consultants Ltd.,
Edmonton, Alberta, Canada

CHAPTER 1

Glacial Geology: A Landsystems Approach

N. Eyles

INTRODUCTION

The extent and importance of sediments (engineering soils) represented by glacial deposits cannot be overemphasized (Figs. 1.1, 9.1, 9.2). Sediment types vary from overconsolidated bouldery silty-clays (tills) and other bouldery aggregates through variably consolidated clays, silts, sands and gravels, to peats and organic soils. A wide spectrum of sediments is represented in very variable associations resting on bedrock that may be a few metres to many tens of metres below the present land surface.

Associated with variable glacial soils are superficial structures, involving both sediments and rock, resulting from glacial and periglacial activity. Solifluction sheets, cambering, valley bulging, frost wedges and glacially-deformed bedrock are just a few examples of superficial structures that pose geotechnical problems. Difficult ground conditions usually arise because of either very rapid transitions from one sediment type to another, the juxtaposition of rock and soil, or the presence of unexpectedly soft horizons. It should be noted that throughout this book the terms ´sediment´ and ´engineering soil´ are used interchangeably.

1.1 Glacial Terrain Evaluation: The Concept of Landsystems.

Terrain evaluation is aimed at understanding the natural features of the landscape and as a process, inevitably involves terrain classification by which the landscape is separated into natural units. Each constituent unit must be internally homogeneous and distinct from the others. Recognition of landscape units implies that there is a genetic relationship between landforms and the processes and materials involved in their development. The processes are mainly surface geological processes that have been active in the recent past, but these may be very different from the processes active at the present time. The materials are the superficial and solid deposits that crop out at the surface and immediately underlie it.

The number of landscape units, or classes, must be reasonably small and three main types can be recognized within a hierarchial classification.

(i) A land element is the simplest part of the landscape and is for practical purposes uniform in form and material and is suited for mapping at large scales e.g. a drumlin or kame.

(ii) A land facet comprises one or more land elements grouped for practical purposes; it is part of a landscape which is reasonably homogeneous and distinct from the surrounding terrain. Land facets are suited to mapping at scales of 1:50,000 to 1:100,000. e.g. a drumlin field or an outwash plain.

1

Fig. 1.1   Maximum extent (irrespective of age) of former Quaternary ice sheets in the Northern Hemisphere. From Denton and Hughes (1981) with modifications.

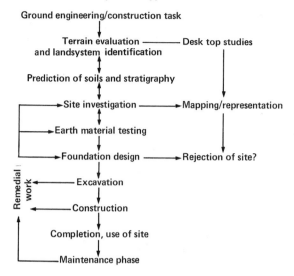

Fig. 1.2 An example of terrain evaluation; idealized sequence of events in the planning and implementation of a typical construction project. In practice many projects commence with a limited site investigation phase and employ foundation designs found to be successful elsewhere (Chapters 11, 12).

(iii) A landsystem is a recurrent pattern of genetically linked land facets, suitable for mapping at scales of 1:250,000 to 1:1m. e.g. subglacial terrain where sediments and landforms of the landscape have been deposited at the base of an ice-sheet. An example would be a drumlin field flanked by an outwash plain and esker system.

Each landsystem can be defined in terms of the sediment complexes underlying and at the same time controlling surface topography. The conditions at rockhead also vary from landsystem to landsystem. Although terrain evaluation is said to be concerned only with the uppermost few metres of the ground (Mitchell, 1973), the glacial landsystems which are considered here extend to bedrock regardless of depth.

The basic premise of the approach followed in this book is that if the land system can be identified from surface landforms, then it is possible to identify in turn the relevant subsurface conditions. As such the approach has an important role in the initial desk-study phase of planning a variety of applied projects with the potential for saving time and hence money (Fig. 1.2). The landsystem, once identified, provides a model not only for planning and the interpretation of the results of site investigation and laboratory testing but focuses attention on those particular features that should be looked for in making a geological assessment of any location for a wide variety of purposes.

1.2 Glacial Landsystems

A complex range and distribution of sediments result from glaciation. As a means of classifying and mapping sediment sequences and landforms at the margins of modern day glaciers, it has been possible to recognize a number of distinct

landsystems each having characteristic topography, subsurface conditions and
sediments (Boulton and Paul, 1976). By the identification of the landforms and
terrain type, the geometry and character of subsurface stratigraphies can be
generalised for large areas. The value of this approach to applied projects in
mid-latitude areas of former Quaternary glaciations lies in its use as a mapping
tool and as a rapid guide to likely subsurface conditions.

Ground conditions resulting from glaciation can be generalised into three
landsystems.

    (i)   subglacial
    (ii)   supraglacial
    (iii) glaciated valley

The subglacial and supraglacial landsystems (Figs. 1.3, 1.4, 1.5) are
characteristic of glaciated lowlands where sediments and landforms were deposited
by large ice sheets. Glaciated valley terrain, on the other hand, results where
the bedrock relief is so marked, as in mountain and highland areas, that bedrock
protrudes through the drift cover and, during glaciation, breaks up ice lobes
into valley glaciers (Fig. 1.6).

Finally, ground conditions and landforms resulting from periglacial, cold climate
conditions may be superimposed on and modify to varying degrees the three
landsystems described above. These periglacially-modified terrains predominate
in a broad zone south of the limit of glaciation; the so-called 'zone of fossil
periglacial landscapes'.

1.2.1 The Subglacial Landsystem

The definition of the subglacial landsystem recognises that landforms and
sediments were deposited at the ice base. The landsystem therefore is the
preserved bed over which the ice sheet moved and landforms are comparable to the
other bedf ms recognized by sedimentologists (e.g. ripples, bars, dunes etc.)
deposited by the movement of a transporting medium (water, wind) over underlying
substrates.

Two subglacial landsystems are recognized. Both are, by definition, deposited at
the ice base but are readily discriminated as they tend to occur in
geographically distinct areas. Thus at the scale of continental ice-sheets,
'shield' lowlands and uplands (e.g. Canadian and Scandinavian shields) of
moderate relief, underlain by hard and complexly structured igneous and
metamorphic rocks, mostly of Precambrian age, can be distinguished from
surrounding lowlands underlain by younger flat-lying sedimentary 'soft' rocks.
Shield areas are characterized by wide exposure of scoured, but not deeply eroded
bedrock having a complex structure and wide areal range in lithological types
with a mamillated knobbly topography of moderate relief. The bedrock topography
has a pronounced effect on the type and distribution of landforms and sediments
deposited subglacially (Fig. 1.3). The activity of subglacial meltstreams and
enhanced ice velocity around rock highs ensures that a thick drift bed does not
always develop to obscure the bedrock surface. Coarse-grained variably
consolidated tills and glaciofluvial gravels and sands predominate in what are
termed 'bedrock-drift complexes' (Fig. 2.13). The absence of a thick drift cover
on shield areas is not the result of enhanced glacial erosion as has been
previously thought but rather is due to ineffective glacial deposition and the
predominance of areal scouring rather than selective linear erosion (Figs. 2.1,
9.2).

see p.40 diagr

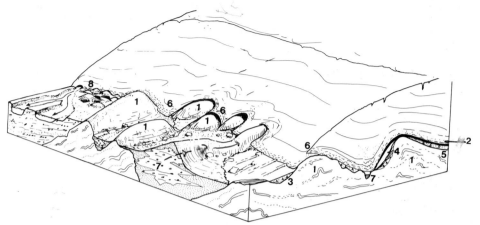

Fig. 1.3a The subglacial landsystem, I (moderate relief ´shield´ beds).

(1) Streamlined granitic rock knobs. Debris in the ice base (2), is deposited as
lodgement till on lower relief bedrock surfaces (3), or dropped particle by
particle into subglacial cavities (4), (Fig. 2.13). Wedges of basal ice lodge on
the stoss (upstream side) of rock highs where they melt down in situ as basal
melt-out till (5), masses of basal debris become detached from the glacier sole
and accumulate as scree-like fans in cavities. Debris melts-out along the
icefront (6), is disaggregated by gravity, and is dumped on the exposed
subglacial surface. (7) Subglacially deposited esker; gravel core in sand.
Subglacial streams flush out abrasion products (silts, clay) which are deposited
in lakes (sediment traps). (8), Rapid build up of outwash fans buries local
areas of the ice front. With ice melt a hummocky kettled outwash surface
develops (Fig. 7.2). The underlying soils serves to distinguish these landforms
from the hummocky forms produced elsewhere by supraglacial deposition (Figs. 1.5,
1.6).

Scale - The width of the ice front shown may vary from several hundred metres to
several kilometres. Esker complexes several tens of kilometres across and
individual rock highs with a relief of over 300 m are not unusual. Subsurface
ground conditions are shown in Fig. 1.3b.

Out from the shield margins and extending radially to the limit of glaciation
(Figs. 9.1, 9.2), lower relief flat-lying or gently dipping sedimentary strata
present different subglacial conditions (Fig. 1.4). The bedrock surface,
commonly disturbed by ice movement (glacitectonized), is buried by thick
finer-grained sequences often with a streamlined slickensided surface in the form
of drumlins. The exposed glacier bed is a shear surface, albeit on an immense
scale, resulting from glacier sliding. The thick soil sequence is the combined
result of repeated glaciation, ineffective erosion towards the margins of ice
sheets, and the presence of deformable matrix rich bedrock units (shales,
mudstones).

Large Quaternary ice sheets flowing radially from shield terrains to areas of
sedimentary strata (Figs. 9.1, 9.2, 9.9) frequently moved into large lake or
marine basins (e.g. the Great Lake Basins and the Baltic) deepened both by
isostatic crustal depression, due to the ice sheet mass, and glacial ponding

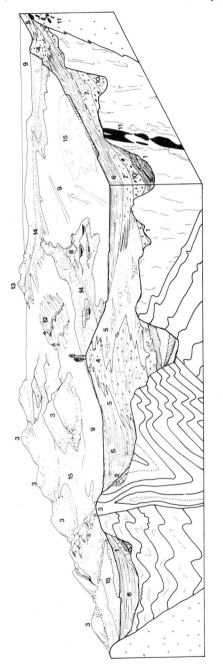

Fig. 1.3b Generalized landforms and sub-surface ground conditions in shield terrains.

(1)   lee-side scree-like accumulations of gravel and melt-out till (see Fig. 2.13)
(2)   stoss-side accumulations of gravel and melt-out till
(3)   streamlined rock knobs resulting from areal scouring
(4)   esker core (cobble facies, grading through sand facies (5), into lacustrine clay (6)
(7)   buried outwash plain ('sandar'; Ch. 7)
(8)   pitted or kettled outwash fan frequently related to eskers (Fig. 7.1b)
(9)   lacustrine or glaciomarine clay plain with keel marks from icebergs
(10) bouldery lodgement till with dispersed mineralised boulders (float) part of
      a dispersal fan down ice of mineralised bedrock (11, Sect. 2.8.1)
(12) inlier of till plain poking through lacustrine clays. Minor washboard moraine ridges
      (Sect. 2.9) or Rogen moraine ridges (Sect. 2.6) lie on its surface
(13) Large (100 m elevation plus) terminal moraine ridge; frequently contains large
      sand and gravel complexes reflecting formation as a delta-moraine in proglacial lake
(14) Proglacial lake beach strands cut in sides of eskers and moraines
(15) peat bog

(Fig. 2.21). In these areas subglacial deposition at the margin of floating and partially floating ice masses is highly significant in determining the type and distribution of soils and landforms (sect. 2.9).

1.2.2 The Supraglacial Landsystem

In many continental areas the streamlined bed of the former ice sheet is obscured by other glacial sediments that were deposited from the ice surface during glacier retreat (Fig. 1.5). These landforms and sediments, superimposed on those deposited subglacially, are collectively referred to as the supraglacial landsystem. Debris may only be transported in large volumes on the surface of an extensive continental ice sheet where the ice margin is thin, slow moving, decelerating and is therefore under strong compression. The ice margin may also be frozen to the substrate. Compressive flow may be aided by large-scale bedrock obstructions such as escarpments within gently dipping sedimentary rock sequences, and by the compression of one large ice lobe against another. As a result of such compression and/or the freezing on of sediment-rich subglacial meltwaters a ´dirty´ ice margin develops where basal debris is exposed on the ice surface. As the ice sheet retreats, it leaves behind large tracts of hummocky supraglacial topography (Fig. 1.5). The geographic distribution of this terrain type reflects its ice marginal formation, frequently forming arcuate belts (moraine complexes) up to several tens of kilometres in width that are frequently associated with bedrock escarpments (Figs. 2.18, 9.1, 9.2).

1.2.3 The Glaciated Valley Landsystem

The third terrain type is encountered in high relief highland or mountain areas where ice sheets are broken up by mountains into many separate glaciers that often coalesce on surrounding lowland margins (Fig. 1.6). Bedrock debris, derived by rockfall from mountainsides above the ice sheet (Fig. 1.7) may be transported far beyond the mountain zone (Fig. 4.6). As the ice sheet thins, such clast rich externally-derived material increases in volume and is deposited from the ice surface in much the same way as occurs on the margins of ice sheets (Fig. 1.5). Hummocky morainic topography along the valley floor is found in association with complex lateral accumulations deposited between the bedrock walls of the valley and the former valley glacier.

1.3 Landsystems As An Aid In Engineering And Earth Science Investigations

In each of the three glacial landsystems, characteristic surface morphologies are associated with distinct sediment types. It follows that indentification of land elements e.g. a drumlin or kame terrace, carries important implications as to likely subsurface stratigraphies and their geometry. However a ´landsystem´ approach to the identification of ground conditions and site characterization will only be effective in areas affected by the last ice sheet as outside this limit any original surface morphology will have been partially destroyed or buried. Nonetheless, with a knowledge of the likely properties and geometry of the principal soils in each landsystem, recognition of old, buried landsystems can be made from either subsurface sediment samples or by consideration of the geometry of certain soil horizons established from bore hole data or limited outcrop exposure.

Thus a cautionary note that should be emphasized is that the landsystems identified here portray the soils and landforms of a single ice advance/retreat cycle, i.e. the latest. Soil sequences deposited during one such cycle at modern ice margins are rarely thicker than a few tens of metres (though lacustrine and marine basins may exhibit thicker sequences) yet, total thicknesses of glacial sediment of several hundred metres are common in many mid-latitude areas. A

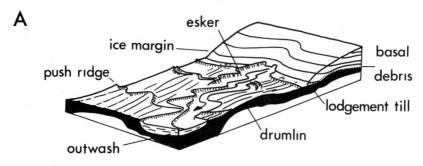

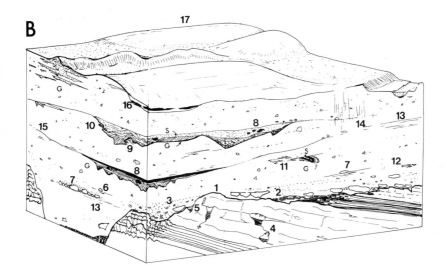

Fig. 1.4 The subglacial landsystem II (low relief bed). Basal transport (A) and deposition (B) by wet-based sliding glaciers.

(1) striated rockhead surface locally overdeepened below sea-level by subglacial erosion (Fig. 2.8).
(2) rock ´rafts´, glacitectonised rockhead and deformation till (Sect. 2.8.2) where alternating competent and incompetent bedrock lithologies are present.
(3) bouldery unit of scree-like debris filling lee-side cavities in (1) above.
(4) ´cold water´ karst from enhanced solution of limestones by subglacial meltwaters.
(5) intrusion of till into joints in rockhead
(6) preferentially orientated clast long axes (pebble fabric).
(7) distinct flat-iron shaping of fine-grained lithologies; coarse grained lithologies produce faceted clasts of higher sphericity, frequently found as boulder pavements (Fig. 2.7).
(8) ´cut and fill´ fluvial sediments deposited as sand (s) and gravels (g) in interconnected subglacial channels or as laminated clays in subglacial ponds (Fig. 2.11). Often contain coherent debris masses dropped from ice roof.

glance at Fig. 9.3 suggests that the number of large-scale climatic glaciations that the mid-latitudes have experienced in the last 700,000 years is close to 20 with a much larger total of ice sheet advance/retreat cycles superimposed. Thus it follows, particularly at the outer limits of glaciation (Fig. 1.1), that subsurface soil sequences record the superimposition of several landsystems interbedded with preglacial and proposition nonglacial deposits.

Many workers have identified the crucial importance of buried geomorphology in engineering site investigations (White, 1974) and it is hoped that the ground condition models portrayed here by reference to block diagrams will enable the student to more easily picture the nature and extent of subsurface sequences. An essential part of any geologic endeavour is the ability to picture sequences in three-dimensions and to make assessments of the lateral changes in strata properties and thickness away from any one data point. The conceptual stratigraphic models can be used therefore in a predictive fashion for many projects (e.g. Fig. 1.2). At a much smaller scale the general distribution of the principal glacial terrain types can be portrayed on maps of the glaciated portions of North America and Britain (Figs. 9.1 and 9.2).

1.4 Terminology and Classification of Tills

A confusing range of terms are used by geologists to describe glacial sediments. This is hardly surprising since the latter cover some 75% of the mid-latitudes (Fig. 1.1). Terminological problems are compounded by both variation from country to country and a long history of reporting such deposits for varied purposes (Charlesworth, 1957). A frequent problem is the use of terms whose meaning is ambiguous and which do not give a clear picture of what is being referred to. Considerable confusion for example surrounds the terms 'boulder clay', 'till', 'moraine' or 'ground moraine' which tend to be used indiscriminately for any poorly sorted admixture of gravel, sand and mud deposited by a glacier. As is demonstrated in chapters 2 to 7 there is a wide range of depositional environments in which poorly sorted sediments are produced.

The question of adequate labelling of poorly sorted sediments is becoming more critical as glacial sedimentological models become more sophisticated, as engineering activity moves into areas of increasingly difficult ground conditions both onshore and offshore, as computerized data banks of site investigation and

(9) till masses diapirically intruded up into base of fluvial channels (Fig. 2.10b).
(10) lenses of resedimented till extending into channel fills resulting from sidewall erosion and collapse.
(11) upper surfaces of cut and fill channels partially eroded by ice flow and resulting in deformed and folded inclusions in overlying till. Smaller channels folded.
(12) shear lamination caused by the shearing out of soft incompetent bedrock lithologies ('smudges')
(13) slickensided bedding plane shears resulting from subglacial shear
(14) near vertical en echelon joints systematically oriented with respect to glacier flow direction, or joint pattern in underlying rock (Sect. 8.3.3).
(15) base of till units may be fluted; orientation consistent with (6)
(16) post-depositional sheared upper surface (Fig. 2.12, Sect. 2.5.1): frequently redeposited by solifluction (Sect. 5.4.1).
(17) drumlinized, streamlined low relief surface; where rockhead is close to surface rock-cored drumlins and 'crag and tail' forms can be mapped. Subglacially engorged eskers are frequently related to (8) at depth.
Scale - variable.

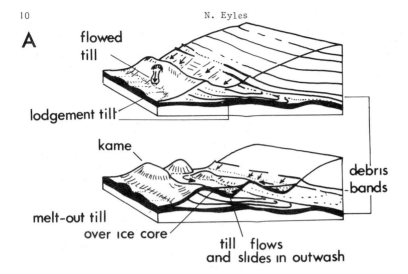

A

flowed till

lodgement till

kame

debris -bands

melt-out till over ice core

till flows and slides in outwash

B

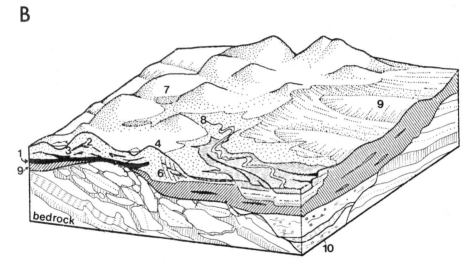

bedrock

Fig. 1.5 The supraglacial landsystem.

Transport (A) and deposition (B) by ice-sheets with frozen or freezing margins of complex thermal regime (after Boulton, 1972). Note that supraglacially deposited flowed tills overlie lodgement till (9) at depth (Figs. 1.4, 1.8). Glacitectonised rockhead results from freezing of the glacier base to rockhead and subsequent deformation by ice movement (Section 2.8.2).

(1) Crudely stratified melt-out till formed by meltdown of sequences of alternating debris-rich and debris-poor basal ice with variable preservation of englacial clast orientation. Best preservation of englacial clast orientation occurs in cold, arid environments where interstitial ice is lost by sublimation

analytical data are built up and as research proceeds into the relationship
between sediment genesis and geotechnical characteristics. There are also good
economic arguments for careful logging of borehole data, exposures, or trial pits
given the high cost of site investigation relative to total construction costs.
A too familiar experience is borehole and mapping data that is virtually useless
in its prime role of imparting a concise and accurate description of sediments
present.

Sediment classifications can either be <u>genetic</u>, where individual terms carry
information as to the genesis and origin of the material, or descriptive where
characteristics of the material are described as objectively as possible though a
genetic term may be employed, in parentheses, at the end of the description e.g.
stiff, fissured grey sandy clay of low plasticity with fine to coarse gravel,
cobbles, and occasional glacial shaped boulders (lodgement till). Schemes
specifically for engineering purposes are reviewed in Section 11.9.

The term ´till´ is argued here to have a specific genetic definition and refers
to ´an aggregate whose particles have been brought into contact by the direct
agency of glacier ice and which, though it may have undergone glacially-induced
flow, has not been significantly disaggregated´ (Boulton, 1972). Knowledge of
the different ways in which debris is deposited by glaciers has developed
considerably in the last fifteen years and genetic classifications of till (and
tillite; lithified equivalent) have evolved rapidly. It is recognized that
textural and mineralogical criteria are inadequate as a basis for the
identification of different till types because of overlap. Increasing emphasis
is instead placed on the use of criteria that emphasize the overall stratigraphy
and surface landforms with which particular till types are found. The presence
or absence of certain landforms or distinguishing characteristics of the
subsurface stratigraphy can be used as a basis for discriminating between true
tills and other ´look alikes´ resulting from non-glacial processes. Thus the
emphasis is on recognition of associated sediment types and landforms,
i.e. landsystems.

The classification of till that has been adopted in this book is restrictive and
intimately related to the three landsystems indentified above. In Fig. 1.7
typical debris transport routes are depicted for a wide range of glacier types.
Two fundamental sources of debris are recognized - that component derived by
erosion as the glacier moves over its bed (subglacial:B) and a component derived
externally from valleysides (supraglacial:A). By following the flow arrows to
the right on Fig. 1.7, sites of till deposition (aggregation) and the respective
landsystem with which each till type is characteristically associated can then be
read off.

(Sect. 2.6). Some flat-iron clasts (Fig. 1.4) may be present.
(2,3) Flowed tills.
(4) Interbedded outwash component may show overall fining up characteristics
recording the filling of the trough and abandonment by melt streams (Fig. 3.6).
(5) Contorted lense-like remnants of flowed tills having moved into outwash
streams (Fig. 3.6).
(6) Strata flexuring as a result of meltdown of adjacent ice-cores (Fig. 3.7).
(7) Hummocky kamiform topography obscuring streamlined surface of lodgement till.
Occurs either as a well-defined morainic complex (or end moraine) or as belts
several tens of kilometers in width (Fig. 9.2).
(8) Lake plains formed by lake ponding between hummocks.
(9) Drumlins on lodgement till surface.
(10) Subglacially cut trench fill or tunnel valley (Fig. 3.13).
Scale: variable.

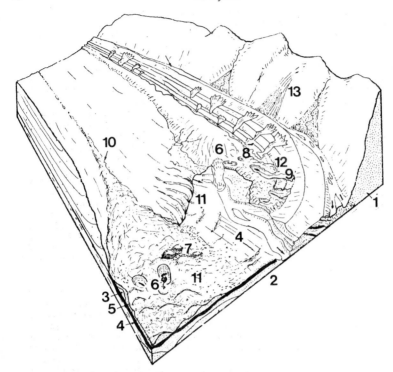

Fig. 1.6 The glaciated valey landsystem

Transport and deposition of valleyside-derived course-grained supraglacial morainic till by valley glaciers.
(1) Bedrock
(2) Buried valley.
(3) Decaying ice-cores with basal melt-out till (Fig. 1.7: Sect. 3.5).
(4) Lodgement till with streamlined drumlinized surface (Fig. 1.4).
(5) Thick hummocky sequences of supraglacial morainic till (Fig. 1.7) along valley floor.
(6) Stratified cores of lateral moraine ridge. Outward dip of both beds and clasts increases the structural stability.
(7) Large angular and far travelled clasts common (Figs. 4.2, 4.6).
(8) Complexly interbedded glaciofluvial sediments and flowed tills deposited in kettle holes or against lateral moraine ridges along valeyside (Fig. 4.4).
(9) Deformation and collapse resulting from melt of ice-cores and glacier recession.
(10) Medial moraine.
(11) Minor ridges of supraglacial morainic till dumped from crevasses or the icefront.
(12) Lateral moraine ridge (with stratigraphy shown in Fig. 4.4 with kame terraces being gullied by the melt of ice-cores.
(13) Valleyside fans which may subsequently breach the lateral moraine ridge resulting in smooth valleyside slopes as part of paraglacial infill processes (Fig. 4.7).
Scale: variable.

More complex variations of Fig. 1.7 have been developed (Dreimanis, 1976) but are difficult to employ as for the most part they are classifications of processes and not sediments as found in the field. Rigorous field criteria whereby such till types can be recognized have not yet been developed.

It is frequently the case that post-depositional mass-movement processes act to redistribute and redeposit tills (e.g. periglacial solifluction, subaqueous slumping). Such processes are particularly common in glaciated valleys with their steep unstable slopes, where ice margins float or front in water and where deposition occurs on the ice surface (Fig. 1.5). These mass movement processes are not specifically glacial and there is increasing unease as to employing the word 'till' to these sediments considering the strict definition to the term (see above and Sect. 3.2). The differentiation of what is original till sensu stricto and what is resedimented using traditional textural and mineralogical approaches is generally unsuccessful (e.g. Falconer 1972, Baker 1976, Shroder and Sewell, 1981). It is in response to considerations such as these that geologists are increasingly abandoning the genetic term 'till' in favour of the non-genetic descriptive term 'diamict' (or diamicton) to refer to any poorly-sorted gravel-sand-mud admixture regardless of origin. In this approach, shorthand codes have been developed which describe the salient characteristics of the diamict (Fig. 1.8) or diamictite (lithified equivalent; Frakes, 1978). The descriptive code can be used, in combination, to identify a large number of diamict types. The code has been thoroughly discussed by Eyles et al, (1983) and the interested reader is referred to that paper for full details.

In addition to objective codes, sediment sequences are also portrayed by constructing a vertical profile of the stratigraphy. The vertical profile is a fundamental starting point in any sedimentological investigation and simply shows the nature of sediment grain-size, sedimentary structures and the type of contacts between individual beds in the section. Excellent recent introductions to the construction of vertical profiles are provided by Thompson and Collinson (1982) and Leeder (1982). The rationale behind the 'vertical profile approach' to glacial stratigraphy is that individual depositional environments give rise to a number of characteristic vertical sediment profiles.

Fig. 1.8 shows representative vertical profiles associated with each of the landsystems identified in Figures 1.4, 1.5 and 1.6, employing both the diamict shorthand codes and genetic till terminology. The width of the vertical profiles in Fig. 1.8 is proportional to matrix grain-size which is estimated in the field; average clast size can also be rapidly estimated and is depicted by varying the size of triangles. Note that codes are also available for glaciofluvial sediments (Tables 7.1, 7.2; Sect. 7.4) and are widely used by sedimentologists for description of sedimentary sequences (see reviews by Miall 1977, 1978). The type of contact between diamicts and associated sediments (erosive, conformable etc.) is also shown by different notation. For comparison, standard descriptive engineering keys in use in North America and Britain for depicting various sediment types are set out in Fig. 10.6 and their rationale discussed in Sect. 11.9. Fig. 1.8 gives some idea of the variation in diamicts and associated sediments found in each landsystem. Note the similarities in diamict character where resedimentation and mass movement are important depositional processes (e.g. during solifluction and supraglacial deposition) emphasizing the need for multiple criteria in order to make safe interpretations of sediment genesis.

It is true to say that with regard to genetic studies, glacial sequences or any other for that matter, have not been fully described until standard sedimentological characterisation outlined above has been completed. The vast majority of glacial stratigraphies have been and are still being described by simply reference to analytical data (i.e. grain-size clay chemistry, bulk

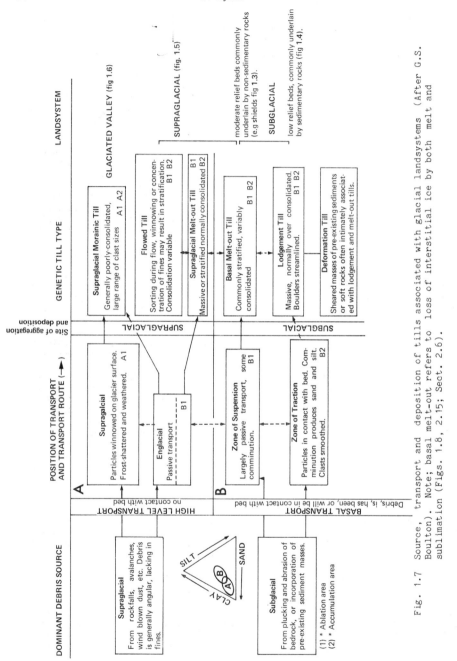

Fig. 1.7 Source, transport and deposition of tills associated with glacial landsystems (After G.S. Boulton). Note; basal melt-out refers to loss of interstitial ice by both melt and sublimation (Figs. 1.8, 2.15; Sect. 2.6).

geochemistry, clast lithology, calcite/dolomite ratios, colour etc. etc.) but the true nature of the stratigraphy remains anonymous.

1.5 The Implications Of The Landsystems Approach To Stratigraphic Practice

The recognition of distinct landsystems composed of genetically related sedimentary sequences (Figs. 1.3-1.6) is not only useful for predicting and understanding subsurface stratigraphies or ground conditions for applied projects but has major implications for stratigraphic investigations aimed at correlating glacial sequences from one area to another. These implications are worth briefly reviewing here.

Traditional stratigraphic practice in the manner formally set-out by Hedberg (1976) is to identify 'type-sites' or reference sections which serve as representative stratigraphies for the geographic area under investigation. The sections are commonly subdivided lithologically into members and beds and subsequently aggregated into formations and groups. Each of these is formally given a name, usually that of the local geographic locality (e.g. Halton Till). This heirarchical litho or rock stratigraphic classification is usually accompanied by mapping in which the members or beds recognized in vertical succession in one area are considered as layer cake mappable units and correlated with those exposed in other map sheets by means of analytical 'fingerprinting' (e.g. texture, clay mineralogy etc. - see above). This can be regarded as a 'correlation chart' approach to stratigraphy because the vertical succession at any one site is matched with successions elsewhere employing horizontal or near horizontal lines on a correlation chart. Examples of this approach to Pleistocene stratigraphy on various scales, can be found in Mitchell, et al, (1973), Mahaney (1976) and Bowen (1978).

By way of contrast, the field of sedimentology is currently undergoing a profound reordering of stratigraphic concepts in which the focus of stratigraphic work has become the 'depositional basin'. The term 'basin study' or 'basin analysis' is an umbrella term for a variety of multi-parameter sedimentological studies aimed at understanding the three-dimensional picture of sediment types (facies) contained within sedimentary basins. The basins may be large regional alluvial basins defined by their plate tectonic setting (e.g. Miall, 1981) or depositional basins found at the margins of modern or Quaternary glaciers. The most detailed 'state-of-the-art' examples of basin analysis have been completed by petroleum geologists against a background of increasingly sophisticated subsurface description employing improved seismic data processing, geophysical well logging and depositional models based on modern environmental analogs (e.g. Mitchum et al, 1977). These data reveal that the subsurface is essentially composed of three-dimensional sedimentary bodies overlapping with each other. Analysis of the shape and internal components of each of these 'depositional systems' and the nature of the relationship between adjacent systems, shows the importance of lateral changes in sediment type (facies) across the basin as a result of changing depositional processes. Many studies show that applying generalised stratigraphic names derived from a single or few 'type-sites' obscures the real nature of regional and local sedimentary sequences.

The 'depositional system' of the petroleum geologist is none other than the 'landsystem' employed in this book. The term landsystem is used because of the importance of landform recognition in interpreting subsurface sequences and expected variation in sediment type. It is a fact that in many glaciated areas, a profusion of formally identified sedimentary units often reflects lateral and vertical variability within a single landsystem, (Sects. 3.5.1, 9.4.1) the number of geologists who have worked in that area and overzealous use of formal stratigraphic nomenclature rather than a long history of deposition.

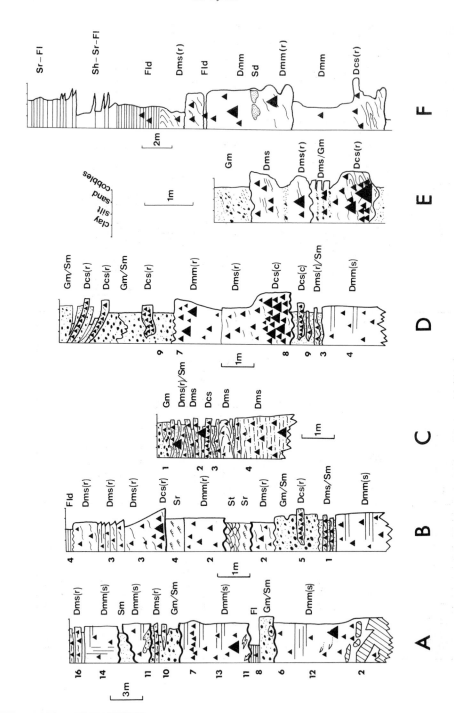

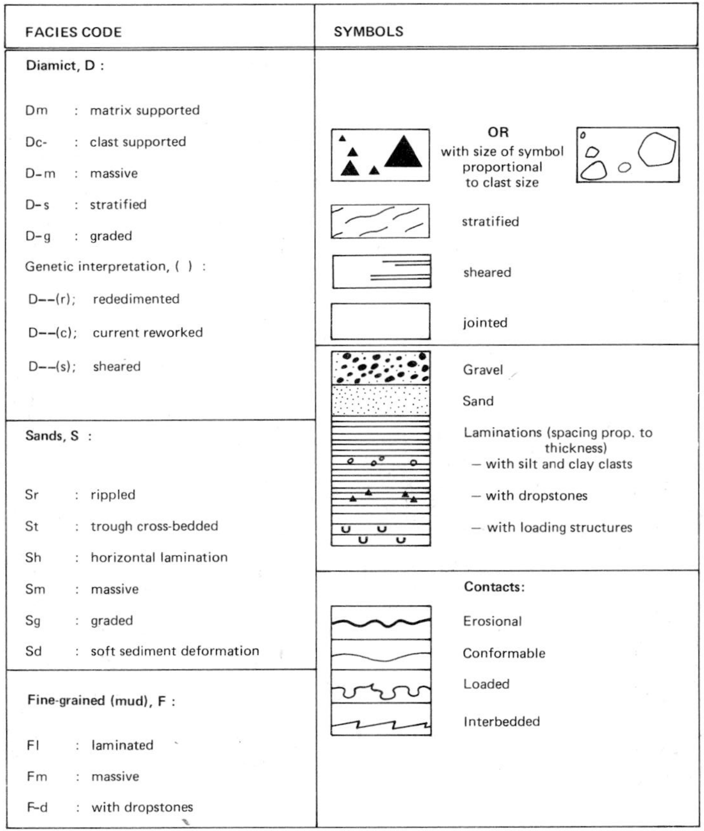

Fig. 1.8 Coded vertical profiles showing diamicts and sediments associated with glacial landsystems and principal till types (Fig. 1.7).

A) Subglacial landsystem: lodgement till (numbers as in Fig. 1.4).

B) Supraglacial landsystem; melt-out and flowed tills (Figs. 1.5, 3.6, 3.7) on top of lodgement till. Note greater preservation of former englacial structures where ice is lost by sublimation, C, (sublimation till; Fig. 2.15) Melt-out generally results in loss structure as a result of water release during ice melt.

D) Coarse-grained supraglacial diamicts (supraglacial morainic till: Fig. 1.7) deposited by valley glaciers (Figs. 1.6, 4.5) may overlie lodgement or basal melt-out till.

E) Sequence of debris flows interbedded with gravels resulting from paraglacial valley infill processes (Sect. 4.6), or solifluction (Sect. 5.4).

F) Glaciolacustrine diamict sequence formed subaqueously as in Fig. 2.20d. Note association with laminated 'varved' (Sect. 6.3.1) silty clays (Sect. 2.9).

Diamict code at left after Eyles et al, 1983. The varying width of the column reflects average grain-size (matrix grain-size in diamicts). For comparison, engineering 'keys' for different glacial sediments are shown in Fig. 10.6 and discussed in Sect. 11.9.

FACIES CODE	SYMBOLS
Diamict, D :	
	OR with size of symbol proportional to clast size
Dm : matrix supported	
Dc- : clast supported	
D–m : massive	
D–s : stratified	stratified
D–g : graded	
Genetic interpretation, ( ) :	sheared
D—(r): rediemented	
D—(c): current reworked	jointed
D—(s): sheared	
Sands, S :	Gravel
	Sand
	Laminations (spacing prop. to thickness)
Sr : rippled	— with silt and clay clasts
St : trough cross-bedded	— with dropstones
Sh : horizontal lamination	— with loading structures
Sm : massive	
Sg : graded	**Contacts:**
Sd : soft sediment deformation	Erosional
	Conformable
Fine-grained (mud), F :	Loaded
Fl : laminated	Interbedded
Fm : massive	
F–d : with dropstones	

The new genetic approach to stratigraphic interpretation and correlation is reviewed at length by Miall (1983). In this approach naming the sediments by reference to formally defined stratigraphic procedure (Hedberg, 1976) is strictly secondary to detailed work aimed at identifying the depositional environment and its three-dimensional sedimentary record. Only then can formal stratigraphic nomenclature be applied. For example, each glacial landsystem could be regarded as a formation with considerable internal variability. More detailed subdivision of the landsystem into members and beds is probably unwarranted given lateral impersistence and facies change.

For many glaciated areas, other than those possessing a simple succession of glacial sediments, which can indeed be physically traced over large areas, pre-existing formal stratigraphic classifications of Pleistocene sediments should be critically evaluated and, if necessary, rebuilt in terms of the stratigraphic variation expected within glacial and non-glacial sedimentary environments (see Reading, 1978; Leeder, 1982).

CHAPTER 2

The Subglacial Landsystem

N. Eyles and J. Menzies

INTRODUCTION

In the glaciated mid-latitudes, landforms and sediments developed at the base of large Quaternary ice sheets fall readily into two distinctive associations. Shield areas, close to the former ice sheet centres (Fig. 9.2) of moderate relief, widespread bedrock exposure and thin covers of coarse-grained sediments (Fig. 1.3) can be contrasted with the low relief lowlands surrounding the shield that are underlain by more fine-grained sedimentary rocks and where thick sequences of glacial sediments (Figs. 1.4, 1.5) extend out to the limit of glaciation (Figs. 9.1, 9.2). Prior to examining the contrasting ground conditions associated with shield and sedimentary lowland zones it is first necessary to review the broad scale controls on patterns of erosion and deposition at the base of large ice sheets.

Quaternary ice sheets were dome-like topographic highs, several kilometres thick (Fig. 2.1) on which snows accumulated, became metamorphosed at depth into glacier ice and then flowed outward in response to gravitational spreading. An equilibrium line, in effect the regional snowline at the end of the summer melt period (Fig. 4.1), defines an upper area on the ice sheet of net mass gain (winter snow accumulation > summer melt) from which ice and snow is transported to the ablation zone of mass loss (winter accumulation < summer melt). The velocity of ice movement is systematically related to distance from the equilibrium line (Fig. 4.1). Gravitational spreading is retarded by frictional drag as ice moves over the subglacial topography and a characteristic ice sheet surface slope profile develops defined by;

$$h = 4.7D^{0.5} \qquad (2.1)$$

where h = elevation in metres and D = distance from the ice sheet margin also in metres. This assumes that ice is a perfect plastic with a yield stress of about 100 kN/m^2 (=1 bar; Appendix II) at a temperature just below 0°C; higher yield stresses at lower temperatures steepen the long profile as does frictional drag against valleysides. Reduction of the yield stress by excessive basal lubrication or deformation of very weak bed materials lowers the profile (Boulton and Jones, 1979). Generally, if the length of a flow line of a former ice sheet can be determined then thickness can be reconstructed; the Laurentide Ice Sheet (Figs. 1.1, 9.9) was about 4.2 km thick as its centre; the last British (Devensian) Ice Sheet was 1.8 km thick (Sugden, 1977; Boulton et al, 1977; Fig. 9.1).

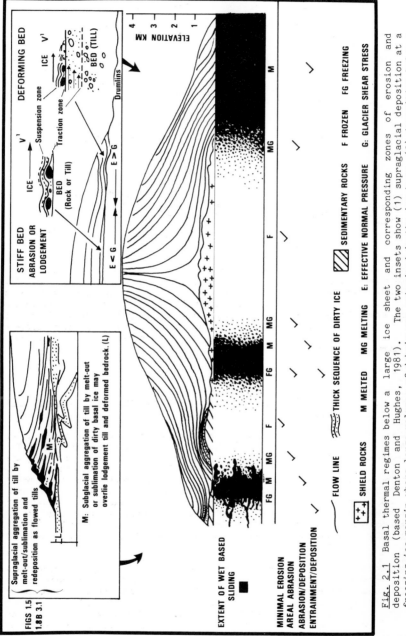

Fig. 2.1 Basal thermal regimes below a large ice sheet and corresponding zones of erosion and deposition (based Denton and Hughes, 1981). The two insets show (1) supraglacial deposition at a freezing ice margin where large volumes of debris are entrained in the ice base, (2) deposition below a melted ice margin with abrasion or lodgement depending on values of effective stress (Fig. 2.3). Deformation of bed materials occurs in areas of low effective stress (Fig. 2.12). Note destruction of ice in melting and melted bed areas and formation in freezing zones.

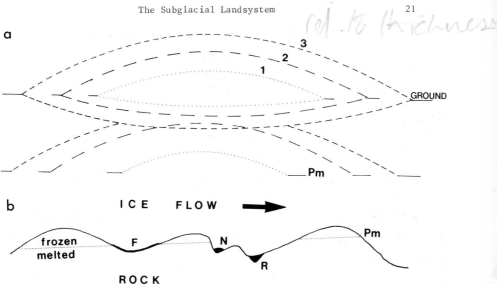

Fig. 2.2a) Growth phases of a frozen based ice-sheet (1——2——3) resting on a ground surface with subsurface permafrost. Note the rise of the pressure melting point (Pm) to meet the ice base giving wet-based sliding conditions and b) a subglacial drainage network composed of either films (F) or channels. The latter may be R (Rothlisberger) channels cut in the bed, or N (Nye) channels cut in ice (Sugden and John, 1976). Note the frozen bed highs and the downglacier gradient on the Pm surface as a result of friction produced by increasing ice velocities. After Denton and Hughes (1981).

2.1 Basal Temperatures, Thermal Conditions And Patterns Of Erosion And Deposition Below Ice Sheets

The most important factor in the erosion, entrainment and deposition of debris at the base of an ice sheet is the thermal condition at the ice-bed interface. Ice masses are usually grouped into temperate or wet-based glaciers that are generally at the pressure-melting point throughout (basal temperatures $= -1°C$ to $-3°C$); polar or cold-based glaciers that are below the pressure melting point throughout and are frozen to their beds (basal temperatures $= -3°C$ to $-18°C$ and sub-polar glaciers that are temperature in their inner regions but possess a polar or cold-based margin. This classification is simplistic however and it is clear that thermal conditions vary both temporally and areally at the base of ice sheets (Boulton, 1972a; Sugden, 1977; Goodman et al, 1979; Denton and Hughes, 1981).

Reconstruction of the basal thermal characteristics of Quaternary ice sheets is a difficult problem and the following review is based upon two recent attempts by Sugden (1977) and Denton and Hughes (1981). Important and complexly interrelated variables are ice thickness and bedrock relief, the position of the ice sheet margins, the rate of accumulation of snow, ice flow vectors, velocity and the position of the equilibrium line. In general, as ice thickness increases so basal temperatures increase in response to pressure-melting and frictional heat developed as the ice mass slides. The average geothermal heat flux (≈ 35 cal/cm /yr) is sufficient to melt 10 cm/yr of basal ice; about the same heat supply results from basal friction at ice velocities of 20 m/yr ($\simeq 46$ cal/cm/yr). Basal temperatures are also affected by the snow accumulation rate and surface temperatures above the equilibrium line. A high rate of accumulation buries cold ice more quickly resulting in a colder ice base than occurs with lower accumulation rates. A further problem is that ice sheet basal thermal regime undergoes complex evolution in time particularly when the ice sheet is growing or thinning. For this reason most modelling refers to the situation when the ice sheet is at its maximum and can be presumed to be in some sort of 'steady state'. Despite this complexity, generalizations are possible as to the nature of the control by basal thermal regime on patterns of erosion and deposition.

For example, if an ice sheet develops by the merging of extensive snow fields lying on permanently frozen ground (permafrost: Sect. 5.1; Williams, 1979), a critical ice thickness is reached where the pressure melting isotherm ($0°C$) at the base of the permafrost layer ascends to the ice base and produces a melting bed interface (Fig. 2.2). Initially, basal melting will commence at low points in the bed. These thawing patches increase in size and an integrated subglacial drainage system develops until only frozen patches remain on bed highs (Fig. 2.2). Melted zones are the focus of enhanced basal ice velocities and abrasion of the bed as a result of water lubrication.

Fig. 2.1 shows a polythermal ice sheet contrasting with the simple mono or bithermal classification of polar, temperate and sub-polar ice masses identified above. Four zones within the ice sheet can be defined and are; melted, melting, frozen and freezing, with complex transition zones existing at the boundaries of each, thereby defining a basal thermal mosaic.

Below the central dome of the ice sheet the bed may be frozen in response to high accumulation rates (rapid burial of cold ice), low flow rates (no frictional heat produced) and the higher elevation of the bedrock plateau on which the ice sheet rests. As ice accelerates away from the central dome to the equilibrium line (Fig. 4.1) frictional heat is developed, the bed progressively thaws and basal ice is destroyed by melting (Fig. 2.1). Beyond the equilibrium line, ice flow decelerates, frictional heat is reduced and a freezing bed develops around still frozen bedrock highs; low points in the bed give rise to melted zones whilst the slow moving ice sheet margin freezes.

Basal flow lines in areas of melting beds are downward toward the bed, acknowledging that ice is lost by melting, and upward away from the bed in areas of freezing where melt waters refreeze to form regelation ice and debris is frozen on to the base (Fig. 2.1). These processes have implications for the distribution of zones of erosion and deposition below the ice sheet. For example, a melting bed results in a continual supply of debris to the ice/bed interface, high sliding velocities and high effective stresses (Fig. 2.3). Such areal abrasion results in the production of typical shield type terrains (Figs. 1.3, 2.16). With lower ice velocities and reduced effective stresses the bed is not swept clear (Fig. 2.3) and debris may be plastered onto the substrate to form lodgement till intercalated with sediments deposited by subglacial melt waters (Figs. 1.4, 2.11, Sect. 2.4.1).

By way of contrast, in freezing or thawing zones where a complex thermal mosaic exists, erosion and deposition are likely to be strongly selective resulting from the streaming of basal ice around frozen high points (Fig. 2.2). Abrasion by basal debris may accentuate frozen highs and debris may be accreted on lee or stoss sides to form drumlins (Sect. 2.5).

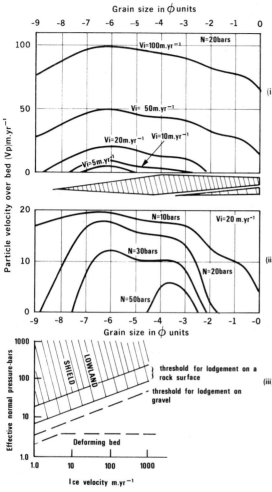

Fig. 2.3

Critical conditions determining abrasion (Vp>Om.yr) and lodgement (Vp=Om.yr) for debris in the traction zone at the base of sliding ice masses (Fig. 2.5).

(i) Particle velocity curves (Vp) are shown (i) for different ice velocities (Vi) at constant effective stress (N=20 bars) and (ii) at different effective stresses at an ice velocity of 20m.yr. The relative volume of particles in the size range -8 to 0 Ø (boulder to very coarse sand) is depicted graphically for basal debris from typical shield areas of granite rocks (double stripe) and lowlands underlain by sedimentary rocks (single stripe). Note that coarse,

(ii) matrix-poor shield debris remains entrained whereas lowland debris is lodgement prone. This may partly explain the predominance of areal abrasion on shield areas (Sect. 9.6.1, Fig. 9.8) with higher effective stresses needed for lodgement on rock surfaces compared with gravel bed (iii). Note conditions under which bed deformation occurs

(iii) i.e. conditions of low effective stress (low bed strength Fig. 2.1, 2.12). Modified from Boulton, (1975). See sect. 8.5 for further discussion.

Lastly, in frozen bed zones, thick stratified debris/ice sequences are built up by continued freezing of meltwaters onto the ice base. These sequences may either be consumed by a change to a melting, sliding bed, survive to be deposited as supraglacial moraine complexes at the ice cap margin (Fig. 1.5, 3.2) or may melt-out or sublimate subglacially (Sect. 2.6, Fig. 2.15). Areas of frozen bed provide loci for extensive bed deformations (Fig. 2.1) in particular the quarrying of bed materials (Section 2.8.2).

2.2 Ice Flow

Ice movement occurs by processes of internal deformation and basal sliding (Fig. 2.14A). In response to the pattern of ice accumulation and loss (Fig. 4.1) flow vectors within the ice mass describe a downward concave flow line in the

Fig. 2.4 Small roche moutonnee with streamlined stoss-end and quarried lee-side.
Direction of ice flow was left to right.

accumulation area and an upward curving flow line beyond the equilibrium line.
Ice deforms as a visco-plastic material with the rate of deformation dependent
upon temperature (Fig. 4.1a) and with a yield strength of about 100 kN/m^2. This
process of deformation is described by Glen's Law (Sect. 4.2.1) such that:

$$\xi = A\tau^n$$

$$(2.2)$$

where ξ is the uniaxial strain or creep rate, τ is the internal shear strength of
polycrystalline ice and A and n are dimensionless constants related to
fluctuations in temperature, particularly in the range 0 to -10°C and confining
pressure.

The value of the factor n varies from 1.5 to 4.2 with a mean of around 3
(Paterson, 1981). Mellor and Testa (1969) have shown that ice deformation
between 0°C and -10°C is highly temperature dependent involving melting of ice
crystal and grain boundaries. Around 0°C, the deformation rate is influenced by
both grain growth and re-orientation induced by shear.

In the central vertical plane of an ice sheet it is assumed that there is no
shear stress and that the bed is frozen. Shear stresses are developed as the ice
deforms outward under its own weight without basal sliding (Fig. 2.1). In the
model developed by Nye (1952) of a sloping parallel-sided slab it can be shown
that shear stresses vary linearly with depth from a value of zero on the upper
surface to a value at the base of:

$$\tau_b = pgH \sin \alpha$$

$$(2.3)$$

where p is the density of ice (0.9gm/cm^3), g is the gravitational constant
where α is the surface slope and H is ice thickness. It is important to note
that basal shear stresses are a function of the surface profile of a thick ice
mass, not bottom topography. It is this driving control on ice flow exerted by

surface slope that allows ice to flow uphill. Basal shear stresses of glaciers are usually found within the range of 0.5 to 1.5 bars (50 to 150 kN/m^2). In simple shear, equation 2.3) reduces to:

$$du/dy = 2A[pg(H-y)\alpha]^n \qquad (2.4)$$

where du is the velocity component in the y direction. From this equation it can be predicted that the shear strain rate varies with depth. With integration of the flow law (Eq. 2.2) and the linear depth variation of shear stress (Eq. 2.3) a relationship between surface and basal ice velocities can be derived where

$$u_s - u_b = [2A/(n+1)](\tau_b)^n H \qquad (2.5)$$

On valley glaciers on the other hand, where movement occurs within confining valley walls, factors such as channel shape, width and side-wall friction become increasingly important (Sect. 4.2).

Movement of an ice mass by basal slip along the ice/bed interface can only occur where ice is at the pressure melting point and water is available for lubrication. Far from being a continuous movement, basal slip consists of a "stick-slip" or jerky movement. Recent studies indicate that micro-seismic 'shocks' are produced by this style of movement which can be attributed to the rupturing of small areas of the ice base frozen onto the bed in lee-side areas of obstacles (Goodman et al, 1979). Ice moves past such obstacles by the mechanisms of pressure melting and enhanced plastic flow. As ice moves against the upstream side of an obstacle high pressures develop, ice melts and meltwaters flow around the obstacle to freeze (regelate) on the downstream side where ice pressures are reduced. This process is fuelled by conduction of latent heat from the downstream (freezing) to the upstream (thawing) side of the obstacle (Fig. 2.5). Such conduction however is only effective for small obstacles up to a centimetrer or so in length. By way of contrast, at obstacles with a size equal to or greater than about 50cm, ice moves around the obstacle by plastic deformation and as obstacle size increases so deformation becomes increasingly effective. At some intermediate obstacle dimension, both regelation flow and plastic deformation do not operate effectively and such 'controlling obstacle sizes' will determine basal velocities. The effect of such obstacles is reduced by water films and subglacial cavities (Fig. 2.13) as both involve decoupling of the glacier from its bed and therefore reduce frictional retardation. The importance of regelation flow and plastic deformation was first established by Weertman (1957) who idealised a glacier bed as consisting of a set of cubic obstacles from which he calculated sliding velocities as:

$$u_b = B\tau_b^{(n+1)/2} S^{(n+1)} \qquad (2.6)$$

where u_b is basal sliding velocity τ_b is basal shear stress, n is Glen's Law flow constant, B is a constant related to the flow law parameter A, and is thus influenced by temperature, and S is bed smoothness (defined as the ratio of distance between each cube to its height). The importance of regelation and deformation flow to processes of glacial debris transport is reviewed in Sect. 2.4.

2.2.1 Ice Velocities

Variation in surface velocity on valley glaciers and ice sheets is considerable but is usually related to position relative to the equilibrium line (Fig. 4.1). Velocity values typically range from 10 to 200 m yr for many valley glaciers to values of 250 m to 1.4 km yr for parts of the Antarctic ice sheet. Velocities of several km yr are found in icefalls where glaciers flow over extremely steep

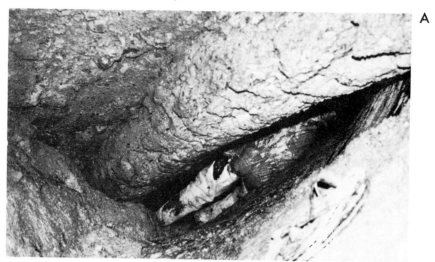

A

B

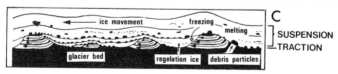

C

Fig. 2.5
A) Subglacial cavity below Glacier d'Argentiere, France. Wet based ice flow
(250 m yr) from right to left. Figure lies on bedrock (abrasion rate = 36 mm
yr (Boulton, 1975). The flute in basal ice reflects shape of bedrock
upglacier. Boulders in the ice base fall out as a cavity fill (Fig. 2.13).
B) Cut section through basal ice at x, showing suspension and traction zones (1
and 2 respectively) at ice/bedrock interface. Suspension layer contains 40%
debris by volume. Crude stratification results from regelation flow (c) and
freezing in of basal debris as ice moves around obstacles (after Boulton, 1974).
For grain-size data see Fig. 2.6.

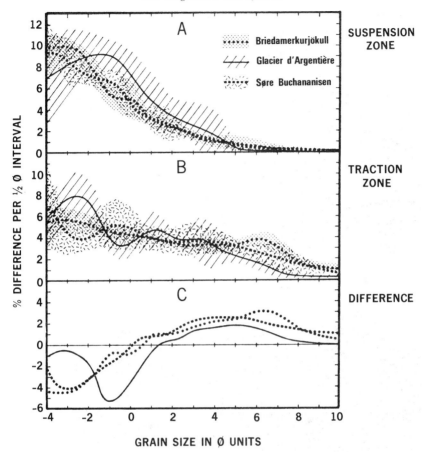

<u>Fig.2.6</u> Grain-size data (mean weight percentages) from basal debris in three sliding glaciers (After Boulton, 1978). (A) portrays data for debris in the suspension zone (Fig. 2.5) above the ice/bed contact, (B) is data for debris in the traction zone at the ice/bed contact and (C) shows the difference in the two curves. Note fines (rock flour) produced by abrasion in the traction zone. Shaded area is the standard deviation from mean.

bedrock gradients. Whereas records of basal sliding velocities are not numerous, typical rates appear to be 10-25% of the surface velocity (Fig. 4.1B). Mid and late melt season velocities are usually highest; diurnal variations in melt, increased rainfall and geothermal activity are factors influencing flow since they reduce frictional retardation at the bed.

Glacial surges are flow instabilities that occur in valley glaciers and individual ice streams (Fig. 4.1) of major ice sheets and valley glaciers (see

Paterson [1981] for a review of surging mechanisms). Typically, a glacier will change from a normal flow regime to one of rapid movement, as much as 100 times faster, over a period of a few months to as long as 3 years, thereafter returning to the "normal" flow pattern. Surging glaciers are peculiar to specific areas of the globe although in those areas not all glaciers surge. In many cases the ice front itself does not advance, the surge consisting of a reactivation of dead ice along the ice margin. Some form of decoupling of the glacier from its bed is necessary for surging to occur through the sequence of events leading to decoupling remains controversial (Clarke, 1976; Paterson, 1981). Surging may result from ice thickness build-up following the transmission of a kinematic wave downglacier (Section 4.2.1) or from excessive ice-bed lubrication and ´drowning´ of controlling obstacles due to increased geothermal heat and the blockage of downglacier meltwater conduits. The influence of surging upon basal processes of erosion and deposition is not well known but surging commonly results in complex supraglacial deposition in response to intense compression of the ice margin during the surge and the freezing on of basal meltwaters to form a thick basal debris zone (Clapperton, 1975, Sect. 3.4).

It has been argued that several major lobes of Quaternary Ice Sheets may have advanced as surges, such as parts of the Superior Lobe of the southern margin of the Laurentide Ice Sheet (Wright, 1973; 1980, Sect. 3.4) and the eastern lobe of the British Devensian Ice Sheet (Boulton et al, 1977). Recent work in North America suggests that several southern lobes of the Laurentide Ice Sheet had low surface profiles indicative of low basal shear stresses at the ice base (Boulton and Jones, 1979).

2.3 Processes of Bedrock Erosion

Major processes involved in glacial erosion of bedrock are quarrying (plucking) and abrasion by debris held within a sliding ice base. Other processes of varying intensity include crushing, fracturing and fissuring.

2.3.1 Quarrying

Quarrying or plucking variably occurs in response to stress unloading at both the micro and macro levels in basal cavities in the lee of bedrock highs, frost penetration and heaving, and freezing-in of weakened areas of bedrock. Boulton (1974) has shown that the stress field set up by ice moving across a bedrock obstruction can be derived as follows:

Assuming no cavitation, the total normal pressure fluctuation

$$\Delta P = \frac{10nVi}{\lambda} \qquad\qquad (2.7)$$

assuming a wavelength to amplitude ratio of 4/1, where Vi is the glacier sliding velocity, n is the ice viscosity and λ is the wavelength of the obstruction. Where

$$\frac{\Delta P}{pgH} > 1 \qquad \text{cavitation will occur (see Fig. 2.13).}$$

Repeated pressure fluctuations across bedrock surfaces in the lee of bedrock hummocks may result from seasonal changes in the size of subglacial cavities. The distinctive plucked lee-end of roche moutonnees probably reflects the fracturing of rock in response to such pressure fluctuations (Fig. 2.4). ´Plucked´ fragments of widely varying dimensions are a major source of abrading tools downglacier and it has been argued that plucking may perhaps be a more important erosional agent, volume for volume, than abrasion (Boulton, 1979a).

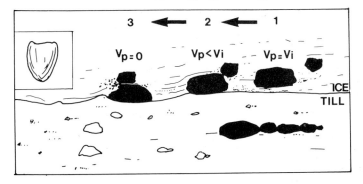

<u>Fig. 2.7a</u> Production of a glacially-shaped (flat-iron) clasts, from pebble to house size, at the base of sliding ice (After Boulton, 1978). 1) Englacial clast with particle velocity (Vp) equal to ice velocity (Vi). 2) Clast meets bed as a result of basal ice melt and velocity is retarded. 3) Clast is lodged, stoss-side becomes streamlined by dirty basal ice; lee-side is plucked (e.g. Fig. 2.4) resulting in a highly distinctive ´bullet´ shape when seen in plan (inset and Fig. 2.7b). Clast collisions on the bed result in boulder pavements which are jams in basal clast traffic preserved for posterity when lodgement till is deposited around the boulder (Fig. 1.4). Large clast clusters may act as nuclei for drumlin formation. As an ice margin recedes englacial boulders melt-out onto the lodgement till surface (Humlum, 1981). These are not shaped but are usually of high sphericity reflecting repeated cycles of abrasion and transport without deposition and shaping.

<u>Fig. 2.7b</u> Bullet-shaped glacial boulder resting on exhumed lodgement till surface of Late Precambrian age (680 million y.b.p.). Jbeliat Group, Western Sahara Desert. Direction of ice flow from right to left. Photo courtesy of C. Eyles.

2.3.2 Abrasion

Abrasion between debris in the sliding glacier sole and underlying bedrock is
expressed in the form of small-scale striae, variably orientated crescentic
gouges and chattermarks, and larger scale grooves of several metres depth and
kilometres in length (Smith, 1965, Goldthwait, 1979; Sugden and John, 1976;
Allen, 1982). Abrasion between debris-free ice and bedrock is extremely limited
(Hallet, 1979). Two processes of rock failure during abrasion are brittle
failure and plastic deformation. Brittle failure of rock is due to tensile
failure along micro-cracks and flaws (Metcalf, 1979) whilst plastic deformation
occurs when the effective pressure on the grinding surface is at least 10^5 bars
(Scholz and Engelder, 1970). In view of the fact that bedrock surfaces consist
of minute asperities, the basal stresses at these points can be expected to be
much higher than the average basal shear stress value of 1 bar and it is
therefore likely that plastic deformation is important in producing small scale
abrasional forms on rock surfaces.

However, many other factors are also involved in the abrasion process. For
example, Boulton (1975) has argued for the importance of effective normal stress
(=weight of glacier on the bed less any water pressure acting to buoy the glacier
up). When stress increases beyond a critical value, abrasion ceases and a phase
of deposition occurs as debris particles moving over the bed become lodged
(Fig. 2.3). In addition Hallet (1981) suggests, just as G.K. Gilbert pointed out
over 70 years previously, that abrasion can also be viewed as a function of the
differential pressure between the abrading particle, the underlying bed and
surrounding bed area. This differential pressure and the "effective contact
force" is argued to be independent of the mean effective pressure at the glacier
bed interface (Hallet, 1979, 1981). Thus, discounting the weight of a debris
particle or aggregate, frictional drag or retardation of velocity in sliding ice
is:

$$F = (pgh - p_w)A_1\varphi \qquad (2.8)$$

where A_1 is the apparent area of contact between particle and bed, φ is a
coefficient of inter-particle friction and p_w is water pressure. Significant
frictional drag due to debris in basal ice is reported by Boulton et al (1979).
The rate of abrasion ($\bar{A}$) as a result of such frictional drag at the glacier bed
interface has been varyingly modelled (Boulton, 1979; Hallet, 1979) and can be
regarded as a function of the following parameters in the form:

$$\bar{A} = \frac{c\ N\ V}{H_R}p \qquad (2.9)$$

where c is the debris concentration within the ice base, N the effective normal
pressure at the particle/bed contact, V_p the particle velocity and H_R the
penetration hardness of the bed. Other things being equal, the abrasion rate
will increase as a function of increasing debris concentration and frictional
drag though it is likely that critical debris concentrations occur at which the
drag imparted by the bed roughness asperities exceed that exerted by glacier
sliding. At this point movement of debris across the bed interface ceases and
the debris assemblage in effect becomes part of the bed. This argument simply
states that at high basal debris concentrations particle inter-reactions may
become sufficiently important that the values of N and V_p become so reduced that
abrasion virtually ceases and deposition of lodgement till occurs (Sect. 2.5).
Thus processes of abrasion and deposition are intimately related (Fig. 2.3).

Average rates of glacial abrasion are approximately 1 mm yr but rates of 2.5 mm
yr have been reported from the Nisqually Glacier (v = 60 m yr ; Metcalf,

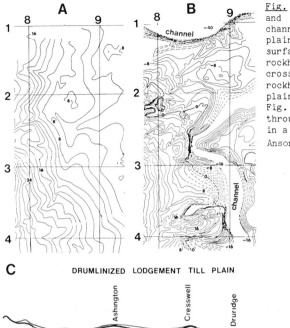

Fig. 2.8 High relief rockhead and subglacially overdeepened channels below a lodgement till plain in northeast England. A; surface contours in metres, B; rockhead contours. C; cross-section showing irregular rockhead and smoothed till plain. Kilometre grid. Fig. 2.11 shows a cross-section through the till plain exposed in a coal strip mine. After Anson and Sharp (1960).

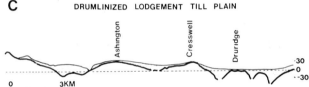

1979), 3.75 mm yr from an Icelandic ice lobe (Breidamerkurjokull, v = 15 m yr) and up to 36 mm yr from a much more rapidly moving valley glacier (Glacier d´Argentiere, v = 250 m yr ; Boulton, 1975; Fig. 2.5).

2.4 Processes of Debris Transport By Sliding Ice Masses

For effective debris transport at an ice base the forces of intertia and frictional resistance caused by the traction zone (Fig. 2.5) moving over the bed have to be overcome. Boulton (1975a) showed that a preferential transport occurs in response to differing debris sizes. With particles smaller than 7ϕ (silt; Appendix I) regelation flow occurs around individual particles resulting in only limited entrainment.

Larger particles greater than -3ϕ in diameter (pebbles), where plastic deformation may begin to operate (Sect. 2.2), will also not be entrained. Thus it has been argued, debris between these two sizes is actively transported (Fig.2.3) leading in the short term to a sorting process (Fig. 2.6).

Particles within the basal zone of suspension (Fig. 2.5) move with the same velocity as basal ice, whereas those in the zone of traction have more complex velocities determined by frictional resistance and particle collisions. The frictional drag of an individual particle or particle aggregates against the bed has been briefly described above (Sect. 2.3.2) and in general as particle size increases so the velocity of particles in the basal zone of traction will be

<u>Fig. 2.9 a,b</u> The subglacial landsystem; air photos and corresponding maps, by
courtesy of R.J. Price, of the surface exposed by the retreat of an Icelandic ice
lobe (Breidamerkurjokull).
Scale: 1 cm = 300 m. a) central margin b) east margin. Heights in metres above
sea-level.

A	Drumlinized lodgement till plain	E	Braided stream and
B	Flutes		outwash fan (Fig. 7.1)
C	Small annual moraine ridges	F	Eskers; note tendency to
D	Large push moraine ridges bulldozed		bifurcate and degrade into fans
	by the ice margin during eposodic	G	Medial moraine ridge on ice
	advances and composed of stacked		margin (Fig. 4.1)
	sediments	H	Ice margin

The development of the ice margin during recession has been thoroughly documented
by R.J. Price.

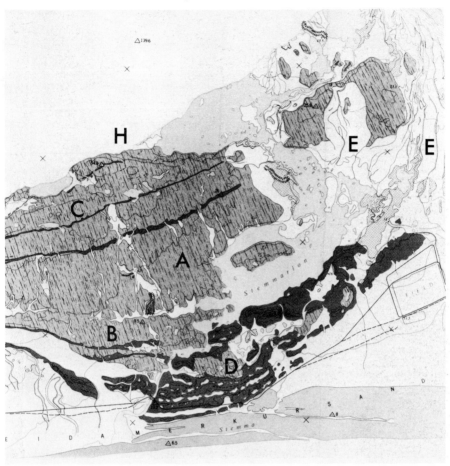

reduced. Similarly, with increasing aggregation of debris at the ice/bed
interface the overall velocity of the tractive debris load will be reduced.
Other variables operate. For example, there is a tendency for platy particles to
slide ‚and to be pushed over the bed whereas more spheroidal particles will tend
to roll. This is normally a function of parent bedrock grain size. Fine-grained
lithologies and those that are well jointed (limestones, basalts etc.,) produce
platy particles that develop a characteristic glacial ´flat-iron´ shaping as they
are pushed along the bed and are abraded by dirty ice moving past them
(Fig. 2.7). On the other hand coarse-grained lithologies (granites) tend to
produce clasts of high sphericity that roll along the bed (Boulton, 1978).

General Comments

The ice-bed interface below large Pleistocene ice sheets has been a major source
of fine sedimentary particles (e.g. loess; Smalley, 1966; Sect. 5.6). The
interface has been likened by many workers to a closed ball mill, or jaw crusher,

Fig. 2.9b Same caption as Fig. 2.9a.

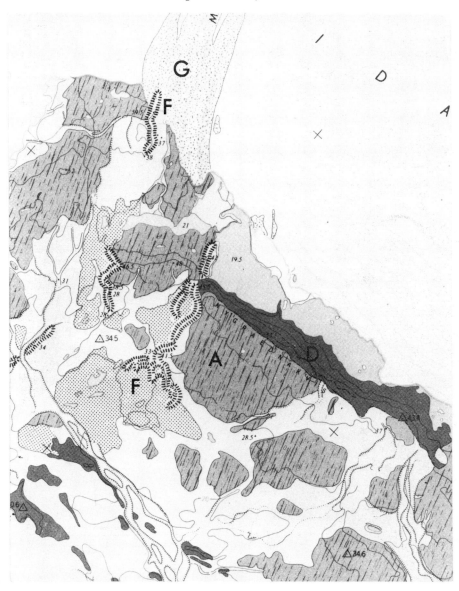

in which fresh bedrock debris is progressively comminuted along a basal flow line
over several hundreds of kilometres in length until characteristic ´terminal´
grain-sizes are reached for each mineral present (Harris, 1968; Dreimanis and
Vagners, 1971; Fig. 2.10a). The analogy of the ball mill is an attractive one,
is widely accepted, but unfortunately does not bear much critical analysis. The

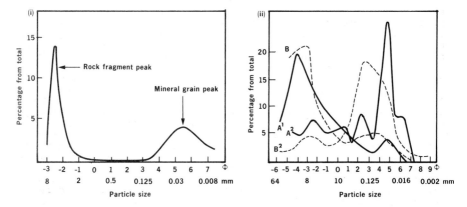

<u>Fig. 2.10a</u> (i) Bimodal grain-size distribution of quartzite debris showing production of matrix mode by crushing in a ball mill.
(ii) Evolution of matrix composed of minerals at 'terminal grade' sizes by long distance glacier transport. A^1-A^2; dolomite after 75 km transport. B^1-B^2; mixed igneous and metamorphics after 1200 km of glacier transport. After Dreimanis and Vagners (1971). See Sect. 2.4 for details.

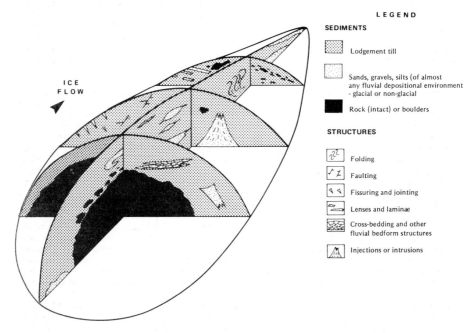

<u>Fig. 2.10b</u> Structure of a rock-cored drumlin; see also Fig. 1.4. Length from 50 to 500 m. Height up to 100 m.

ice-bed interface is a dynamic open system, ball mills are closed, and significant size-sorting processes operate at a glacier bed that do not in a ball mill (Fig. 2.3). In addition to removal of fines from the system by water (subglacial wash-out), subglacial geochemical processes are highly significant particularly on carbonate terrains (Hallet, 1976). The importance of continued reworking of pre-existing sediments and pedogenic soils, subglacial meltwater processes and inheritance of grain-sizes from parent sedimentary rocks (e.g. multi-cyclic quartz grains) is also stressed by Gravenor (1957), Slatt (1971), Gillberg (1977), Haldorsen (1978, 1981), Kemmis (1981), Slatt and Eyles (1981) and by studies of quartz sand grain surfaces using electron microscopy (Bull, 1981). Above all, long distance transport and comminution of single debris populations (a la ball mill) is strictly prohibited by wet based sliding conditions because debris in the basal suspension zone (Fig. 2.5) is repeatedly moved through the traction zone as a result of basal ice melt and is continually sorted and dispersed. As a result, a wide range of grain-size populations are found in basal ice of the suspension layer in modern glaciers (Fig. 2.6; Lawson, 1981) with no progressive fining of these populations with increasing downglacier transport (Slatt, 1971). It used to be thought that this lack of increasing 'maturity' downglacier was due to the short transport distances available in modern valley glaciers compared to Pleistocene ice sheets (Slatt, 1972; Eyles and Rogerson, 1978). However, it can be demonstrated that 'terminal grade' abrasion products are produced even during a single abrasion event at the bed of short valley glaciers. The amount varies according to bed roughness in that smooth beds (polished rock surfaces) result in efficient comminution and enhanced abrasion whereas rougher beds result in grain crushing and the production of reduced fine-grained products (Boulton, 1978). However, regardless of bed-type and glacier length, fines produced by abrasion ('rock flour') are washed out by subglacial waters and do not progressively accumulate within the basal suspension layer in a downglacier direction as is implicit in the 'ball mill' model. Fine abrasion products are also much more likely to be lodged against the bed as a result of regelation processes and thus be removed from the debris transport system (Figs. 2.3, 2.5). Coarser debris particles are similarily subject to significant size sorting processes (Sect. 2.4).

If there is no progressive fining of basal debris down the flow lines of modern wet based sliding glaciers (and we presume that pleistocene ice sheets were similar in character) how then are the large matrix modes of Pleistocene tills in mid-continent areas (Figs. 2.10a; 8.1b) produced? One possible answer appears to be the repeated reworking of subglacial sediments and incorporation of lacustrine sediments (former subglacial washout) in which minerals are present in abundant terminal grade sizes (Fig. 8.1b). Studies of clay mineralogy also indicate that the incorporation of fine-grained sedimentary rocks (mudstones, shales) is critically important. A further consideration is that certain fine-grained glacial units have been incorrectly identified as tills and are in fact glaciolacustrine in origin (Figs. 2.20, 2.21).

A further problem in estimating the characteristics of basal debris transported by Pleistocene ice sheets is apparent. This is that different grain-size populations in the ice base (resulting from the subglacial sorting processes discussed above) are mixed during lodgement deposition. It can be argued that the textural homogeneity and typical poor sorting of lodgement tills (Fig. 8.1a) results from the lodgement process by which former heterogenous basal debris populations are eliminated and does not result from transport and abrasion processes per se. Post-depositional processes such as subglacial shearing (Sect. 2.5.1) are also very important in modifying the particle-size characteristics of lodgement till (Boulton, et al, 1974). In short, the particle size distribution of a Pleistocene lodgement till indicates very little of the size characteristics of debris formerly transported at the base of former ice

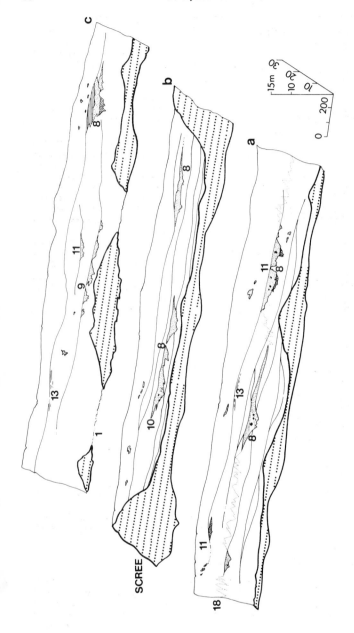

Fig. 2.11 Successive sections (a,b,c), cut during mining operations, through the lodgement till plain shown in Fig. 2.8. V-shaped channel fills cut by subglacial meltstreams, where present, run parallel with former ice flow direction. Same numbers as in Figs. 1.4, 1.8. Multiple cross-cutting till units record erosional episodes during the course of lodgement till deposition rather than multiple glaciation; erosional interfaces fade out laterally within a single till unit. The dotted line is the base of a postglacial weathering profile; note that base coincides with sands and gravels as a result of underdrainage with an irregular base elsewhere (from Eyles, et al, 1982). For geotechnical properties, see Section 8.2.4.

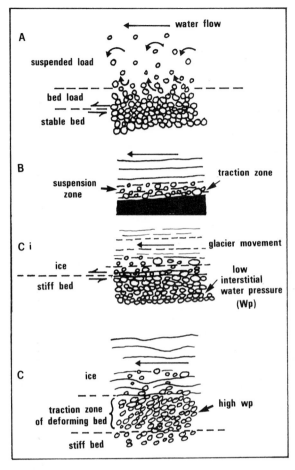

Fig. 2.12 Particle transport by water contrasted with that by sliding ice. A) Particle transport at the base of flowing water.

B) Suspension zone and traction zone at the base of ice sliding on rock.

Ci) Suspension and traction zones with ice sliding over till, gravel or other sediments in which pore water pressures are low and effective normal pressures exceed glacier shear stress (Fig. 2.1) resulting in a stiff bed.
Cii) Development of thick traction zone of deforming bed materials where effective normal pressure is exceeded by glacier shear stress (Fig. 2.1).
Modified from Boulton, 1975

sheets. Basal melt-out and sublimation tills by way of contrast, as they have never undergone lodgement, may preserve large textural variations (Shaw, 1977, 1979; Lawson, 1981) and offer more scope for reconstruction of the characteristics of basal debris, and mechanisms of debris transport, at the base of former ice sheets.

2.4.1 Processes of Deposition By Sliding Ice Masses

High debris concentrations in the basal traction layer of sliding ice and high effective stresses result in considerable frictional drag against the bed (Sect. 2.3.2). Dense overconsolidated lodgement till results from the successive frictional retardation and pressure melt-out of individual bedrock particles and/or debris aggregates against the bed. Modern sedimentation rates vary between 0.5 and 3 cm yr (see discussion in Mickelson, 1973, and Sugden and John, 1976) but recent data indicates that till deposition is not continuous. Erosional episodes may occur in which the till bed is eroded resulting in a

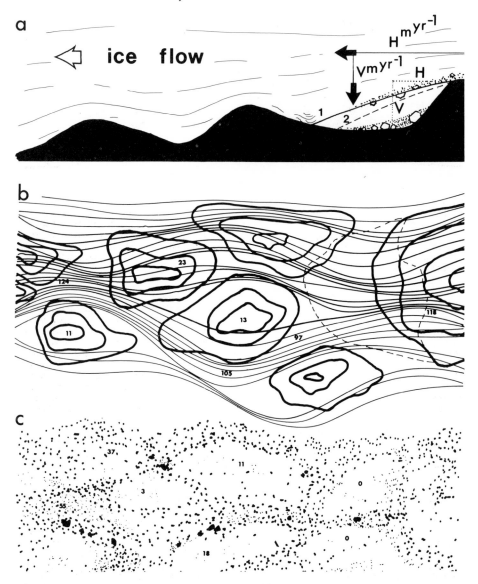

<u>Fig. 2.13</u> a) Flow over bedrock hummocks with lee-side decoupling (cavitation). Hm=horizontal ice velocity. Vm=vertical. Cavitation occurs when Vm/Hm   V/H. Winter cavity size (2) is smaller than summer (1) because of reduction in Hm. b) Basal flow lines and thickness of basal suspension layer (in cm) around contoured rock highs. c) Streaming of basal debris. Figures are debris concentrations (% volume). These streams of dirty basal ice can cut large erosional grooves (Goldthwait, 1979) and help maintain bedrock hummocks.

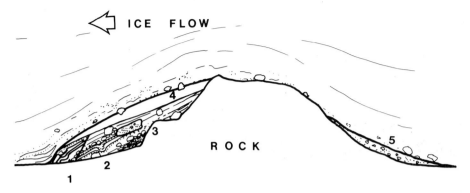

Fig. 2.13 d
a) Bedrock-drift complex consisting of rock high, lee-side cavity and subglacial
scree debris derived from the ice base. Note stoss-side wedge of over-ridden
basal ice (5) melting-out in situ. The subglacial scree may be glacitectonized
(1) at point of cavity closure. Voids and deformed bedding (2) result from decay
of buried ice blocks that at (3) contain basal debris. At (4) boulders are being
expelled from the ice roof into the cavity.

cross-cutting erosive relationship between till units (Fig. 2.11). The lodgement
process requires beds composed of unlithified sediments to be stiff, i.e. well
drained under high effective stresses (Figs. 2.1, 2.3). Elson (1961) suggested
that heat produced by abrasion during the lodgement process results in basal ice
melt and the production of optimum water content for maximum compaction of the
till.   Indeed in any sediment optimum water contents are needed for achievement
of maximum consolidation;   too much or too little water results in under
compaction (Sect. 13.4, Fig. 13.2).   At low effective stresses (poor drainage)
unlithified sediments or already deposited lodgement till will deform, in effect
fluidize, and be retransported by the glacier (Fig. 2.12). In general, higher
effective stresses are required for lodgement till deposition to commence on rock
beds given their reduced roughness compared to a gravel bed (Fig. 2.3). Rock
knobs, however, give rise to complex fluctuations in effective normal pressure
resulting in asymmetric patterns of till deposition around bedrock highs
producing drumlinoid forms (Boulton, 1975).

The essential difference between lodgement and basal melt-out tills (Fig. 1.7) is
that the former are deposited by pressure melt-out of glacial debris particles or
aggregates, in the base of sliding ice whereas melt-out tills per se are
deposited below stagnant ice that is melting in situ (Figs. 2.14, 3.7,
Sect. 3.5). It is likely that a complete transition exists between these two end
members but this awaits detailed documentation in the field. A crude layering or
fissility in lodgement till has been reported and interpreted as former basal
suspension zones (Fig. 2.5) that have become detached from the ice base and have
melted-out in situ (Dreimanis 1976 for refs.). However fissility and horizontal
discontinuities are produced by many processes including vertical stress relief
or the formation of periglacial ground-ice lenses. Discontinuous sub-horizontal
silt and sand ´stringers´ in lodgement tills reported by Rose (1974) and Menzies
(1979a) may be lag horizons resulting from water released during pressure
melt-out.

There is increasing recognition of the importance of both subglacial meltwater deposition and the engineering and stratigraphic importance of sand and gravel units intercalated within lodgement till sequences (Fig. 2.11). Examination of large-scale exposures through lodgement till sequences along coastal cliffs and in strip mines, for the purpose of mapping the distribution of 'buried' subglacial sands and gravels, often shows that sand and gravel units may be traced for several kilometres where the line of section (or boreholes, trial pits, etc.) is aligned close to the direction of ice flow (i.e. direction of former subglacial stream flow). Where the section line lies transverse to ice flow more simple lensoid channel cross-sections are seen up to 15 m thick but generally thinner than 8 m. These channels record impersistent subglacial drainage. With changes in flow direction of basal ice and/or winter reduction in subglacial meltstream discharge, deposition of lodgement till recommences over subglacial sands and gravels whilst active melt streams may reform elsewhere. Persistent subglacial streams that survive until deglaciation form eskers (Figs. 1.4, 2.9). The upper surfaces of fluvially deposited beds are sheared off as till deposition recommences thereby generating intense slickensiding and folding in underlying beds (Fig. 1.4). These units rapidly soften when exposed in cutting slopes and promote slope failure (Sauer, 1979; Sect. 13.5); remedial work requires drainage of the slope by deep wells, drainage galleries or sand drains and regrading of the slope. The exposure of water bearing units during excavations is particularly problematical as artesian pressures are frequently developed.

As an ice margin thins and recedes so effective stresses at the bed are substantially reduced and the strength of previously deposited lodgement till may correspondingly drop. The upper part of the till bed may deform under the influence of shear by the moving glacier bed (Sect. 2.5.1). As the ice margin recedes, dispersed boulders and debris in the suspension zone melt-out and are lowered onto the emerging subglacial surface (Fig. 2.7; Humlum, 1981).

The characteristics of lodgement till are reasonably well known and its stratigraphy (Fig. 2.11), associated landforms (Figs. 1.4, 1.8) and engineering properties (Chapter 8) are a norm against which diamicts of dubious origin can be compared. Lodgement till plains typically cover bedrock surfaces of higher relief amplitude (Fig. 2.8). One of the most written about aspects of lodgement tills is the orientation of clasts ('till fabric', e.g. reviews by Dreimanis, 1976; Embleton and King, 1978). Clasts generally have their long axis parallel to the direction of former ice flow, with a slight upglacier dip. The measurement of clast orientation in the field is a lengthy and uninspiring task - the estimation of flow direction or sedimentary fabric by paleomagnetic means (Sect. 2.6.1) or by reference to a few _in situ_ glacially shaped 'flat-iron' boulders (Figs. 2.7, 1.4) is more rapid (Eyles, _et al_, 1982). Distinct clast fabrics typical of other diamicts have been discussed by Walker (1975), Middleton and Hampton (1976), Benedict (1976) and Collinson and Thompson (1982).

A note of caution is that 'fabric', in an engineering context, refers to structural discontinuities within a sediment that may affect sampling, testing foundation properties and slope performance (Sects. 8.3.8, 12.2; 13.5). The fabric of lodgement tills is distinct and is dominated by fissure systems orientated consistently to direction of former ice flow, topography (McGown and Radwan, 1975) or underlying fissures in bedrock (Cox and Harrrison, 1979).

2.5.1 Subglacial Landforms Deposited By Sliding Ice Masses

Low Relief Till Plains

Subglacial landforms commonly associated with basal deposition by sliding ice masses are streamlined with the resultant till plain resembling a large-scale

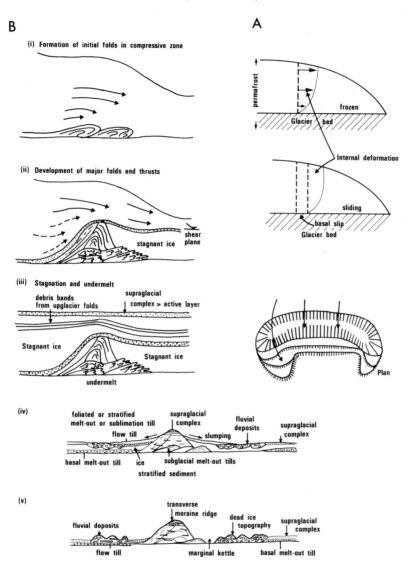

B

(i) Formation of initial folds in compressive zone

(ii) Development of major folds and thrusts

stagnant ice

shear plane

(iii) Stagnation and undermelt

debris bands from upglacier folds

supraglacial complex > active layer

Stagnant ice

Stagnant ice

undermelt

(iv)

foliated or stratified melt-out or sublimation till

supraglacial complex

fluvial deposits

supraglacial complex

flow till

slumping

basal melt-out till

ice

subglacial melt-out tills

stratified sediment

(v)

transverse moraine ridge

dead ice topography

supraglacial complex

fluvial deposits

flow till

marginal kettle

basal melt-out till

A

permafrost

frozen

Glacier bed

Internal deformation

sliding

basal slip

Glacier bed

Plan

Fig. 2.14 A) Differences in flow mechanisms between cold based (frozen) and warm based (sliding) ice masses.
B) Genesis of Rogen moraines at the base of glaciers in permafrost zones (Figs. 2.1, 2.2). After Shaw (1979).

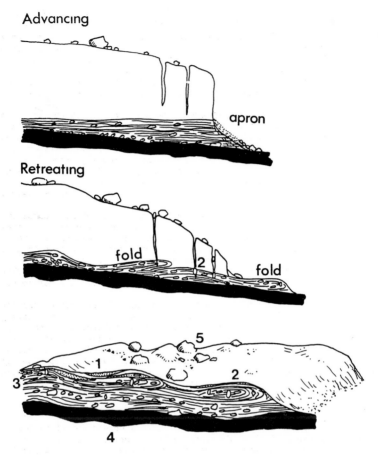

Fig. 2.15 Deposition of sublimation till. Englacial debris, folded as a result of deformation at the base of the cold based polar glacier (2.14A), is lowered onto the substrate as interstitial ice is progressively lost by sublimation. A typical vertical profile is shown in Fig. 1.8c. (1) thin flowed till and outwash units as a result of fluvial reworking; (2) vertical stones truncating stratification; (3) stratified diamict interbedded with thin sand partings recording sublimation of debris rich and debris poor ice; (4) thick stratified diamict unit with abraded boulders from episodes of traction at the frozen ice base; (5) Supraglacial debris. After Shaw (1977a). (Sect. 2.6). After Shaw (1977a).

slickenside (Figs. 2.9, 2.16). The simplest and commonest form of streamlined landform are flutes which are elongate ridges rarely exceeding 1 m in height and 1 - 2 m in width but which may be 500 m to 1 km long. Other ridges termed 'megaflutes' may extend up to 20 km and in dimensions range from 10 to 25 m high and up to 200 m wide (Gravenor and Meneley, 1958). Theories of fluting formation emphasize the intrusion of basal sediments as plastic pastes (Galloway, 1956) or as water-soaked slurries (Hoppe & Schytt, 1953; Andersen & Sollid, 1971) into cavities beneath the moving ice (Paul and Evans, 1974). These cavities are most commonly formed in the lee of large boulders (Boulton, 1976; Fig. 2.9), rock knobs or patches of frozen sediment at the ice base. Strictly speaking flutes are not primary depositional landforms since they result from the past-depositional deformation of subglacial sediments. They must also be carefully distinguished from erosional grooves cut subglacially across bedrock or sediment by basal ice 'streams' (Fig. 2.13b; Smith, 1948; Goldthwait, 1979).

No other glacial landform has attracted so much attention and research as have drumlins. The elliptical plan view and asymmetric long-section with its steep stoss-side and gentler lee-side defines a landform found throughout the glaciated mid-latitudes (Figs. 1.4, 2.10, 9.1, 9.2) and at the margins of many present day glaciers, i.e. in Iceland (Fig. 2.9). Dimensions vary enormously, from 3 to 70 m in height, 100 m to several kilometres in length and 3 to 500 m in width. Many studies have dealt with drumlin shape, drumlin spacing and orientation, clast fabrics within drumlin till, and drumlin structure (Aario, 1977; Menzies, 1979).

Drumlins typically occur in large groups (fields or swarms) sometimes numbering several thousand individual forms as, for example, in central-western New York State (<10,000), in western Nova Scotia (2,300 - 2,500) and in western Pomerania (<3,000). However, drumlins also occur in small numbers of 10 or less in many localities. Several morphological characteristics are worth noting, namely the consistent relationship between drumlin size, spacing and shape in any particular drumlin field or part thereof (Menzies, 1979). Genetic hypotheses are numerous.

'Erosional' hypotheses, although varying in individual detail, all hinge upon the concept that drumlin formation results from the erosion of an already existing sediment. It is therefore necessary, but often extremely problematical, to show that the internal drumlin sediment is preglacial, of a previous glacial phase, proglacial or a contemporaneous subglacial unit subsequently eroded. No matter what sediment type is identified it must be shown that it has been deposited in situ prior to erosion and has not suffered reworking or transport and inclusion into the subglacial environment. Many drumlins are clearly of an erosional origin (Gravenor, 1953; Menzies, 1979; Whittecar and Mickelson, 1979).

Other 'depositional' hypotheses emphasize drumlin initiation by accretion of basal debris around 'proto-drumlins' composed of rock or boulder obstacles or 'clast clusters' (Boulton, 1975; Menzies, 1979; Moran, et al, 1980; Fig. 2.10b). In addition, the significance of bed deformation below sliding ice is receiving increasing attention since the paper of Smalley and Unwin (1968) who identified the importance of changes in subglacial bed strength with time. The deformation of unlithifield subglacial bed materials is dependent on the changing balance between normal pressure on the bed exerted by the glacier, less any subglacial water pressures, and the shear stress exerted on the bed by ice movement (Boulton et al, 1974; Fig. 2.1, Sect. 8.5). The two are independent, but with low average values of glacier shear stress (approx. 1 bar), bed materials will only deform at low effective stresses such as may occur at the ice margin where ice thickness is reduced (Fig. 2.1) or further upglacier where beds of low hydraulic transmissivity pump up local porewater pressures and the bed in effect fluidises and becomes dilatant (Smalley and Unwin, 1968). Deformation occurs in the form shown in Fig. 2.12 with a vertical velocity gradient and an upper velocity slightly below that of basal ice. The areal extent, thickness and nature of the 'remoulded' traction zone remains problematical but can be regarded as having a thickness that is related to porewater pressure.

For example with increasing porewater pressure the upper surface of the till bed will be in failure and the remoulded traction zone will thicken (Fig. 2.12).

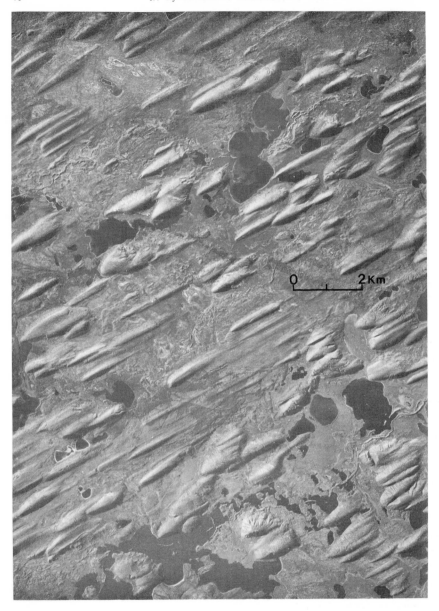

Fig. 2.16 Drumlin swarm; slickensided subglacial surface developed below
           ice flowing from top right to bottom left. Note eskers
           meandering between drumlins. Photo: A14509-5, Energy, Mines
           and Resources, Ottawa.

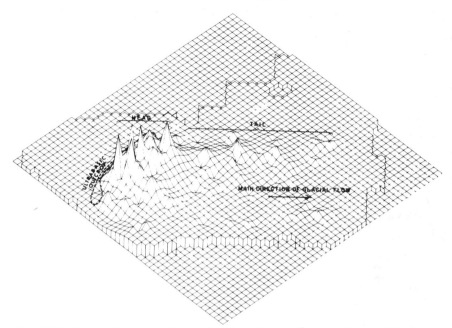

Fig. 2.17 Perspective plot of nickel values in the 64μ fraction of till in a dispersal fan emanating from an ultrabasic bedrock outcrop. Individual squares on the grid are 1 km^2. Note the 'head' and 'tail' of the fan (from Shilts, 1976).

With reduction in internal shear strength, the material, in approaching the state of fluidisation, will deform and thin-out. Similarly where either a stress reduction or a reduction in porewater occurs the thickness and rate of deformation of the traction zone will decrease. Lodgement till, deposited upglacier on a stiff bed under high effective pressures, may start to deform as the ice margin retreats and effective stresses decline. Furthermore, if the bed consists of a mosaic of different hydraulic transmissivities then drained (or frozen) stable areas of the bed may become streamlined as surrounding zones of low strength are remoulded and become part of the traction zone. If basal debris is then deposited around these nucleii of frozen or better drained bed material and if the mass does not subsequently fail under glacier shear stress then a debris hummock or ´proto-drumlin´ may result (Menzies, 1979a, 1982). This process could result in a drumlinized or fluted till surface though the production of large subglacial landforms is not likely. The question of post-depositional deformation in lodgement tills is discussed further in Sect. 8.5. Other drumlin hypotheses emphasize kinematic fluting and local secondary flow patterns set up within basal ice (Shaw and Freschauf, 1973; Aario, 1977; Shaw, 1980) by analogy with bedforms caused by fluvial erosion of cohesive sediment (e.g. Allen, 1971; 1982).

Hypotheses in which depositional processes are thought to dominate have gained considerable support perhaps for no other reason than drumlins tend to occur most frequently within the broad depositional zones of Quaternary continental ice

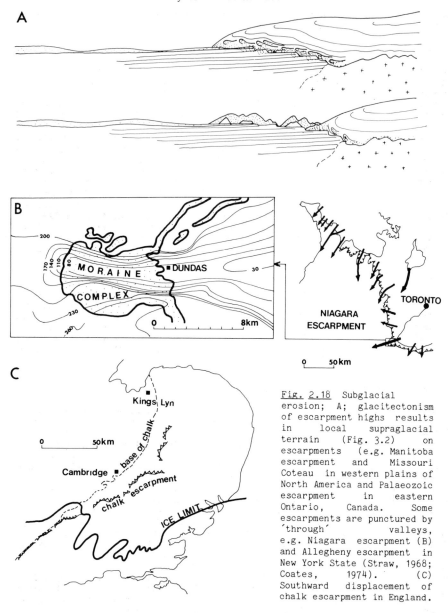

Fig. 2.18 Subglacial erosion; A; glacitectonism of escarpment highs results in local supraglacial terrain (Fig. 3.2) on escarpments (e.g. Manitoba escarpment and Missouri Coteau in western plains of North America and Palaeozoic escarpment in eastern Ontario, Canada. Some escarpments are punctured by 'through' valleys, e.g. Niagara escarpment (B) and Allegheny escarpment in New York State (Straw, 1968; Coates, 1974). (C) Southward displacement of chalk escarpment in England.

sheets (Figs. 9.1, 9.2). Depositional hypotheses attempt to explain the cause of initiation prior to development. It would seem likely that drumlins "evolve"and migrate in much the same way as other bedforms (e.g. ripples or dunes) under certain boundary conditions at the ice-glacier bed interface rather than form "en masse" at any one point in time or space..

2.5.2 Moderate Relief Rock Beds; Shield Areas

The term bedrock-drift complex refers to a characteristic assemblage of bare rock highs and a surrounding ´skirt´ of glacial sediments that tend to be concentrated in stoss and lee-side positions (Fig. 1.3). In areas where high erosive ice velocities were maintained during glaciation an extensive lodgement till surface does not develop, areal abrasion dominates and deposition is strongly controlled by fluctuations in normal effective pressure and enhanced ice velocities as ice moves around bedrock highs (Sect. 2.3.1). The combined action of subglacial meltwater streams, high velocity ice-streaming (Fig. 2.13) and the generally coarse character of basal debris on shield areas results in a rock bed that is generally swept clear (Fig. 2.3). During glacier retreat and ice thinning lee-side cavities develop (Fig. 1.3) and basal debris collects as a crudely statified subglacial scree. Ice blocks fall from the cavity roof and included debris melts-out within the scree; cavity closure as a result of transient changes in ice velocity deform the scree stratigraphy (Fig. 2.13; Hillefors, 1973). Basal ice may also stagnate or freeze against the stoss-side of bedrock highs and slowly release its debris by melt or sublimation; till units, wedging out against bedrock highs, coupled with scree-like lee-side accumulations constitute drumlinoid ´bedrock-drift complexes´ or ´crag and tails´.

The ´streaming´ of debris that occurs as basal ice deforms around bedrock highs tends to emphasize and maintain the irregular bedrock relief. Folding between converging debris streams in lee-side areas acts to thicken the normally thin and continuouly-melting basal suspension zone and as the ice margin retreats so basal debris melts-out and is dumped as an irregular bouldery cover over the emerging rock highs (Haldorsen, 1982). Delta-fans form in the numerous lakes that characterise such topography, receiving debris washed off bedrock highs (Fig. 1.3).

2.6 Debris Transport And Deposition By Cold Based Ice Masses

In areas of either freezing or frozen beds, basal sliding does not occur (Fig. 2.14A) and in contrast with the thin debris suspension zones found in areas of basal melting (Fig. 2.5), thick stratified sequences of ice and debris are built up by the regelation of meltwaters draining into cold bed zones (Fig. 2.1). This process may allow the ice mass to entrain large volumes of debris. The basal ice sequence may either remain frozen to the substrate as a rigid carapace protecting the underlying bed with clean ice moving overhead by internal deformation (Fig. 2.14B), or may move downglacier undergoing deformation and shear attenuation. In this latter case limited abrasion may occur at the ice base. As the sequence is transported downglacier it may move into a melting bed zone and be deposited as lodgement till or, in areas of high sliding velocity, may simply scour the bed. Where basal debris sequences survive, and are not thinned out by transport, deposition can occur supraglacially as the retreating ice margin exposes the sequence (Figs. 1.5, 2.1). Under humid periglacial climatic conditions of discontinuous permafrost (Sect. 5.2) basal debris is aggregated in situ by supraglacial melt-out and redeposited by flow (Figs. 1.5, 1.7, 1.8, 3.2) with little preservation of original englacial debris/ice stratification and structure. Other tills form at depth below overlying sediments as basal melt-out tills (Figs. 1.7, 1.8, 3.7; Sect. 3.2).

Under arid cold periglacial conditions with continuous permafrost, such is represented today in certain areas of the high Arctic and Antarctica, interstitial ice is lost by sublimation where ice passes directly into the

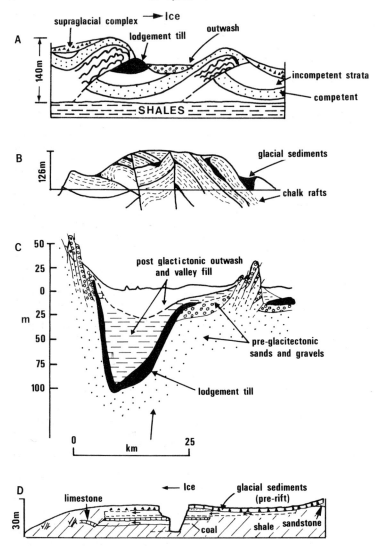

Fig. 2.19 Glacitectonic Settings A) Escarpment type with ice margin against bedrock high (Missouri coteau; U.S.A.). Note supraglacial complexes, as in Fig. 2.18, resulting from ice compression. B) Island type; stacking of chalk rafts on Mons Klint, Denmark. C) Valley type with tectonism of valley side walls in Netherlands. D) Pull-apart dislocation in lee-side position. After Banham (1975).

gaseous phase; aggregation rates are extremely slow however. Rates of c. 20 cm/1000 years are reported for sublimating glaciers in the Dry Valleys of Antarctica (Shaw, 1980). Englacial stratification and structure may survive intact producing a crudely stratified coarse-grained sublimation till often showing folds inherited from deformation at the ice base (Figs. 1.8c, 2.15). The very slow rate at which sublimation tills are lowered onto the substrate may suggest that Pleistocene sequences thicker than a few tens of centimetres are unlikely.

By way of contrast, the preservation of large englacial fold complexes, by being more rapidly melted-out in situ onto the substrate, provides a model for the formation of large moraine ridges that are common to shield areas in Finland, Sweden and Canada. These ridges, often with a fluted, streamlined surface, lie transverse to former ice flow directions and impart a ´rippled´ or ´ribbed´ appearance to the landscape. These Rogen moraine ridges, named after the classic locality in northern Sweden, are typically found where local topography promoted spreading of former ice margins are found closely associated with drumlins (Lundquist, 1969). Diverging flow lines may result in reduced longitudinal ice velocities (Section 4.2.1) and corresponding compressive flow. It can be suggested that englacial ridge complexes form by the stacking and piling-up of the basal debris sequence in which bedrock masses may also be incorporated. The upper surface of this large transverse bed obstacle is then streamlined by clean ice moving overhead (Fig. 2.14). Recent work on shield areas (Fig. 1.3) indicates the importance of a moderate bed relief with frozen zones on bed highs (Fig. 2.2) and the stacking of thick englacial debris sequences on their upstream margins. Cold permafrost conditions prohibit much surface melt and the entire basal sequence is lowered onto the substrate by slow basal melt. Original basal stratification is preserved. The differential compaction that occurs around large clasts is particularly diagnostic of such lowered sequences (Shaw, 1979). Limited meltwater activity on the ice surface coupled with supraglacial melt-out and flowed till deposition may lead to superimposition of typical supraglacial landforms and sediments (Fig. 3.2) on and around the basal ridges (Fig. 2.14).

2.6.1 Paleomagnetic Investigations of Subglacial Deposition

The investigation of magnetic characteristics of diamicts appears to provide a useful genetic tool when considered in context with other sedimentary features and also has several practical applications. Diamicts that are deposited by pressure melt-out at the base of an actively sliding glacier (lodgement tills) generally have coarse silt-sized magnetic particles that are poorly aligned to the present earth´s magnetic field or palaeofield(s) as a result of shear dispersion and distortion during deposition by lodgement. Diamicts formed as glaciomarine or glaciolacustrine muds with ice rafted clasts (Fig. 2.20d) and which are frequently confused with tills (Sect. 2.9) show a high degree of precision of natural remanent magnetism (NRM) that results from the unhindered ability of magnetic particles to align with the earth´s field with regard to azimuth and inclination (Dreimanis, 1976). The orientation of magnetic particles is rapidly measured on a spinner or cryogenic magnetometer of which several types are in use. Samples (small cubes or cylinders) are usually magnetically ´cleaned´ beforehand by alternating field demagnetisation.

Stupavsky and Gravenor (1974) were the first to recognize that remanent magnetization characteristics of certain fine-grained tills in the Lake Ontario and Erie Basins of North America (Fig. 2.21) were inconsistent with shearing and deposition by lodgement processes at an ice base. The suggestion was made that basal deposition occurred when debris was released through a slurry layer below the ice base as a result of subglacial water release (Gravenor, et al, 1973; Stupavsky, et al, 1974). It has been recently argued however that some of the

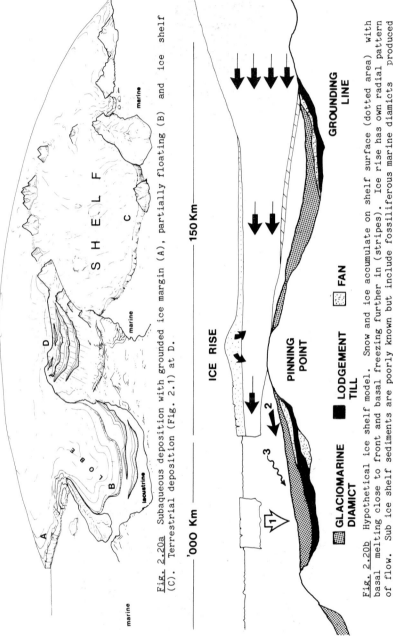

Fig. 2.20a Subaqueous deposition with grounded ice margin (A), partially floating (B) and ice shelf (C). Terrestrial deposition (Fig. 2.1) at D.

Fig. 2.20b Hypothetical ice shelf model. Snow and ice accumulate on shelf surface (dotted area) with basal melting close to front and basal freezing further in (stripes). Ice rise has own radial pattern of flow. Sub ice shelf sediments are poorly known but include fossiliferous marine diamicts produced as in Fig. 2.20d. Lodgement tills and fan sediments are from periods of thicker and more extensive ice cover. Area of marine diamict deposition is very large (Anderson et al, 1980) and offers greatest potential for preservation in the rock record.

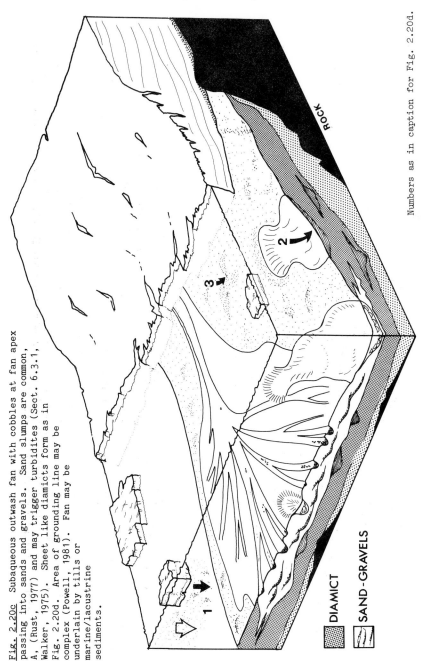

Fig. 2.20c Subaqueous outwash fan with cobbles at fan apex passing into sands and gravels. Sand slumps are common, A, (Rust, 1977) and may trigger turbidites (Sect. 6.3.1, Walker, 1975). Sheet like diamicts form as in Fig. 2.20d. Area of grounding line may be complex (Powell, 1981). Fan may be underlain by tills or marine/lacustrine sediments.

ROCK

DIAMICT

SAND-GRAVELS

Numbers as in caption for Fig. 2.20d.

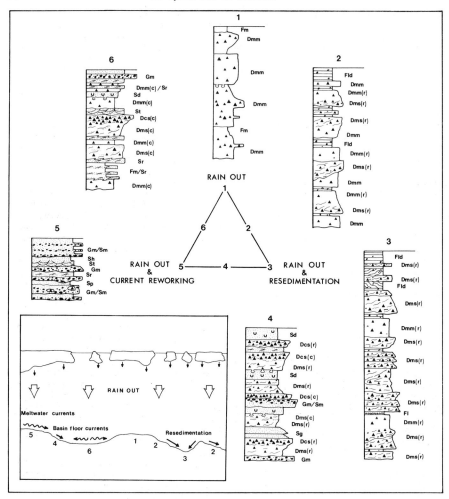

Fig. 2.20d Depositional model for diamict sequences deposited below floating ice of variable type, in glaciated basins subject to substantial influxes of fine-grained suspended sediment (from Eyles and Eyles, 1983). Idealized vertical profiles evolve where the following processes operate either singly or in comination; (1) rain-out of fines from suspension (large arrow) with a variable ice-rafted component (small arrow), (2) downslope resedimentation of diamict and (3) current re-working. Lithofacies coding as in Fig. 1.8. Vertical scale; various. Reproduced by permission of the Geological Society of America.

tills sampled by the above workers were in fact deposited on lake floors as muds with additional coarser material rafted in by floating ice (Eyles, et al, 1983a). Nonetheless the scenario pictured by Stupavsky et al, (1974, 1979) is likely to have occurred at the base of large Pleistocene ice sheets advancing into large lake or marine water bodies (Fig. 2.21). More work is clearly warranted on the use of paleomagnetism as a genetic tool in particular the magnetic characteristics of melt-out tills and lodgement till that has been post-depositionally sheared below the ice base (Fig. 2.12; Sect. 2.5.1). The application of paleomagnetism to problems of drumlin formation and other subglacial landforms would therefore appear to be a valuable research area.

On a more practical note the technique also appears to offer, in proper conjunction with other sedimentological and geotechnical data, a rapid means of discriminating lodgement tills from other diamicts encountered in cores recovered during offshore drilling of marine basins (Sect. 2.9.3) though the effects of coring and drilling need to be ascertained. Measurement of natural remanent magnetism has also been used as a 'way up' criteria for expensive cores whose correct orientation has been lost. Such cases are not well publicized however!

Paleomagnetism offers a relatively rapid way of estimating sedimentary fabric (preferred alignment of clast or matrix particles). The technique of measuring magnetic fabric is referred to as anisotropy of susceptibility. This simply recognizes that in a sample where there is a preferred particle orientation, it is easier to magnetize the particles parallel with their long axis rather than across it (Rees, 1965; Puranen, 1977). Thus the technique requires measurements of susceptibility (the ability of a sample to become magnetized) for different sample orientations. Anisotropy of susceptibility can be rapidly measured for a large number of samples (some field portable equipment is available) thereby overcoming traditional objections to estimation of clast fabric in the field by time consuming, inherently biased and tedious means of a compass. Thus magnetic techniques offer a cheap and statistically valid means of measuring former ice flow directions from lodgement tills and of studying sedimentary fabrics in other diamicts. The rapidity of these techniques is of increasing attraction to industry with regard to mineral exploration projects in glaciated terrains (Sect. 2.8.1).

Bulk magnetic susceptibility has also been employed as a means of 'fingerprinting' till units derived from different source areas (Gravenor and Stupavsky, 1974; see also Sect. 2.8.1). Another use of paleomagnetism is cross-correlation and dating of old glacial sequences in continental interiors with regard to the 'standard' geomagnetic timescale pieced together from deep sea cores and volcanic rocks (Easterbrook and Boellstorff, 1978; Bowen, 1978). For periods up to 10^7 years the earth's field will be of a single polarity and will then reversed for the following polarity epoch. During each epoch, short-lived reversals (events; 10^3 to 10^5 years) and changes in paleofield strength and position (excursions; 10^3 years) occur. In addition, cyclic variation around the mean pole positions (secular variation) also occurs on a timescale of about 1000 to 3000 years (Stupavsky and Gravenor, 1983 for review).

Attempts have been made to determine deposition rates of Pleistocene tills by reference to the numbers of secular variation cycles preserved within a sedimentary sequence (Gravenor, et al, 1973). The identification of geomagnetic reversals and excursions in diamicts or any sediment for that matter, is complicated by sampling techniques and post-depositional sedimentary processes (Verosub, 1977; Eyles, et al, 1983a), and emphasizes the need for very careful sediment description at the time of sampling.

2.7 Subglacial Meltwater Erosion

Major components of the subglacial landsystem are sediments and landforms resulting from subglacial and proglacial (ice marginal) meltwater activity (Figs. 1.3, 1.4, 2.11; Price, 1973; Sugden and John, 1976; Ch. 7). The

importance of bedrock erosion by subglacial streams draining wet-based zones is
stressed by Vivian (1970). Impact fracturing, as large boulders are dislodged,
and abrasion by suspended and bed load transport are significant erosional
processes aided by high hydrostatic pressures. Erosional landforms that result
range in scale from small-scale sculpting and potholing of individual rock
outcrops (Sugden and John, 1976) to large linear trenches (or 'tunnel valleys')
cut across present-day topography and drainage routes (Woodland, 1970,
Sect. 4.7). These trenches may survive deglaciation without being infilled;
hidden buried channels are a major problem and their true character is frequently
only revealed during large excavations for dams and mines (Fig. 2.8, Chap. 14).
For the same reasons that they often provide good shallow aquifers so excavations
through buried channels are frequently beset by drainage problems (Gravenor and
Bayrock, 1956; Norris and White, 1961).

Fig. 2.9 shows a typical proglacial margin exposed by the retreat of an Icelandic
ice lobe, showing fans of outwash cutting across drumlinized lodgement till
plains, with eskers emerging from sub and englacial conduits at the ice front.

2.8 The Subglacial Landsystem: Shields And Sedimentary Lowlands

Broad contrasts can now be established between the subglacial landforms and
sediments of crystalline shield areas and those characteristic of lowland areas
underlain by sedimentary strata (Figs. 9.1, 9.2).

2.8.1 Shield Areas And Drift Exploration

The term 'shield' refers to a subdued scoured bedrock topography consisting of
rock knob lowlands and uplands with poorly integrated drainage, numerous lakes
and little drift cover. The underlying rocks are predominatly crystalline
enclosing troughs of highly metamorphosed sedimentary and volcanic sequences and
are complexly structured. The Canadian Shield, at the core of North America is
an oval area stretching from Greenland to central Canada and extending as far
south as New York State and Wisconsin (Figs. 9.2, 9.8). As a simplification, the
shield zone largely coincides with the extent of wet-based warm thermal
conditions below the Laurentide Ice Sheet (Figs. 2.1, 9.9, Sect. 9.6.1). Similar
remarks as to thermal regime apply to the shield zone in Scandinavia and more
restricted shield-type areas in Britain (Fig. 9.1). In Canada, areal scouring
and abrasion predominated with silt- and clay-sized abrasion products flushed out
and trapped in proglacial lakes forming extensive clay plains following lake
drainage (Fig. 2.21, Ch. 6).

Mosaics of bedrock-drift complexes, 'rippled' till plains composed of Rogen
moraines, lakes, outwash plains, extensive esker systems up to several hundred
kilometres in length and lacustrine clay plains form a recurring shield landscape
(Prest, 1970; Lundquist, 1977; Figs. 1.3, 9.2). On shield areas, there is a
direct economic interest in the way in which glaciers erode and deposit material
aside from the engineering or geotechnical considerations described in
Sect. 13.4.

Drift Exploration

Because shield areas contain complexly structured metamorphic and igneous bedrock
complexes they are major foci for mining activity and mineral exploration.
Direct access to bedrock is prohibited in many areas however by glacial sediment
cover. The term 'drift exploration' refers to the multidisciplinary practice of
locating buried potential ore deposits by the application of glacial geology,
geochemistry, geophysics and drilling technology. A full review of such activity
is out of place here and the following is an introduction to the most recent

literature. Detailed case histories of the successful application of combined
techniques to drift exploration are given by Lee (1971), Jones (1975), Kauranne
(1976) and Kirwan (1978) and a short review of drilling technology is given in
Sect. 11.2.

The starting concept of drift exploration is that as a result of former ice sheet
flow over mineralized bedrock there is a primary dispersal fan, frequently
widening in the direction of ice flow away from the source, composed of
mineralized bedrock debris (Fig. 1.3b).   The three-dimensional shape of these
fans can be thought of as a smoke plume drifting downwind but also subject to
lateral drift in response both to basal ice processes (Sect. 2.4) and shifts in
ice flow direction. Shilts (1976) emphasized that glaciers disperse mineral
'float' from a point source in the form of a negative exponential curve
(Fig. 2.17). The concentration of minerals in till peaks within a few kilometres
of the source followed by a rapid decline thereafter with a more gradual decrease
with increasing distance. The terms 'head' and 'tail' are applied to this
distribution (Fig. 2.17).   The tail is normally the target of reconnaissance
sampling with more detailed work needed to identify the head.   Examples of
reconnaissance till sampling programs are given by Peacock and Mc.L. Mitchie
(1975), Stigzelius (1977) and Shilts (1980). Bearing in mind the high cost of
geochemical laboratory $_2$work review of the extensive literature suggests that
about 0.5 samples per km^2 are needed for reconnaisance sampling programs whereas
between 5 and 10 samples per km^2 are required for resolving the head of a
dispersal fan.

Dispersal fans close to the head can in many cases be identified by reference to
samples taken from mineralised surface boulders or 'float' (Dreimanis, 1958).
The earliest recorded use of such classical boulder tracing is in Finland in 1740
by Daniel Tilas and in North America by Stephen Reed working in the early
nineteenth century. Mineralized boulders were frequently found by farmers
clearing fields (see also Gunn, 1967); modern day practice is to employ sniffer
dogs (for sulphide ores), geiger counters, scintillometers, magnetometers and
other portable hand-held electromagnetic devices (Hyvarinen, et al, 1973).
Sampling is usually carried out on a systematic grid basis designed to identify
the limits of the dispersal fan or train of erratic boulders and to focus in on
possible source areas at the head of the fan where more detailed drilling work
can be started. A major problem is that surface boulders and bedrock debris in
the surface weathered soil zone (i.e. those sediments most easily sampled by
simple 'dirt-bagging') are prone to weathering and dispersion.   Weathered
products may however be scavenged by clay minerals or secondary iron and
manganese oxides further down the pedogenic soil column and geochemical analyses
are frequently based on these fine-grained scavengers (Shilts, 1975).   However,
anomalous geochemical values of targetted metals (geochemical anomalies are
defined as normal background value plus twice the standard deviation) derived by
analysis of the clay-sized fraction are not always trustworthy and weathered
debris is best avoided by test pits or trenches designed to penetrate through
weathered surface material. The size fraction generally considered most useful
is 0.180 to 0.063 mm (fine sand-very fine sand; Appendix 1) which avoids metals
bound in clays, organic material, carbonates and secondary oxides (Alley and
Slatt, 1976). In general, texture is an important variable in geochemical
studies of glacial sediments (Shilts, 1973). Post depositional 'epigenetic'
geochemical processes affecting primary glacial dispersion of metals such as
occur as a result of circulating groundwater movement, (hydromorphic dispersion)
are complex and are exhaustively reviewed by Bolviken and Gleeson (1979).

Apart from surface boulder tracing the most common situation is that where
dispersal fans radiate from buried bedrock sources and are restricted to sediment
horizons buried within multiple glacial stratigraphies. As a result, mineralised

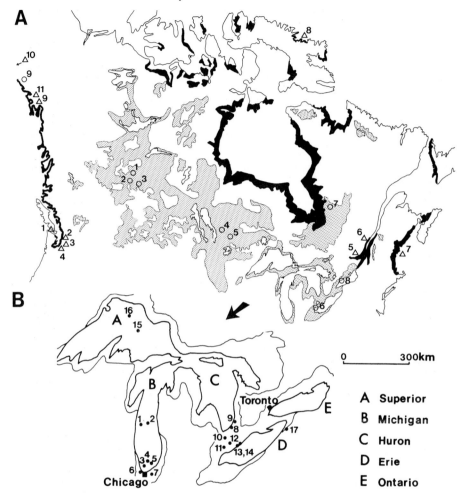

Fig. 2.21   A) Maximum last glaciation extent of glaciolacustrine water
bodies (stipple) and marine overlap (black) in North America (after various
sources). Circles; location of references to glaciolacustrine diamicts formed
by deposition of suspended fines and ice rafting (Fig. 2.20d); 1) Bayrock, 1969;
2) Shaw, 1982; 3) May, 1977; 4) Nielsen, 1982, 5) Teller, 1976; 6) Quigley,
1980, 7) Hillaire-Marcel, et al, 1981; 8) Eyles and Eyles, 1983; 9) Ferrians,
1963, Triangles; location of selected references describing representative
glaciomarine stratigraphies; 1) Hicock et al, 1981 2) Armstrong and Brown, 1954;
Armstrong, 1981 3) Domack, 1982 4) Easterbrook, 1963 5) Rust, 1977; Cheel and
Rust, 1982 6) Gadd, 1971 7) Larson and Stone, 1982 8) Nelson, 1981 9) Miller,
1973; 10) Miller, 1953 11) Powell, 1981.
B) References to 'silty-clay tills' associated with lacustrine facies in the
Great Lakes area and which are likely to be or include glaciolacustrine diamict
facies: 1) Wickham, et al, 1978; 2) Lineback, et al, 1974; 3) Lineback and

debris is not found either at the surface or even in overlying and underlying sediments. The location of mineral deposits buried below barren overburden is a frequently encountered problem (Bradshaw, 1975). Deep sampling of coarse-grained shield sediments was not economically feasible until the late nineteen-fifties with the development of new drilling techniques (Sect. 11.2). Burial of dispersal fans occurs most commonly when outwash or fine-grained glaciolacustrine sediments are deposited over lodgement till thereby sealing the dispersal fan within the subsurface stratigraphy. This overlying cover may be sufficient to prevent subsequent glacial incorporation and secondary dispersal and dilution of erratic material during later glaciations. Secondary pick up of mineralised debris from tills by later ice incursions is a particular problem in shield areas of moderate bedrock relief where ice flow directions are largely a function of ice sheet thickness. The latest ice flow direction determined by (1) the orientation of surface landforms on the ground or on air photography or (2) fabric studies from shallow till exposures or trenches is frequently misleading as to previous flow histories and careful examination of multiple stratigraphies is therefore required in order to identify secondary dispersal of mineralised debris during later glaciations (Alley and Slatt, 1976; Kujansuu, 1976). Flow directions recorded even in a single lodgement till unit can diverge considerably from a mean flow direction (Shilts, 1978; Shakesby, 1979; Eyles, et al, 1982) and, as a result of non-continuous deposition of lodgement till (Fig. 2.11; Sect. 2.5), mineralized zones may occur as discrete lenses with abrupt vertical changes in metal contents through the till sequence (e.g. Shilts, 1975a).

In thick subsurface sequences where direct access to large exposures is impossible the accurate determination of former ice flow histories is extremely difficult. Where good quality core is available sedimentary fabric resulting from ice flow can be identified by reference to the anisotropy of magnetic susceptibility (magnetic fabric; Sect. 2.6.1). If oriented core is available pebble fabrics are possible by direct measurement or by x-ray photography. Core recovery is not however common in most drift exploration projects where reverse circulation drilling is usually employed and which simply returns disaggregated debris to the surface (Fig. 11.3). In this case bulk magnetic susceptibility can be used to determine former ice flow directions (Puranen, 1977; Day and Morris, 1982). The bulk susceptibility of a sample, whether loose particles or a cohesive sediment, is dependent upon the content of magnetic minerals (most commonly silt-sized magnetite in the matrix together with any clast inclusions). If the location of large magnetite-rich bedrock sources is known in the area (for example as an aeromagnetic anomaly) then the changing bulk susceptibility through sediment sequences around the anomaly can be used to derive former ice flow directions. The technique is also very useful, in addition to its use as an index of ice flow directional variability, if the occurrence of magnetite to a targetted metal is understood. In this case bulk susceptibility can be employed as a rapid and inexpensive geochemical indicator of associated targetted metals. Thus for example magnetic minerals are frequently found in association with lead-zinc or sulphide ores. The technique also provides data as to distance of transport from source since increasing distance will result in reduced susceptibility as magnetite contents decrease.

Recent work on ore deposits below barren overburden also emphasizes the value of measuring vapour geochemistry of soil air ´exhaled´ from buried mineralization. This is a long-used and successful technique in the Soviet Union but has only

Gross, 1972; 4) Lineback, et al, 1971; 5) Vitorello and Van der Voo, 1977; 6) Peck and Reed, 1954; 7) Wu, 1958; 8) Broster, 1977; 9) Gravenor and Stupavsky, 1976; 10) Quigley and Ogunbadejo, 1976; 11) Soderman, et al, 1961; Soderman and Kim, 1970; 12) Desaulniers, et al, 1981; 13) Quigley, et al, 1977; 14) Gibbard, 1980; 15) Farrand, 1969; 16) Dell, 1976; 17) Calkin, et al, 1982).

recently been employed in the West (Lovell, et al, 1980). Peat bogs, vegetation and soil humus has also been employed as geochemical indicators of subsurface mineralization (see Bolviken and Gleeson, 1979, for review).

A major, industry-wide problem is one of recognizing subsurface sediment types and stratigraphic relationships from reverse circulation drilling (Fig.11.3). The advantage of this drill type is primarily that it is robust and is capable of sampling both very coarse-grained shield sediments (e.g. bouldery outwash or till) and bedrock (Skinner, 1972; Thompson, 1979). Its use however, effectively prohibits analysis of sequence context as a guide to sediment genesis and has probably retarded the application of sophisticated glacial depositional models in the exploration industry. Identification of sediment type is particularly difficult as only water-flushed debris is returned to the surface. In many cases only the material directly on the bedrock surface is sampled and this is the meaning of the term ´basal till´ as widely used in the industry. The term till is in many cases inappropriate; ´till-like´ diamicts are produced in a wide range of environments (Sect. 1.4, Fig. 2.20, Chs. 3 to 7) but only lodgement and melt-out till have not been reworked by fluvial processes. Other diamicts are compositionally complex and will have secondarily dispersed metal contents as a result of glaciofluvial or hydromorphic dispersion of primary ore fragments (Lee, 1968; Shilts, 1975a, 1976; Shilts and McDonald, 1975). In particular coarse-grained supraglacial diamicts on valley glaciers (Fig. 1.6, Ch. 4) and flowed and supraglacial melt-out tills (Fig. 1.5, 3.2) contain far travelled bedrock debris reflecting the absence of glacial comminution during englacial and supraglacial transport (Fig. 3.1, Boulton, 1970a; Eyles and Rogerson, 1978; Evenson et al, 1979). Thus it is not unusual to find in a thick lodgement till sequence progressively further travelled debris upwards in section associated with an upper capping of stratified melt-out till (Fig. 1.8B) containing low background metal contents in effect the tail of a dispersal fan from a source some distance away (Fig. 2.17).

In shield areas with their knobbly bedrock relief subglacial deposition is strongly controlled by bedrock hummocks in the form of drumlinoid crag and tails and lee-side bedrock-drift complexes (Sect. 2.5.1, Fig. 2.13; Hillefors, 1973). Under thick covers of glaciolacustrine clay or outwash the distribution of glacial sediments that are likely to contain primary dispersal fans can be determined by (1) construction of a bedrock topography map by various means (geophysical, gravimetric and drill data), (2) determination of ice flow directions and (3) target drilling in lee-side areas of bedrock highs thereby aiming for tills deposited in lee-side cavities. These are likely to give an accurate picture of the distribution of mineralized bedrock debris in the former ice base. Lee-side areas may also yield tills preserved from former flow episodes (Alley and Slatt, 1976). Highly overconsolidated melt-out tills pressed on the stoss-side of bedrock highs are also valuable targets but their preservation potential is very low.

In general, genetic identification of diamict type is critical and without it so many drift exploration drilling projects are simply expensive excercises in vertical dirt-bagging where bulk samples are simply reduced to heavy mineral concentrates for geochemical analysis regardless of stratigraphy or sedimentology. The development of cheaper drilling technologies giving continuous core recovery (e.g. resonant drilling; Sect. 11.2; Fig. 11.3c) is crucial as access to good quality core material significantly enhances the chances of finding mineralized sediments and allows such mineralization to be placed in a more sophisticated sedimentological and stratigraphic context. Detailed integrated studies using this approach have barely begun. However it is likely that, given the correct economic and research climate and allowing for proprietary restrictions, such studies coupled with recent advances in airborne

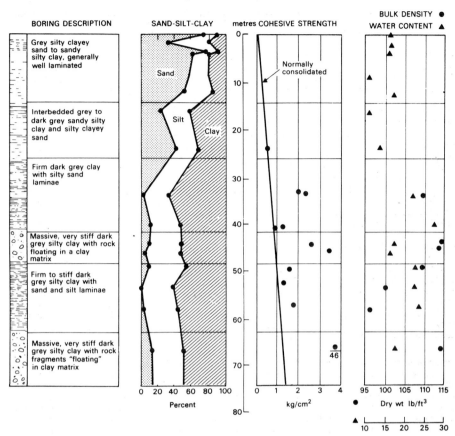

Fig. 2.22 Geotechnical profiles through North Sea Quaternary sediments lying in 130m of water. (Modified from Edwards, 1978, see also Fig. 11.7A). The overconsolidation and high bulk density of the massive diamicts between 40 and 50m depth and below 60m suggests that they may be lodgement tills but note the normal consolidation of adjacent marine silty clays. For a detailed genetic study the nature of contacts between each sediment type would be useful data (e.g. Fig. 1.8) and can be derived by very careful core logging.

and field portable instrumentation will significantly enhance the success rate of drift exploration in shield areas.

2.8.2 Sedimentary Lowlands

Sedimentary lowlands on the margins of shield zones in North America (Fig. 9.2) and Europe (Fig. 9.1) are underlain by bedrock successions of variable competency ranging from mudstones to limestones (Figs. 9.6, 9.7). The presence of incompetent units of low permeability is recognised as a critical factor in the evolution of bedrock slopes (Section 5.4.4) and these successions have often been severely deformed by ice sheets (glacitectonism). It is possible to recognize, in general, a deformable glacier bed in contrast to the rigid beds offered by hard competent shield rocks (Boulton and Jones, 1979). Lodgement till deposited on soft sedimentary rocks may overlie a thin complexly deformed and brecciated layer of rock that shows evidence of limited ice transport. This has been referred to in the literature as 'deformation till' or 'local till' (Dreimanis, 1976) and may be associated with extensive glacitectonised rock zones (Figs. 2.18, 2.19).

In a selected review and classification of large-scale glacitectonic structures, Banham (1975) stressed the association of deformed bedrock strata with distinct topographic settings that result in severe compression of the ice margin. Compressive flow is also a prerequisite for development of the supraglacial landsystem (Figs. 2.18, 3.2) and in many areas the two are intimately associated (Moran, 1971; Sauer, 1978) in the form of moraine complexes draped over thrust bedrock (Fig. 1.5). The extent of glacier thrusting and the size of typical areas involved needs emphasis. Individual thrust blocks may be 200 m thick and cover areas of 13 km^2 or more and, given limited subsurface outcrop in lowland areas, the recognition of repeated or reversed geological sequences in borehole logs have been crucial in identifying thrust terrain (Christiansen, 1971; Christiansen and Whitaker, 1976). The importance of these thrusted, often fragmented masses, lies in the adoption of suitable shear strength characteristics for cut slope design in large open pit mines, disturbance of hydrogeological routes and slope failures where transport routes or pipelines cross postglacial rivers cut in thrust terrain (Sauer, 1977, 1978).

Sites susceptible to glacitectonic disturbance include valleysides, where ice lobes moved down or across deeply incised stream channels resulting in tectonized interfluves (Fig. 2.19) and scarp faces, islands and peninsulas that lay across the direction of ice flow. Tensional 'pull-apart' rifting is also reported in the lee of upland margins facing down the direction of ice flow (Banham, 1975).

As shear strengths of all but the weakest rocks exceeds by several times the available shear stresses that glaciers can normally exert on their beds (c. 25 - 150 kN/m^2) for deformation to occur rock strengths must be reduced to low values. The elevation of porewater pressures is crucial and the development of high hydrostatic uplift forces in aquifers overpressured by wet-based glaciers is critical (Clayton and Moran, 1974; Moran et al, 1980). Other glacitectonic mechanisms involve decoupling of permafrost layers from the unfrozen substrate, the decapitation of bed highs frozen into the ice base (Figs. 2.1, 2.2) and shear through soft fine-grained sediments or rocks (Brodzikowski and Van Loon, 1980; Schwan and Van Loon, 1981).

Ground conditions in thrust terrain range from stacked, overthrust slices of bedrock with a hummocky surface relief and frequent lakes. The large size of these thrust blocks has already been commented on. On the other hand, 'rubble terrain' consists of smaller fragmented rafts ripped up and transported several kilometres down the flow line. These blocks decrease in size away from source areas, reflection basal comminution, and are intimately associated with basal melt-out tills as a result of the in situ stagnation of the debris-charged glacier. Rubble terrain is therefore commonly associated with the supraglacial landsystem.

The importance of a frozen outer ice margin (Fig. 2.1) in blocking subglacial drainage along buried aquifers, thereby elevating subglacial porewater pressures, is stressed by Moran (1971) and Moran et al (1980) in reviews of glacier thrust terrain in North America. Thrust blocks may be streamlined and covered with a veneer of lodgement till deposited by wet based ice a few kilometres upglacier of the advancing ice front. The limits of glacier advances are commonly marked by unstreamlined thrust blocks often with source sites occupied by lakes. With regard to the classic glacitectonized landscapes of Holland, North Germany and Poland younger thrust slices and ridges commonly overlie ridges indicating multiple phases of deformation in successive glaciations (Maarleveld, 1981) but this has not so far been reported in North America.

2.9 Glacial Deposition in Water

The subaqueous glacial environment (marine or lacustrine) ranges from sub ice-shelf locations, to high energy areas close to partially floating or grounded ice margins, to more extensive distal localities of predominantly marine or lacustrine deposition with background ice-rafting of coarser debris (Fig. 2.20). A large number of subenvironments exist reflected in a substantial but mostly descriptive literature of Quaternary and Pre-Quaternary sequences (Andrews and Matsch, 1983). Few detailed depositional models based on observations of modern processes have been published that can be widely used for stratigraphic interpretation or for applied projects. As industrial development of formerly glaciated marine shelves and lake basins accelerates in response to urban, military and energy pressures so the development of subaqueous glacial depositional models is of great significance.

On low relief, isostatically-depressed continental surfaces and coasts in North America and Europe, ice sheets advanced and retreated in association with large frontal lakes and marine embayments (Figs. 2.21, 9.2) in which ice margins were either fully or partially afloat. Whilst space prohibits a detailed review it is in these environments that there is greatest opportunity for the deposition of diamicts commonly labelled as massive or stratified ´waterlaid tills´ (Dreimanis, 1976, 1979) but which were not deposited directly at a grounded ice base and which as a result exhibit anomalous and sometimes extreme ranges in geotechnical properties. The sedimentary sequences and topographic settings in which these diamicts occur are distinctive however and it is by the analysis of associated sediments that their origin can be identified. Sedimentologists refer to this either as sedimentary association or sequence context (Sect. 1.4).

An important corollary is that the glacial sediments most widely encountered in the Pre-Quaternary rock record are those of glaciomarine and, to a lesser extent, glaciolacustrine origin (Hambrey and Harland, 1981) given the greater preservation potential of sediments deposited in continental margin basins (Sect. 9.1). Terrestrial glacial sediments have a low preservation potential whereas the ocean area affected by ice-rafting around Antarctica is many times greater than that of the glaciated area on that continent (e.g. Fig. 2.20b). Consideration of Fig. 1.1 also suggests that nearly 40% of the area of the world´s continental shelves have been affected by Quaternary glaciations (Fig. 9.3) so the ability to distinguish terrestrial from marine glacial sediments is important not only for the accurate reconstruction of geologic history but for a variety of applied projects along continental margins and in offshore areas.

2.9.1 Subaqueous Depositional Processes

Subaqueous environments close to floating or partially grounded ice margins are complex and the application of Walthers Law (which states that a continuous vertical sequence of sediments records the lateral succession and overlap of former depositional environments) allows the conclusion that simple packages of lodgement till sitting within marine or lacustrine sediments are improbable. Apart from the fact that the prerequisite need for a stiff substrate on which lodgement may operate is unlikely to be satisfied on soft subaqueous sediments (Fig. 2.3, Sect. 2.5), the entry of glacier ice into water produces a wide range of subenvironments and therefore complex depositional sequences.

Recent descriptions of sedimentation adjacent to or below Antarctic ice shelves are available (Orheim and Elverhoi, 1981; Drewry and Cooper, 1981; Domack, 1982a) whilst Powell (1981) has described several tidewater ice margins in Alaska. In general the possible combinations of water depths, icefront characteristics thermal regime, retreat rates and meltwater discharges is very wide and attention has focussed on ice-contact deposition at grounding lines or ice fronts (proximal glaciomarine) and the nature of deposition further away from the ice margin (distal glaciomarine).

Proximal subaqueous environments are characterised by the rapid buildup of chaotic morainal banks and ridges containing subaqueous outwash, reworked basal debris and dropped supraglacial and englacial debris. Substantial moraine ridges up to several hundreds of kilometres in length have been formed at the grounding line of large floating ice margins (Hillaire-Marcel et al, 1981) whilst smaller moraine ridges are generated as ice frontal ramps hinge upwards and float off (Barnett and Holdsworth, 1974). Ridges formed in this fashion below floating ice are variably referred to as De Geer ridges, cross-valley moraines and washboard moraine (Sugden and John, 1976). Resedimentation and mass movement processes are very important downslope agents of transport in the subaqueous ice-contact environment and are associated with the triggering of turbidites (Sect. 6.3.1, Fig. 6.7) producing laminated silty clays in deeper water (Fig. 6.8).

A sedimentary context in which such downslope mass movement occurs is frequently that of subaqueous outwash fans or cones at the exit points of sediment laden subglacial meltstreams (Fig. 2.20c). The fans may have apices above water level analogous to subaerial braded stream fans (Fig. 2.9, 7.1) and the fan shape may also be closely controlled by inlets in the ice front. Whereas esker-like ridges composed of superimposed outwash cones are reported (Rust and Romanelli, 1975; Cheel and Rust, 1982), more extensive sheet-like sedimentary bodies, several kilometres wide result where fans are closely spaced and mutually overlap or where a broad fan surface migrates laterally. Sediments range from coarse cobble sizes, in feeder channels close to the former fan apex (and may be associated with glacitectonic deformation and subaqueous moraine ridges), to sheet-like or channelled massive sand bodies recording distal stream deposition and the importance of gravitational slumping (Rust, 1977).

Diamicts found within these fans are of several types - those produced by resedimentation and others formed in situ in quiet water areas as a result of suspended sediment deposition and ice rafting. Mass movement processes are particularly common in fine-grained and deltaic sediments as a result of rapid deposition and oversteepening of slopes, tectonic adjustments, floods and iceberg calving (Middleton and Hampton, 1976; Coleman and Garrison, 1977, Lowe, 1979). Mixing of fine and coarse-grained deltaic sediment by flow down a delta front for example produces crudely stratified diamicts easily mistaken for subaqueous ´flowed tills´ (Fig. 1.7) supposedly released directly from an ice margin (Powell, 1983).

Elsewhere in more distal glaciomarine and glaciolacustrine environments diamicts formed by suspended sediment deposition and ice-rafting occur as drape-like bodies infilling lows in the basin floor, and are subsequently buried as active fan or delta lobes build out and migrate over areas of formerly quiet water. Repeated switching of active delta lobes into interdistributary areas of quiet water suspension deposition may result in alternations of glaciolacustrine diamicts and sands (Fig. 2.20c).

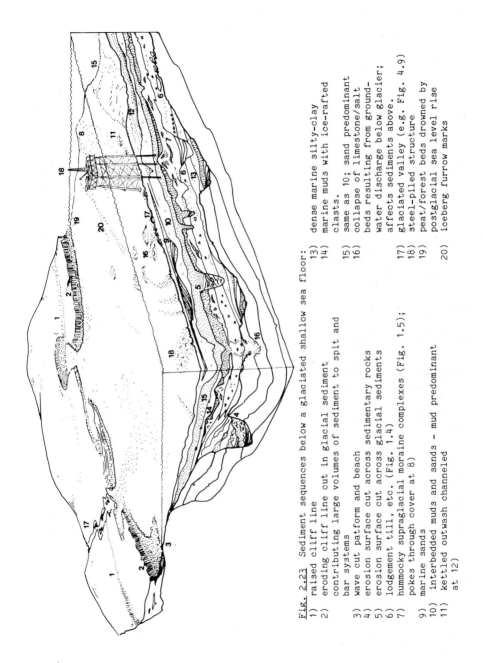

Fig. 2.23   Sediment sequences below a glaciated shallow sea floor:

1) raised cliff line
2) eroding cliff line cut in glacial sediment
   contributing large volumes of sediment to spit and
   bar systems
3) wave cut patform and beach
4) erosion surface cut across sedimentary rocks
5) erosion surface cut across glacial sediments
6) lodgement till, etc. (Fig. 1.4)
7) hummocky supraglacial moraine complexes (Fig. 1.5);
   pokes through cover at 8)
9) marine sands
10) interbedded muds and sands – mud predominant
11) kettled outwash channeled
    at 12)

13) dense marine silty-clay
14) marine muds with ice-rafted
    clasts.
15) same as 10; sand predominant
16) collapse of limestone/salt
    beds resulting from ground-
    water discharge below glacier;
    affects sediments above.
17) glaciated valley (e.g. Fig. 4.9)
18) steel-piled structure
19) peat/forest beds drowned by
    postglacial sea level rise
20) iceberg furrow marks

It is possible to make a generalised model for diamict deposition in either the distal glaciomarine or glaciolacustrine environment (Fig. 2.20d) by reference to the varying relationship at any one depositional site between (1) ice rafting and deposition of suspended fines (rain-out), (2) current activity and (3) downslope mass movement. Interaction between these three processes results in a wide range of possible diamict facies and associated sediments. In areas of quiet water and low bottom relief massive fine-grained diamicts accumulate with textural variation resulting from changes in suspended sediments or in the volume of ice-rafted debris (Fig. 2.20d). In quiet water areas adjacent to floating ice fronts, rates of suspended sediment deposition may be so high so as to produce largely clast free muds (Orheim and Elverhoi, 1981; Drewry and Cooper, 1981). In marine environments diamicts commonly contain in situ molluscs and foraminifera (Boulton, et al, 1982; Domack, 1982a).

In other slope-dominated environments, downslope resedimentation of diamict produces stratified diamicts interbedded with laminated silty-clays (turbidites) reflecting the genetic control on turbidite deposition by sudden slumping (Sect. 6.3). Elsewhere in areas of variable current activity, perhaps close to deltas or fans, suspended sediment deposition is reduced and sandy diamicts are deposited as part of a continuum of sediments from sands to pebbly sands and stratified and massive diamicts with complex gradational contacts between them (Eyles and Eyles, 1983a). In distal areas several tens to hundreds of kilometres away from ice margins 'residual' stratified diamict facies (Anderson, et al, 1980) record continued background ice rafting, and the interrupted deposition of suspended fines by bottom currents. Deposition rates may be extremely low (e.g. Clark et al, 1980).

Particularly useful diagnostic criteria for recognition of glaciomarine and glaciolacustrine diamicts deposited by the 'triad' of rain-out, current activity and resedimentation are absence of evidence for direct glacier traction over the site, frequent soft sediment-deformation of the diamicts (indicating low strengths at time of deposition), an intimate association with 'normal' marine or lacustrine sediments an absence of subglacial landforms, and distinct patterns of natural remanent magnetism compared to lodgement tills (Sect. 2.6.1). Glaciomarine diamicts also appear to possess a distinct geochemistry compared with terrestrial tills (Frakes and Crowell, 1975).

2.9.2 The Engineering Significance Of Discriminating Waterlaid Diamicts From Tills.

Complexes of fine-grained diamicts, containing up to 80% silt and clay, and interbedded with lacustrine or marine sediments are found across a broad area of North America that saw either the formation of enlarged lakes or experienced marine overlap during Late Pleistocene glaciations (Fig. 2.21). These fine-grained diamicts (referred to widely as silty-clay tills) have generated much discussion by virtue of a variable state of consolidation and high void ratio which are not consistent with their supposed origin as subglacially deposited tills (Easterbrook, 1964; Milligan, 1976). Recent work shows that they are of glaciomarine or glaciolacustrine origin in the manner identified in Fig. 2.20d (Domack, 1982a; Quigley, 1980; Eyles and Eyles, 1983). These units sit characteristically within deltaic or subaqueous outwash sequences. The large frontal lakes and marine embayments that developed during the entry and retreat of Quaternary ice sheet margins in continental North America (Fig. 2.21), acted as efficient sediment traps and inhibited direct glacial deposition. Within these basins, floating ice (whether ice shelves, bergs or seasonal floe ice) introduced clasts into bottom muds and clays resulting in a poorly-sorted till-like material with high mud fractions (Fig. 1.8F).

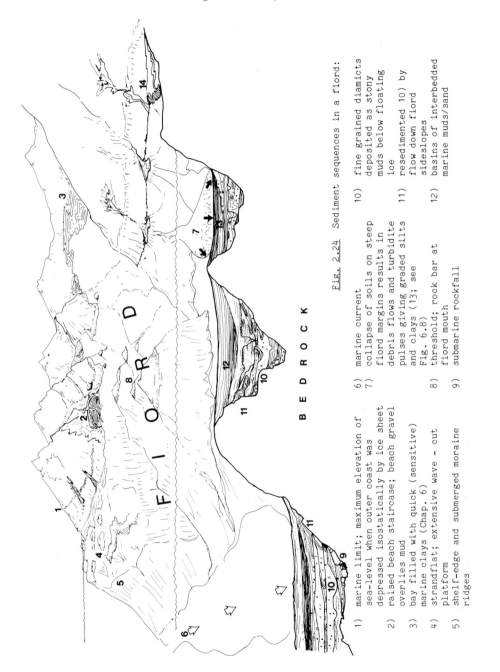

Fig. 2.24 Sediment sequences in a fiord:

1) marine limit; maximum elevation of
   sea-level when outer coast was
   depressed isostatically by ice sheet
2) raised beach staircase; beach gravel
   overlies mud
3) bay filled with quick (sensitive)
   marine clays (Chap. 6)
4) strandflat; extensive wave - cut
   platform
5) shelf-edge and submerged moraine
   ridges

6) marine current
7) collapse of soils on steep
   fiord margins results in
   debris flows and turbidite
   pulses giving graded silts
   and clays (13; see
   Fig. 6.8)
8) threshold; rock bar at
   fiord mouth
9) submarine rockfall

10) fine grained diamicts
    deposited as stony
    muds below floating
    ice
11) resedimented 10) by
    flow down fiord
    sideslopes
12) basins of interbedded
    marine muds/sand

It is difficult at present to isolate the engineering properties and problems of such waterlaid diamicts as hitherto they have been described as ´tills´ in site-investigation reports and thus hidden from more critical analysis. However, recent industry discussion focusses on the problem of ´soft-zones´ recognized increasingly within silty clay tills in the areas shown on Fig. 2.21. This concern stems from the traditional description of any diamict as a basal till and the belief that this general category of sediments provided homogenous problem-free foundation materials. Soft zones exist precisely because the diamicts are not tills (Sect. 1.4) and can be generally predicted by more critical sedimentological description of the sedimentary sequences associated landforms and general depositional setting under investigation. For example, tunnelling for a light rapid transit line in Edmonton in an area of former glacial lakes (Fig. 6.6) has presented longer than usual exposure of substrate units previously considered to be two superimposed lodgement tills. Soft zones, identified by anomalously low SPT blow counts (Sect. 11.6.1, Table 12.3, Sect. 12.2) and high fracture permeability, ascribed to glacial shearing at the upper boundary of the lower lodgement till, were subsequently shown to be thicker and more widespread than previously identified by normal borehole numbers and arrays. These soft zones present the danger of differential settlements under high loads and require reevaluation of existing use of spread footing foundations and changes to either end-bearing piles resting on competent sub-till sediments, or rafts (Ch. 12, Thompson, et al, 1980). The vertical association of laminated lacustrine clays, silts and sands with irregular beds of diamict suggests deposition of melt-out and flowed tills in a shallow lake body coupled with deposition of glaciolacustrine diamicts as stoney lake muds (May and Thompson, 1978). These units are variably over-consolidated by groundwater lowering (May, 1977) and are well-jointed. The latter, coupled with frequent interfaces between different sediment types results in tunnel roof failures and water entry into workings (Matheson, 1970). Variable consolidation in these diamicts is also achieved by carbonate cementation, suction stress as the muds dewater after lake lowering and/or extinction, and erosion of overburden. Lateral and vertical variability, in particular the existence of ´hard´ and ´soft´ zones can often be explained by reference to internal sand drainage blankets and tree roots (Soderman and Kim, 1970; Quigley and Ogunbadejo, 1976; Table 12.3) and the presence of laminated silty clays (Fig. 2.20d). A stiff surface crust up to 10 m deep overlying relatively low shear strength material is frequently encountered.

Typical topographic settings in which such notorious silty-clay ´tills´ are found are low relief lake or marine plains with no evidence of streamlined subglacial landforms (drumlins, flutes, etc.). Older glacial landforms may however protrude through the overlying glaciolacustrine sediments. A variable cover of beach sediments and blown sand is normally present. Subsurface ground conditions typically show strong vertical associations of muddy diamicts with laminated (´varved´) silty clays with or without dropstones (e.g. Fig. 1.8F) or marine silts and mud, with mollusc faunas, within an overall layer-cake stratigraphy emphasizing that the sequences are the preserved sedimentary infillings of old lake or marine basins. Their layer cake stratigraphy contrasts with the more complex stratigraphies developed where an ice margin fronts in a lake and is involved in the depositional process (Evenson et al, 1977; Gibbard, 1980).

A further example shows the importance of careful examination of the structure and stratigraphy of ´massive´ silty-clay diamicts below lake plains. These diamicts have attracted attention as possible sites for toxic waste storage on the basis of their very low permeability (Sect. 15.5). A recent commercial investigation report for one site in Ontario, Canada describes ´glacial till´, below glaciolacustrine clays, as a grey silty-clay with variable pebbles, massive, compact to dense and plastic. Closer examination of the ´till´ shows an S.P.T. resistance (Sect. 11.6.1) of less than 12 blows/ft., lower in fact than for the overlying clays that have been over-consolidated by desiccation ($\simeq 17$ blows/ft.). Furthermore, detailed examination of borehole samples shows that diamict units frequently show thin sand 'stringers' and lamellae that in some cases can be correlated from one borehole to another over many tens of metres, suggesting possible hydraulic continuity, despite the conclusions of the final site investigation report which emphasized the impermeability of the 'till' substrate and the acceptability of the site for toxic waste storage. In this case the so-called 'tills' were deposited as glaciolacustrine diamicts, with the sands deposited by episodic current activity on the lake floor derived from nearby deltas. The presence of hydraulically integrated horizons within the stratigraphy would clearly be a major disadvantage to any toxic waste scheme and indicates the need for a thorough sedimentological approach to the problem of assessing silty-clay 'tills' for such schemes.

On a related theme, Donovan (1978) has emphasized the role of sand units, deposited as delta lobes and later buried by marine clays (Champlain Sea Clays: Sect. 6.2) in determining sites of large retrogressive slope failures in raised marine sequences in eastern Canada. Transient high porewater pressures developed in the sands during snow melt trigger valleyside flow-slides in the soft clays.

2.9.3 Future Prospects

Glaciated mid and high latitude marine shelves are sites of frontier offshore development in connection with hydrocarbon exploration and production. The description, sampling and testing of thick glacial diamict sequences below sea floors (Figs. 2.22, 2.23, 2.24) offers a challenge to traditional geological description and genetic interpretation of diamict sequences and associated sediments. Smith (1975) has highlighted the importance of geophysical methods to overcome difficulties in traditional geotechnical sampling. Use of the electric cone penetrometer (Sect. 11.6.2, Fig. 11.7) is increasing as considerable problems are associated with recovery of undisturbed samples in deep water as a result of gas and porewater expansion. At the same time many uncertainties surround in situ sea floor geotechnical testing given variable test precision and data interpretation. There is increasing interest in testing of reconstituted sediment samples under simulated depositional and post-depositional histories for the design of large oil and gas platform foundations (Focht and Kraft, 1977; Hight, et al, 1979; Carneiro, et al, 1982). Substantial data from drilling programmes demonstrate the importance of subaqueous glacial deposition on many shelves (Nelson et al, 1974; Holmes, 1977; Thompson and Eden, 1977; Eden, et al, 1977; Boulton, et al, 1982) and a good case can be made for integrated geotechnical and sedimentological studies of raised glaciomarine sediments in order to more closely define subsurface offshore ground conditions.

The area of former glaciomarine conditions in North American can be delimited with some precision (Fig. 2.21). It is difficult to be as firm with regard to the British Isles even though high-level 'shelly drifts' are widespread (e.g. Mitchell et al, 1972). To date these have been commonly explained by reference to glacial incorporation and shorewards rafting of marine sediments by ice sheets. It is argued here that particularly along the East Anglian coast, in Northeast Scotland and the coasts around the Irish Sea Basin the importance of glaciomarine deposition has been underestimated and that in situ glaciomarine sequences have been as confused with tills deposited by grounded ice sheets (Eyles and Eyles, 1983a). Further research around the British coasts, drawing on the broader glaciomarine literature and data that is available from offshore projects will not only result in a more detailed understanding of former glacial environments but offers a means of correlating terrestrial sequences with the thicker offshore Quaternary and provides a firm basis for engineering purposes.

Figure 2.23 summarizes typical ground conditions found in shallow offshore marine basins where water depths are less than 150 m and complex buried geomorphologies result from repeated incursions of grounded and partially floating ice (Fig. 2.20). Deep water glaciomarine sequences are not well known apart from shallow cores and sporadic drilling; Fig. 2.24 should therefore be regarded as a hypothetical model for further testing.

CHAPTER 3

The Supraglacial Landsystem

M. A. Paul

INTRODUCTION

The highly characteristic supraglacial landsystem (Fig. 1.5) develops when sediment is deposited on the surface of a glacier. This sediment is for the most part subglacial in origin, and is exposed on the glacier surface as a result of the pattern of ice flow. As a consequence of this supraglacial deposition, bodies of glacier ice (dead ice) become buried within the sedimentary sequence. The final irregular pattern of landforms and sediments is the result of deposition above and between dead ice bodies that subsequently melt-down and disturb overlying strata.

The engineer who is carrying out a project within supraglacial terrains will wish to know the likely surface and subsurface distribution of sediments, so that not only may the best site or route be chosen, but also so that suitable sources of fill or aggregate may be found and exploited. Difficulties arise particularly at the site investigation stage, when it is required to assemble an overall picture of the three dimensional layout of the sediments from a number of sources such as trial pits, borings and surface exposures. This stage is often rendered extremely difficult by the variable nature of glacial sediments; supraglacial deposits are especially difficult in this respect, and so it is a great advantage to have a conceptual model to guide interpretation. The landform and sediment model that is described here also provides an understanding of the geotechnical properties that the sediments possess and the ways in which they vary.

3.1 Previous Research

The recognition of the supraglacial landsystem has occurred only relatively recently. It was previously thought that these deposits were entirely subglacial, despite the complex glacial histories that were needed to interpret them. Hartshorn (1952; 1958), Gravenor and Kupsch (1959), Kaye (1960) and Clayton (1967) used the supraglacial model in Pleistocene terrains whilst Boulton (1972) dealt with the occurrence and sedimentology of supraglacial deposits in Spitsbergen, highlighting applications to Pleistocene sequences. Later, systematic accounts of the geotechnical aspects of the landsystem were provided by Boulton and Paul (1976). In recent years several sedimentological and geomorphological studies on Pleistocene deposits have invoked the supraglacial model (e.g. Clayton and Moran, 1974; Marcussen, 1977; Boulton, 1977; Eyles and Slatt, 1977). The recognition of a supraglacial origin for glacial sequences often has repercussions for the local and regional stratigraphy of an area (see below) since concepts such as type sections, upper tills, middle sands, etc. cease to be valid basis for either stratigraphic correlation or as a framework for geotechnical sampling and testing.

71

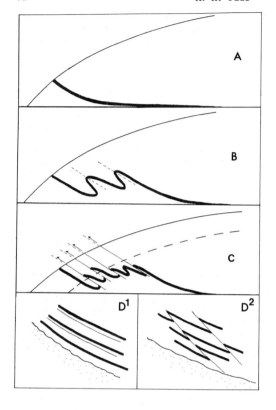

Fig. 3.1. Development of a multiple debris band sequence. (A) Idealised path taken by the basal debris band during compressive flow. (B) Folds form on the debris band, which are drawn out by flow into tight structures whose axes lie parallel to flow lines. (C) Thrust faults dissect the folds along lines parallel to flow. Ablation of the glacier to dashed line leads to a typical structural sequence at the glacier terminus. In plan, the thrust faults may lie parallel to the strike of the debris bands (D_1) or oblique to them (D_2). Ice melt, under a cover of till derived from the debris, leads to a topography of ridges and troughs (Fig. 3.2).

3.2 Processes Of Supraglacial Deposition

Basal debris is exposed on the ice surface as a result of compression of the ice front (Fig. 3.1). This behaviour is characteristic of glaciers of freezing or frozen thermal regime where thinning of the basal debris layer in response to basal melting is either slow or non-existent (Fig. 2.1). Once englacial debris is exposed on the ice surface, it is released by melting, and a series of debris flows form at the points of outcrop and spread over the ice surface, so producing a cover over much of the glacier snout. Melting continues beneath this cover (at a reduced rate) and its thickness is increased both by the continued release of debris at the lower surface and by the superposition of fresh flows from higher up the glacier. Debris that flows into position is termed flowed till and those that are released at depth from melting ice, and which does not subsequently move, is termed basal melt-out till (Figs. 1.7, 1.8). The criterion of movement is an important one, since it is believed that the structure and properties of the sediment are altered during the flow process. In some cases the sediment is variably disaggregated during flow. There is general agreement that such materials are not true tills (Sect. 1.4; Lawson, 1979, 1981, 1982, Eyles et al, 1983) but a specific name for them has not yet been established. In this chapter the term flowed till (Fig. 1.7) is used throughout, recognizing however the wide range of diamicts that result from processes of resedimentation by gravity sliding and flow (Fig. 1.8; Lowe, 1979).

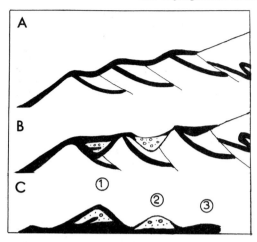

Fig. 3.2 The development of the supraglacial landsystem by relief inversion. (A) shows the initial system of ridges and troughs in which outwash and flowed till accumulate (B). After deglaciation (C) the topography is 'inverted', with the formation of ridges from the trough fillings. Some ridges (1) contain both outwash and flowed till; others (2) contain mostly outwash, and others (3) mostly flowed tills. (See Figure 3.6 for details of the sequences).

The local thickness of the till cover depends on the irregularities in the underlying ice. Since the rate of flow of atmospheric heat through the till depends on its thickness, the rate of melting of the buried ice surface will vary from place to place, being greatest beneath those areas of least till thickness. This differential melting leads to the production of an irregular surface topography of ice-cored ridges separated by ice-floored troughs and hollows (Fig. 3.2). It is in these troughs and hollows that subsequent sedimentation occurs, and so the initial arrangement of the dead ice is central to the overall distribution of sediments and the final landforms.

3.3 Local Patterns Of Landform Distribution

Deposition in the supraglacial environment leads eventually to one of two landform patterns. These are termed the hummocky supraglacial assemblage and the supraglacial till-plain assemblage. The production of either the hummocky or the till plain assemblage depends on the interplay of two factors. The first concerns the character of the sediment that covers and fills the ice-cored topography; the second concerns the distribution of englacial debris bands within the ice front.

The type of sediment that fills the ice-cored topography determines the extent of redistribution during the melting of the ice-cores. If only limited redistribution occurs, the trough fillings will remain upstanding as distinct ridges after deglaciation, whereas if the fillings collapse when their lateral support is removed, the resulting topography is more uniform. It is evident that the one case leads to the production of the hummocky assemblage, whereas the other leads to the production of a till plain. As a result the variety of landforms that results from supraglacial deposition is very wide ranging from the extremes of very hummocky terrain underlain by thick supraglacially deposited sequences and basal melt-out tills to thin flowed till plains through which streamlined elements of the underlying subglacial surface will protrude (Fig. 1.5). Correspondingly a diverse morphological terminology has evolved in areas of supraglacial terrain (Gravenor et al, 1960).

The distribution of englacial debris bands outcropping along the ice front controls the distribution of ice-cores. Thus, when the bands crop out in a

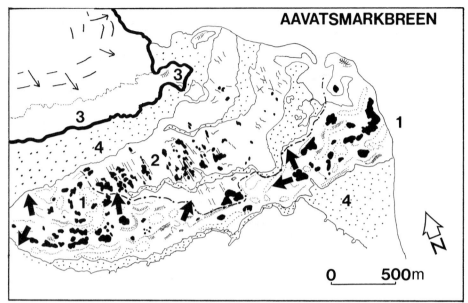

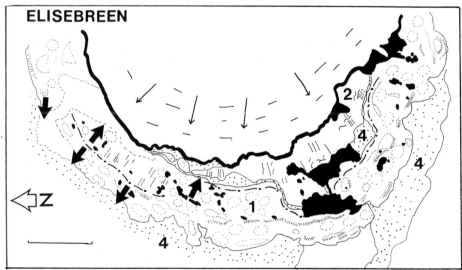

Fig. 3.3. Landforms and sediments at the margins of two glaciers in northwest
Spitsbergen.   Supraglacially deposited flowed tills (broad arrows) and ice-cored
ridges and troughs (1) overlie and obscure fluted lodgement till (2).   Flow
directions at the ice margin (3) are shown by slim arrows. Outwash deposition
(4) is limited to areas between controlled ice-cored ridges. The scale bar in
all cases is 500 m.

regular pattern, the ice-cored ridges that form also have a regular pattern, and
vice versa. The regularity or otherwise reflects the tectonic structure that was
acquired during the compressive flow that brought the debris to the surface.
Figure 3.1 shows two possible arrangements. In arrangement (D_1) the bands lie on
limbs of tight, isoclinal folds whose axes strike parallel to the glacier margin.
Overthrust strike faulting has occurred also, both processes leading to
repetition of the stratigraphic sequence in the direction of ice flow. Surface
ablation then reveals the debris bands as a series of outcrops that strike
parallel to the ice margin. The carpet of flowed till that forms from these
outcrops thus lies as a series of stripes that are transverse to ice movement,
and are relatively continuous. The ice-cored ridges that form are also
continuous in this direction, perhaps with occasional breaches. Such a
situation, in which the positive topographic elements are parallel to the local
ice margin, has been termed ´controlled´ (Gravenor and Kupsch, 1959).

In arrangement (D_2) the debris bands have been folded isoclinally as before, but
the thrust faults lie oblique to the strike of the outcrops and so the resultant
ice-cored ridges are discontinuous. The term ´uncontrolled´ has been used for
this irregular ridge pattern. ´In principle both the hummocky and the till plain
assemblages can be controlled or uncontrolled: in practice, the controlled
pattern is usually much more evident in the case of hummocky topography.

3.4 Regional Patterns of Landform Distribution

When considered on a larger scale, areas of supraglacial landforms have a
consistent relationship to other landform assemblages. Compressive flow is
required for supraglacial deposition to occur and commonly results when the ice
decelerates, as may result from meeting a large bedrock obstruction such as an
escarpment, compression between other ice lobes (forming an interlobate moraine
complex) or at the outer limits of glaciation when the ice margin is frozen, very
slow moving and dams more active ice upglacier. The passage of kinematic waves
downglacier (Sect. 4.2.1) in response to changes in mass balance may also be
important. The supraglacial landsystem is commonly developed on the flanks of
bedrock obstructions (Fig. 2.18) while it most frequently occurs as an outer
fringe at the limit of a glacial advance.

In upglacier areas where compressive flow is less severe and extending flow
occurs, debris is carried near to the glacier bed and will be released by
subglacial processes. In these areas the supraglacial landsystem will be absent,
and the subglacial landsystem (Fig. 1.4) will be revealed when the ice withdraws.
There is thus a standard pattern developed, in which the supraglacial landsystem
is found at the outer limits of the subglacial landsystem. This pattern can be
recognised around present day glaciers and also in areas of Pleistocene glacial
terrain (Fig. 9.2). Figure 3.3 shows landforms at the margins of two glaciers in
Spitsbergen. At Aavatsmarkbreen there is an outer belt of large supraglacially
formed moraine ridges, which marks the limit of the most recent glacial advance.
Inside this belt there is streamlined topography of subglacial origin formed
partly in lodgement till and partly in bedrock. Over this subglacial landsystem
is draped a thick veneer of discontinuous flowed till. At Elisebreen a similar
pattern can be seen, with the exception that the subglacial surface (lodgement
till) is fully revealed. Both these examples are considered to have formed at
the limit of glacier advance since there is no visible obstruction to flow. At
other glaciers in the area, ice-cored ridges are developed on the upstream flank
of major bedrock obstructions, otherwise they too show the same pattern with
fluted lodgement till exposed further inside the glaciated area (Paul and Evans,
1974).

In Britain, analogous systems of supraglacial and subglacial landforms can be
recognised. One of the largest occurs on the western flanks of the Pennines in

Fig. 3.4. Schematic picture of landforms and sediments of the Shropshire basin. A broad arcuate belt of hummocky flowed till draped over outwash ridges abuts the western Pennine edge. The till plain to the north is composed of lodgement till and is locally obscured by lacustrine sediments and flowed till. This combination of bedrock escarpment and supraglacial terrain (Fig. 2.18) is a common feature of glaciated terrain. W; Woore. C; Crewe, N; Nantwich.

Cheshire and on the north flank of the Staffordshire coalfield. In this region there is a belt of hummocky topography some ten kilometres wide which extends in a sweeping arc from Macclesfield to the vicinity of Whitchurch (Worsley 1969). This is considered to be a complex of supraglacial deposits that formed when southward moving 'Irish Sea' ice, derived from the Irish Sea Basin, impinged against the western Pennine escarpment. Figure 3.4 shows the regional geology and geomorphology. The sedimentology of so large an area is complicated but in general the main topographic elements are built of outwash, with tills forming a capping and discontinuous subsurface interbeds. Inside the belt of 'supraglacial' terrain a streamlined lodgement till plain emerges (Poole and Whiteman, 1966; Peake, 1981).

In North America the supraglacial landform and sediment association is spectacularly developed over some 500,000 km^2 on the Albertan, Saskatchewan and Manitoban Plains, in adjacent parts of North and South Dakota and in the western Arctic coastal lowlands (Fig. 9.2). Two prominent escarpments, the Missouri Coteau and the Manitoba/Missouri escarpment, up to 130 m high, run counter to former iceflow direction of the Laurentide Ice Sheet, thereby inducing intense compression of the ice margin and widespread elevation of basal debris to a supraglacial position. In the Dakotas the width of the supraglacial zone is 25-90 km widening to several hundred in Canada. Within this terrain a hummocky relief of up to 100 m is not unusual. Once initiated an ice margin, such as that in Fig. 1.5 and 3.2, becomes a major obstacle to ice flow and generates repeated compression as more active ice upglacier flows into the stagnating margin. 'Incremental' stagnation then occurs whereby the zone of compression migrates

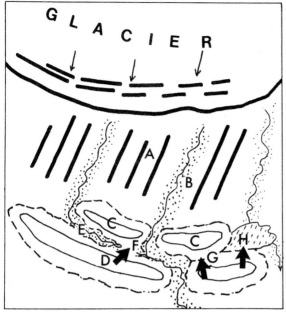

Fig. 3.5. Idealised pattern of sedimentation in the supraglacial landsystem. A) Streamlined forms of the subglacial landsystem (Fig. 1.4). B) Outwash sedimentation in channels cut radially into subglacial deposits. C) Ice-cored ridges that impede the radial flow of melt streams. D) Carpet of flowed till that originated on the flanks of the ridges. Types of trough environment: E) Trough receiving mostly outwash sediment since flowed tills do not penetrate to centre of channel. F) Elsewhere flowed tills penetrate to centre of channel, producing laterally continuous interbeds of flow deposit. G) Trough not occupied by meltstreams, thus sedimentation is dominated by flowed tills. H) Flows moving into lacustrine basin.

upglacier and the supraglacial landsystem is finally deposited over a very large area.

Clayton and Moran (1974), who initiated a landsystem approach to the glacigenic sequences of the western plains, have described the great similarities between supraglacial landforms in Dakota and those currently forming on large modern glaciers in coastal Alaska. These glaciers are of the ´surging´ type that is they exhibit episodically very high flow rates triggered by unknown processes that may include water at the bed, earthquakes, kinematic waves and changing thermal regime (Sect. 2.2). These ice masses experience intense compression at the terminus where active ice encounters decaying ice-cored moraines resulting from previous surges and are thus good models for the evolution of Pleistocene supraglacial landscapes and sediments.

Wright (1980) also argued that the surging Klutlan glacier in interior southern Alaska was a suitable analog for the development of 20-40 km wide arcuate ´controlled´ moraine complexes of supraglacial origin in Minnesota. The ´controlled´ arcuate form of the moraine complexes is thought to result from repeated compression of more active ice against a terminal dam. Klutlan glacier lies in a dry continental climate and therefore is argued to provide a better analog for the development of Pleistocene supraglacial landforms in continental interiors than glaciers in more temperate coastal locations. It appears that in Minnesota, ice-cores took 7-9000 years to melt-down; much more rapid melt-rates are exhibited by glaciers in more temperate coastal areas. Shaw (1980) has argued that the style of deposition at cold polar glaciers in Antarctica may be an even more appropriate analog for Pleistocene continental interiors. These ice-masses contain a well-stratified basal debris zone which is deposited as the ice mass downwastes in situ; ice is lost by sublimation and englacial structures are preserved (Figs. 1.8c, 2.15) in the final deposited sediment sequence. Deposition rates are however extremely slow (Sect. 2.6).

Two points should be noted. First it is not being argued that Pleistocene ice masses in Minnesota and the western plains were of the surge type; modern surging glaciers however provide a good model for landform evolution. Second on surging glaciers, which are valley glaciers, the till which they deposit is coarse and derived principally from the valley sides (=supraglacial morainic till;   Fig. 1.7, Ch. 4).   In Pleistocene supraglacial terrains till has been derived from the ice base (Fig. 2.1) and is thus more fine-grained.

The association of supraglacial terrain and major bedrock escarpments is repeatedly encountered (Fig. 3.4).   In southern Ontario, Straw (1968) has demonstrated the association between arcuate moraine complexes of the supraglacial landsystem along the Niagara escarpment which again provided a major transverse obstacle to ice flow (Fig. 2.18). The escarpment has been severely modified by glacial erosion in the form of deep trough-like indentations akin to the 'finger lakes' of New York State (Clayton, 1965). Within and on the hummocky surfaces of the morainic complexes large joint-guided bedrock masses also testify to glacial quarrying of the underlying escarpment.   Enhanced quarrying of flat-bedded sedimentary strata along the margin of the Canadian Shield is reflected in the pattern of lakes (Fig. 9.2);   the sedimentaries present north-facing scarps transverse to ice flow direction. The Dummer 'moraine' in southern Ontario mapped by Chapman and Putnam (1973) is a classic example of the development of the supraglacial landsystem in such a position. In this example much of the material underlying the hummocky terrain is exceptionally coarse-grained and includes glacially transported bedrock masses. Palmquist and Bible (1974) also describe hummocky supraglacial moraine complexes (end moraines) overlying bedrock slopes opposed to former glacial flow directions in Iowa. Moran (1971) and Moran et al (1980) have clearly demonstrated that highly disturbed, glacitectonized rockhead can be expected in areas of supraglacial terrain (Fig. 1.5). This association with thrusted bedrock highs is related to the severe compression that occurs at the ice margin and may also involve the freezing on of the ice base to underlying frozen highs (Fig. 2.2a;   Sect. 2.8.2) or overpressuring of subsurface aquifers (Sect. 2.8.2).

The origin of the extensive tracts of ice-cored terrain in the western Arctic coastal lowlands (Fig. 9.2) has been debated for some time (French, 1976). Massive ice underlies sequences of till and outwash with ground conditions similar to that shown in Fig. 1.5 (Rampton, 1974). Some of this buried ice may be glacier ice as it shows tectonic structures similar to that found with a modern glaciers.   Mackay (1972) however has argued that the ice-cored hummocky terrain is the result of the thermokarstic disintegration of periglacial ground ice (Sect. 5.2.3) that had developed earlier in coastal plain sediments exposed by lowered sea levels.   Processes of thermokarst decay in the periglacial environment are described in detail by French (1976) and are similar to the processes of glacier ice melt-down below a cover of melt-out and flowed tills (Fig. 3.2).

3.5 Typical Sedimentary Sequences

Once differential surface melting has established a ridge-trough system (Fig. 3.2), the troughs then act as loci of further sedimentation. There are two sources from which they can receive sediment; material may repeatedly flow in from the slopes of the ridges (Fig. 3.10) or the trough may be occupied by the meltwater drainage system, with the possible deposition of outwash or glaciolacustrine sediment (Fig. 3.11).   In many cases, both sources are active simultaneously and together produce complex sequences of sediments.

The ice-cored ridges act as barriers to the outward flow of meltwater from the ice margin;   for this reason the major component of hummocky morainic complexes

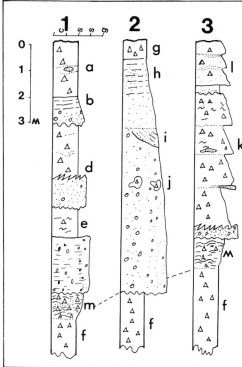

TILL

SAND

GRAVEL

EROSIVE

SHARP

INTERMIXED

Fig. 3.6. Typical sequences and contacts, shown on vertical profiles, due to deposition in supraglacial troughs. The numbers (1) to (3) correspond with the troughs shown in figure 3.2. Varying width of column is proportional to grain-size (e.g. Figs. 1.8, 7.8).

Sequence (1) - a trough receiving both flowed tills and outwash

(a) Capping of flowed till, introduced when the stream had abandoned the trough, consisting of one or more flow units, which may be separated by lag surfaces or minor current deposited horizons.

(b) Fining-up sequence of outwash, formed as stream activity diminished.

(c) Eroded upper surface of flowed till unit deposited during outwash deposition. This flow unit has survived disaggregation.

(d) Intermixed basal contact of flow unit and outwash.

(e) Flow unit, similar in its relationships and origin to the unit above; note different basal contact.

(f) Lodgement till of the subglacial landsystem underlying the supraglacial sequence (Figs. 1.4, 1.5, 1.8b).

Sequence (2) - a trough receiving mainly outwash

(g) Flowed till cap:- as in (a) above. If the trough is large, or the accumulation of sediment has overtopped the adjacent ice core, the cap may be absent, since the flow can only spread a limited distance from its source.

(h) Fining-up sequence of outwash.

(i) Main body of outwash. Typically complex, with lateral and vertical alterations of grading. Sedimentary structures typical of channel flow are usually well developed. Erosional episodes may be frequent.

(j) Remnants of till flows, such as till balls, lenses, stringers or boulder lags.

Sequence (3) - a trough receiving mainly flows

(k) Individual units separated by minor erosional (lags) horizons, or minor depositional horizons.

(l) Composite unit of multiple flowed tills. The upper part may itself be composed of several flows; these may show layered or banded structures.

(m) Basal melt-out till overlying lodgement till.

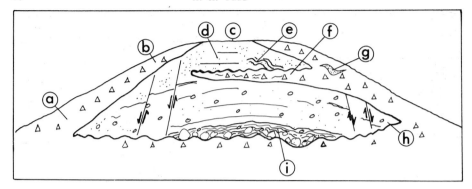

<u>Fig 3.7</u>. Cross section in a trough filling after deglaciation. The width of the
hummock may range from ten to several hundred metres.
(a) Flowed till, thickening towards the flanks of the trough, due to the
increasing proximity to the former source.
(b) The lunate hummock shape is the result of the removal of confining ice
walls.
(c)) In the central part of the trough deposition occurred directly onto bedrock
or lodgement till; there is no subsidence in this region resulting in the flat
topped form and undisturbed bedding (d).
(e) If the entire trough were underlain by ice, a rounded form with arched and
faulted sediments would be the result.
(f) Tongues of flowed till that intrude into the trough are deformed conformably
with the other sediments.
(g) Small outwash bodies show secondary contortion as a result of movement on
deglaciation. These are often useful indicators of previous movements of the
trough fillings.
(h) Greatest deformation in the form of warping and normal faulting is
experienced on the flanks of the ridge, since here the greatest thickness of ice
has melted (Fig. 3.2).
(i) Crudely stratified basal melt-out till (Fig. 1.8B).

is outwash and as such constitutes major aggregate resources. Melt streams are
forced to flow parallel to the ridges and along the troughs (Fig. 3.5). These
meltwater systems change rapidly in discharge and spatial location, and so
individual troughs are quickly occupied and abandoned. During the period of
occupancy, the stream may be erosive, or may deposit outwash ranging from cobbles
to fine sand and silt. In addition, some of the major basins may contain lakes
which act as stilling pools in which silts and clays are deposited, together with
deltaic sediments at the point where the stream enters the lake. These lake
basins may attain large dimensions; with the subsequent melt down of surrounding
ice a raised lake plain is left (Fig. 3.11; Mollard, 1973; Schwan <u>et al</u>, 1980).

There will be other troughs which do not contain melt streams, either because
they have never been occupied, or because they have been abandoned. These
troughs will be infilled by flowed tills from the adjacent ridges. The overall
pattern of these flows will be of a continuous sheet that fills the trough, but
in detail each flow will be a distinct unit, with its own set of structures, and
each having its own orientation. It may be noted that these flows as a whole are
orientated transverse to the troughs, in contrast to the meltwater sediments,
whose depositional structures will indicate paleocurrents along the trough.

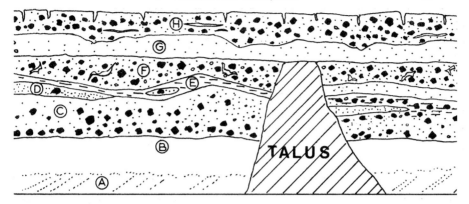

Fig. 3.8. Glacial succession at Taxmere Quarry, Arclid, Cheshire, England
(A) Bedrock.
(B) Sands.
(C) Lodgement till containing sand stringers, occasionally tightly folded with the fold noses trending eastwards (Fig. 1.4)
(D) Sand, no apparent bedding, containing clots of flowed till.
(E) Clay, containing streaks of sand and silt and till inclusions.
(F) Flowed till containing stringers of sand, irregularly flexured, but not overfolded. The unit infills an irregular topography on top of unit (E); the upper contact is planar over the observed length of section.
(G) Sand, cross-bedded over the length of section.
(H) Flowed till with frequent sand layers.

The pattern of sediment distribution, since it is dictated by the orientation of the ridges, may be either controlled or uncontrolled (Figs. 3.11, 3.12). In both cases the sediments will tend to occur in belts that lie parallel to the former ice margin. In a controlled distribution, these belts are nearly continuous and relatively narrow, whereas in an uncontrolled distribution they are discontinuous, and often consist of discrete patches or spreads of sediment that are displaced by the offsets of the ridges. On the larger scale however, they are still organised into belts that run transverse to the direction of ice movement. Figure 3.5 shows an idealised distribution of such sediments around an ice margin.

Three general types of geological sequence can be recognised within the supraglacial landsystem. They derive their character from the relative importance of the contributions of the meltwater system, flowed tills from the ridge flanks and the thickness of basal melt-out tills. The sequences are as follows:-

(a) Interbedded flowed till and outwash sequences (Sequence 1, Fig. 3.6). In channels where meltwater activity is periodic, or which are abandoned by the stream after a period of residence, flowed tills move into the channel and will not be destroyed. This leads to sequences in which major flowed units are either interbedded with outwash or which are sealed by flowed units. In the latter case the outwash may become finer upwards as the stream activity in the channel progressively diminished. Lacustrine deposits also occur in such sequences due to the blocking of the drainage system by flowed tills which the streams cannot remove.

M. A. Paul

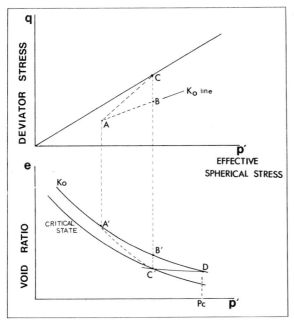

Fig. 3.9. Critical state model for the stress history of flowed till. Details are discussed in the text.

Fig. 3.10. Hummocky supraglacial deposition; unstable flowed till lobe moving over outwash sediments in the trough bottom. The back scarp is 10 m. high. Outwash plain in the background; Spitsbergen.

(b) Dominantly outwash (Sequence 2, Fig. 3.6). In channels where meltwater streams are active, flowed tills derived from the flanking ridges (Fig. 3.10) will be disaggregated by the stream. Deposits in these channels tend therefore to be mostly composed of outwash, and range from gravel and cobbles down to sand or finer (Ch. 7). Since the finer sediments are deposited from the less energetic periods of flow, they may be accompanied by bodies of till that have survived complete disaggregation. Such bodies may be recognisable flow units, or may be eroded remnants. Since stream activity in such channels is strongly episodic, major channeled erosion surfaces will normally be present in the sequence.

(c) Dominantly flowed tills (Sequence 3, Fig. 3.6). These form in troughs in which there is little meltwater activity and so contain mainly flow deposits derived from the ridge flanks. The sequence overlies basal melt-out till. Flowed till units are often separated by washed horizons, gravel lags or depositional laminae of sand, silt or clay. These interfaces may show evidence of subaerial soil processes such as drying or frost action in the form of infilled dessication cracks, mud flakes and ice-wedge casts. The latter form rapidly and do not necessarily indicate lengthy breaks in deposition and multiple glaciation in a stratigraphic sequence (Boulton, 1972, 1977). A typical sequence originates when the trough is filled by a series of flowed tills that range from till slumps or slurries. A series of such events leads to the superposition of several units (Fig. 1.8B). Intervening quiescent periods allow the development of minor stream activity (the washed horizons and laminae) and the development of subaerial pedogenic soil structures.

Figure 3.7 shows a typical cross-section through a hummock and one point is worthy of further discussion. The flowed tills often show a variety of basal contacts with the underlying sediments (Fig. 1.8B, 3.6). Basal contacts may include sharp, planar bases that have eroded the underlying deposit; intermixed contacts where tongues or streaks of the one sediment are drawn into the other, often with complex local folding; depositional contacts in which structures on the surface of the lower bed are faithfully preserved by infilling around them, and, finally, gradational contacts where no distinct boundary can be drawn, but where instead the one deposit replaces the other over a distance of a few centimetres.

The characteristics of basal melt-out tills that have accumulated _in situ_ at the base of overlying stagnant dirty glacier ice are not well known but have been discussed by Lawson (1979, 1981), Kemmis _et al_, (1981) and Haldorsen and Shaw (1982). For the most part they are crudely stratified as a result of the lowering and compaction of alternating dirty ice layers in the lower ice base and may retain original englacial structures and clast orientations (Fig. 2.14). Such preservation is even more marked where deposition occurs by sublimation (Figs. 1.8C, 2.15). These sequences overlie either lodgement tills of varying thickness or deformation tills, that record the active flow or glacier ice prior to _in situ_ stagnation (Shaw, 1982a; Eyles _et al_, 1983). The formation of thick multiple sequences of melt-out till and interbedded lenses of subglaciofluvial sediments over areas of many square kilometres as argued by Carruthers (1939, 1947, 1953) is problematical (Boulton, 1972; Eyles _et al_, 1982). Extensive till plains underlain by planar superimposed sheets of melt-out till recording the _in situ_ decay of an ice sheet bed over a large area are unlikely given the tendency for hummocky ice-cored topography to develop as debris is revealed on the ice surface (Fig. 3.2) which then control the style of subsequent deposition (Fig. 3.5). Thus the most likely occurence of basal melt-out tills is as ´cores´ to hummocks and ridges within large morainic complexes (Sect. 3.4; Figs. 3.6, 3.7, 9.1, 9.2; Kemmis _et al_, 1981).

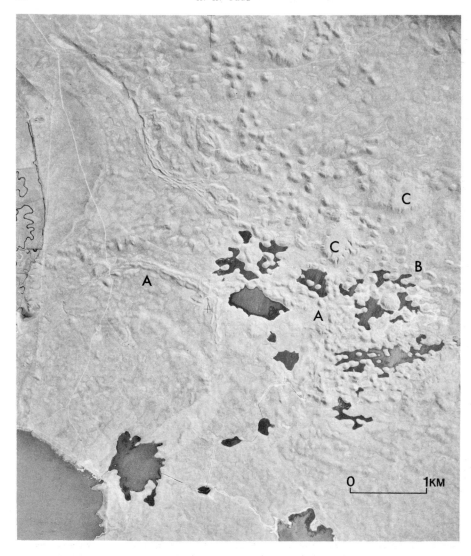

Fig. 3.11. Supraglacial Landsystem. Hummocky terrain composed of former trough fillings (Fig. 3.2) now 'inverted' to form hummocks. A controlled series of hummocks at A records the infilling of an elongate trough contrasting with uncontrolled hummocks at B. Note the plateau of lake sediments at C, formed by the melt-down of surrounding ice. Photo: A155- 11-16, National Air Photo Library, Ottawa.

3.5.1 Stratigraphic Investigations

It can easily be appreciated how difficult it is to construct a representative stratigraphy based on sequences in supraglacial terrain (Figs. 1.5, 3.7). With poor exposure between the few available sections, it is perhaps not surprising that stratigraphic work in supraglacial deposits has usually resulted in controversy. In addition to such depositional complexity, the effects of post-depositional bulldozing and thrusting by ice advances, reactivation of dead-ice and reversed density gradients set up by rapid deposition of contrasting sediments must also be considered (Brodzikowski and Van Loon, 1980; Schwan et al, 1980). This approach sees multiple tills as the normal expectation in glacial sediments. If the entire supraglacial landsystem is viewed as a stratigraphic entity, albeit on a large scale, it can be mapped as such, and a stratigraphy of landsystems can be produced (Sect. 1.5). The uppermost association is commonly supraglacial or glaciated valley origin and overlies lodgement till of the subglacial landsystem (Figs. 1.5, 1.8).

3.6 The Supraglacial Landsystem of the Cheshire-Shropshire Basin, Britain

It was stated earlier that a wide belt of hummocky topography flanking the Staffordshire escarpment is considered to be of supraglacial origin. The relationship of the hummocky landforms to the rising bedrock floor is directly analogous to that seen at several modern glaciers where compressive flow has brought debris to the glacier surface and to similar settings in North America (see above). The sedimentary model developed above can now be ued to illustrate the subsurface ground conditions in this area. Drift maps of the area (Fig. 3.4) show hummocky terrain running parallel to the 65 km long bedrock escarpment. The distribution of outwash suggests the infilling of troughs between ice-cored ridges aligned transverse to ice flow suggestive of a 'controlled' pattern of deposition.

The stratigraphy within the hummocky belt also provides evidence for a supraglacial origin. Many sections show complex sequences of sand and silt, interdigitated with flowed tills and often capped by one or more major flowed till units. One such sequence is shown in Figure 3.8. This sequence shows in descending order a multiple flowed till capping resting on an extensive outwash body, which in turn is underlain by a complex of outwash and flowed till units. The upper part of the sequence has formed in a trough that received a mixture of flows and outwash, analogous to sequence 3 in figure 3.6. Subglacially deposited lodgement till forms the base for the whole sequence and contains interbedded channel fills of sand and gravels resulting from subglacial meltwater activity in the manner depicted in Fig. 2.11. In other areas, basal melt-out tills overlie the lodgement till 'basement' (Fig. 1.8B).

Stratigraphic sequences such as these appear to be common throughout the belt of hummocky topography. There is general agreement on the complexity of the glacial sequence (McQuillan et al, 1964; Worsley, 1967, 1969, 1977), and on the frequent presence of flowed tills. The area provides a good example of the supraglacial landform and sediment association in terms of its location, landform patterns, and stratigraphic sequences.

3.7 Changes in Landforms During Deglaciation

Subsequent to the phase of trough sedimentation, there is a deglaciation phase during which ice-cores finally melt. Compared with the period over which sedimentation occurs (typically a few years) the deglaciation period is long, perhaps of the order of hundreds of years or more (Clayton and Moran, 1974). This period is controlled by the rate of heat flow through the sediment cover.

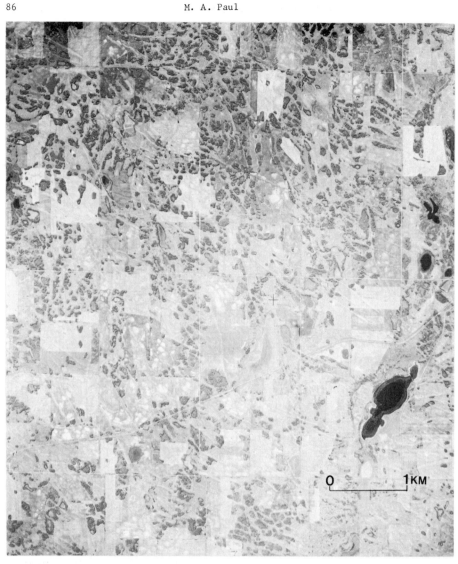

<u>Fig. 3.12</u>. Supraglacial Landsystem. Controlled distribution of elongate trough
fills orientated top left to bottom right probably based on transverse crevasses
in the ice front (Gravenor <u>et</u> <u>al</u>, 1960).   Photo: A12391-115, 160-1368-38.
Alberta Department of Lands and Forests.

If this cover equals the thickness of the active layer of the permafrost (Fig. 5.4), melting will cease. Any remaining ice is then effectively part of the permafrost, and will survive indefinitely until such time as the external conditions change, for example by erosion of the sediment cover or an amelioration in climate.

During deglaciation, lateral support is removed from the trough fillings. In some cases this leads to minor faulting or flexuring (Fig. 3.7), whereas in other cases the filling collapses completely and the landform is destroyed. Observation shows that fillings that are largely outwash will usually display the former behaviour. Those fillings that are of largely flowed tills or fine grained sediments tend to display the latter. Survival of the topographic form thus depends on the bulk sedimentology of the filling.

The overall stability of the filling can be explained by considering the stability of the newly formed slope under thaw conditions discussed by McRoberts and Morgenstern (1974). When frozen sediment thaws, an excess pore pressure is generated. This pressure depends on the rate of thaw and on the coefficient of consolidation of the soil; in fine-grained soils it may reach substantial proportion of the overburden pressure. As a result, slopes in such sediments often have an angle of stability of only a few degrees. It follows that when deglaciation removes the lateral support from a fine-grained filling, the contemporaneous thawing of the permafrost within the filling itself will lead to adoption of a shallow slope angle. For a granular filling resting on melt-out till, on the other hand, the angle of stability is about half the angle of internal friction, and slope angles of around 15° are displayed by many ridges that are composed of outwash. Thus the final form of the landsystem will be dominated by steep ridges that are composed mainly of outwash; former channels filled by flow deposits will now be represented by subdued, undulating landforms. Those flow-based fillings that survive as ridges usually contain small but significant outwash bodies that act as internal drains and so allow the dissipation of excess pore pressure. The proportion of one against the other in the overall morphology leads to the contrast of the hummocky and supraglacial till plains identified in Sect. 3.3.

3.8 The Relationships Between Depositional Processes And Geotechnical Properties

The supraglacial landsystem contains three principle types of sediment. These are melt-out tills (or sublimation tills; Figs. 1.7, 1.8), which have formed by the direct in situ release of englacial debris; flowed tills, which have moved downslope under gravity with consequent disturbance of their structure, and outwash sediments (Ch. 7).

The process of flow is a complex one, in which the grain support mechanism ranges from complete support by a continuous soil skeleton, through support by intergranular collisions, to support by fluid turbulence alone. The operation of these mechanisms is related to the water content of the flowing sediment, which usually lies between fifteen and thirty percent (corresponding to a liquidity index of 0 to 1.5; Sect. 8.2.3). At lower water contents the sediment behaves as a coherent slide which moves principally on a basal shear surface, whereas at higher water contents it behaves as a thick fluid in which relative movement occurs throughout the soil mass. As a result, an important distinction can be made concerning the constraints on interparticle movement, in the former case particles are heavily constrained by their neighbours, while in the latter there is little or no constraint. This lack allows the dispersion of different sized particles during flow, and so allows the development of internal structures gradations in particle size and clast orientations that characterize the more fluid sediment flows (Middleton and Hampton, 1976; Lowe, 1979; Nardin et al, 1979). Similar remarks appertain to solifluction sheets (Sect. 5.4.1).

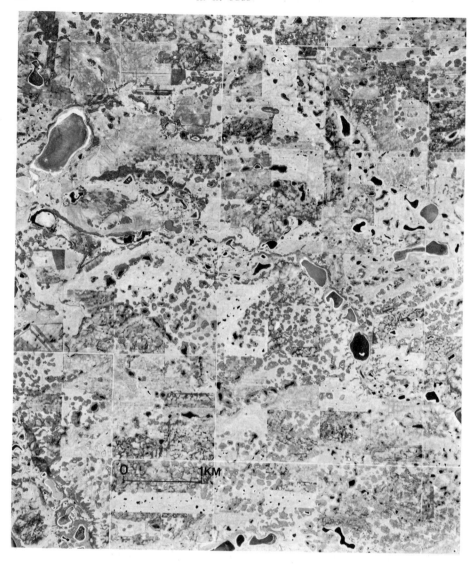

Fig. 3.13. Supraglacial Landsystem. Linear stream trenches; formerly open
channel meltstream trenches draining through supraglacial ice-cored terrain and
infilled with flowed tills and outwash. Infilling results in blurring of the
original linear form (Gravenor and Bayrock, 1956; Norrish and White, 1961).
Photo: 160-5303, 1333, National Air Photo Library, Ottawa.

It should be noted, however, that in any one flow and hence in any one flow unit, several grain support mechanisms may have operated simultaneously; for example, the lower part of a unit may have moved into position as a coherent body under the influence of intergranular support, whiles the upper part of the same unit may have been simultaneously emplaced as a slurry. Consequently the lower part will lack sedimentary structures, except perhaps a basal gravel layer, whereas the upper part will contain structures that result from sorting during flow. This 'couplet' of a lower massive and upper structured zone in a flowed till unit has been reported by Boulton (1971), and a more detailed discussion of the relationship between grain support mechanism and sedimentary structures has been given by Lawson (1981, 1982).

The depositional processes described above can be invoked to explain certain geotechnical properties of supraglacial tills and their related sediments. The properties that will be considered here are the grain size distribution and associated classification indices, since there are in widespread use and correlate well with several other parameters (See Ch. 8). The stress histories and state of consolidation of the sediments will also be discussed, since depositional processes can throw some light on a complex situation. The arguments presented here are partly based on those given in Boulton and Paul (1976) and the reader is referred to that paper for full discussion.

The grain size distribution of englacial debris is acquired during transport. During supraglacial deposition, the distribution may be modified, and within a given sedimentary unit there may be considerable grain-size variation due to the processes that produced that unit. Melt-out till, since it is formed directly from englacial debris, has a similar but more variable grading compared with lodgement till (Sect. 2.4). The grading curves of flowed tills often show loss or gain of finer particles, settling out of coarse particles, etc. and may show rapid vertical and lateral variation within one flow unit (Figs. 1.8, 3.6, 8.1).

In a study of supraglacial flowed tills exposed below hummocky topography deposited by the Des Moines ice lobe in Iowa, Kemmis et al (1981) found that flowed tills had generally lower bulk densities (mean 1.62 g/cc) than lodgement tills (1.9 g/cc). With regard to Index properties, similar Atterberg indices will be exhibited by melt-out and flowed tills to those of lodgement tills. On a plasticity chart melt-out tills will fall close to the T-line whereas flowed tills will exhibit much greater variation (Fig. 8.2).

The state of consolidation of a flowed till is related to both its depositional and post-depositional history. In general, many flowed tills are overconsolidated, with past maximum stresses of around 150-300 kN/m^2. This is often the result of post-depositional drying, which tends to affect flowed tills due to their exposed positions on the crests of, and within local high points of the topography. It is clear that glacial over-riding has played no part in the loading of a supraglacial sediment. However, compression tests on flowed tills from modern glaciers, which have not dried, also show a light (about 100 kN/m^2) degree of over-consolidation.This is clearly not a post-depositional effect and can be explained instead as a result of the flow process itself.

The position is best explained by the use of critical state theory (Atkinson and Bransby, 1978). Figure 3.11 shows the stress history of a small element of supraglacial till that is first released as melt-out till. Initially this till is located at points A and A'. If there is superposition of sediment from above, it will compress along a K_o path to point B'. However, if there is also an increase in shear stress due to changes in the underlying ice slope, the element will move instead towards point C on the failure surface, and will thus also move off the K_o line and onto the critical state line which it reaches at C'. At this point the sample has flowed and is thus a flowed till (hence illustrating in mechanical terms the conversion of melt-out into flowed till). Once flow ceases, the element of till finds itself removed from the virgin compression line, and can only return along a recompression path (such as C′D where D corresponds to a ′preconsolidation′ pressure P_c) and it is thus overconsolidated. This overconsolidation is quite distinct from any post-depositional effect and is solely the result of shear during flow.

It should be stressed that the above argument only applies to flowed tills that have moved as slides and where a continuous soil skeleton is maintained by intergranular contacts. In slurries where grains are supported by fluid turbulence and intergranular collisions and where effective stress is insignificant, geotechnical properties cannot be investigated by stress paths of the type discussed here. A full discussion of the geotechnical properties of flowed and melt-out tills can be found in Paul (1981).

3.9 Prediction Of Subsurface Stratigraphies

The supraglacial model is a valuable aid to site investigations in glaciated areas, particularly at the reconnasissance stage (Fig. 1.2). The essential step is to recognize that positive elements in the topography are in the main, former sedimentary troughs and that negative elements are sites of ice-cores that formerly bounded them. The type of trough filling can also be identified (Fig. 3.6). Thus the subsurface conditions that may be anticipated on site or along a route will be related to the positive or negative nature of the terrain, and to the lines of former drainage that operated between the ice-cores. It must of course be stressed that local conditions can vary considerably even within one depositional trough particularly with regard to the presence or absence of melt-out tills, and so a generalised model can never replace a detailed site investigation; however, such a model can provide a conceptual framework within which the results of the detailed investigation may be evaluated. Positive topographic elements with steep sides (provided these are not due to post-depositional erosion) are usually modelled on granular materials. Within these materials there may, however, be beds or lenses of flowed till or fine grained outwash. Other hummocks may be modelled on basal melt-out tills. Thus the engineer should be wary if the intention is to exploit the deposit for fill or aggregate, and it will be necessary to investigate the quality of the material further in this respect. Notwithstanding, these topographic features are a good starting point in the search for borrow materials.

Positive elements with subdued slopes represent former trough fillings that have collapsed when the surrounding ice walls melted (Fig. 3.11, 3.12, 3.13). They are usually fine-grained, although they commonly also contain beds of coarser material. These beds are normally thin and discontinuous.

Negative topographic elements represent the sites of former ice bodies. They may be based on a variety of possible sediments, including;

(i)   basal melt-out tills derived from debris that was contained in the stagnating ice,
(ii)   flowed tills that have partially refilled the area from adjacent trough fillings as the ice-cores melted, or
(iii) nothing, in the sense that there has been little or no supraglacial sedimentation in the area of the ice-core, and its site now provides a window into underlying subglacial or pre-glacial deposits.

This latter possibility should be borne in mind when putting together stratigraphic sequences from borehole evidence or open exposures in such areas.

CHAPTER 4

The Glaciated Valley Landsystem

N. Eyles

INTRODUCTION

The term glaciated valley is used here in a broad sense and includes both high mountain valleys (e.g. Rockies, Alps), and upland areas where the pattern of ice sheet flow was locally dictated by valleys (e.g. Appalachians, English Pennines, Highlands of Scotland, etc.). Glaciated valleys frequently contain some of the most difficult engineering ground conditions, a situation compounded along valley floors both by severe constraints on location and intense competition between different land uses (Anderson and Trigg, 1970; Fookes et al, 1975; Newberry and Subramaniam, 1977; Voight, 1978).

In this chapter the general bedrock form, rockhead conditions and typical sedimentary sequences of glaciated valleys are reviewed. Postglacial modification of glaciated valleys by glaciofluvial and paraglacial processes will be stressed.

4.1 The Bedrock Form Of Glaciated Valleys

The bedrock floor of glaciated valleys commonly shows evidence of enhanced erosion in the form of a stepped long-profile and oversteepened valleysides. Enclosed sections overdeepened by tens or hundreds of metres (Figs. 2.8, 9.4) bounded by rock bars (reigels) and occupied by thick sediment sequences, or deep water bodies, are typical of glaciated valleys and are the product of either enhanced erosion where powerful tributary ice streams joined the main valley or differential resistance to glacial erosion of variable bedrock. The erosional activity of subglacial meltwaters is also important. The term overdeepened recognises that normal fluvial erosion produces a graded, or ´asymptotic´, long profile that cannot be reduced below base-level; ultimately sea-level. The classic example of overdeepened valleys are the fiord basins of the western sea boards of Scotland, Norway, British Columbia, New Zealand and Chile, locally overdeepened to as much as 1700 m below sea level, a response to the high precipitation of the coastal maritime environment and resultant large ice discharges through the basins. An up-and-down long profile is a recurring feature of glaciated valleys (Fig. 9.4).

Glacially modified valleys exhibit a wide range of cross-sections at rockhead. The classic U-shape held to be a characteristic of glaciated valleys is frequently the result of postglacial slope processes that erode sediments along the valleysides and redeposit them as coalescing fans. Below these fans strongly asymmetric rockhead cross-profiles with narrow gorge-like sections are common (Sect. 14.4).

4.2 The Dynamics Of Valley Glaciers

Debris transport and deposition by valley glaciers is intimately related to the mass budget of the glacier system. This budget (or mass balance) identifies the relationship between mass gain (snow) and mass loss (melt) and its dimensions are dictated by local and regional climate. The amount of potential mechanical work which any glacier can realise is related to the volume of snow and ice added each year in the accumulation basin(s) at high elevation and transferred downglacier. Glaciers of high ´mass transfer´ where large inputs of mass are balanced by large exports downvalley, to be lost by melt at lower elevations, are powerful abrasive agents (Andrews, 1972a). Good examples are temperate valley glaciers in wet coastal mountain environments (e.g. British Columbia, Norway, Chile) and, in conjunction with repeated valley glaciation, helps to explain the occurrence of overdeepened fiord basins in these areas.

The volume of winter snow input to a glacier can be assessed by various means that include probing the snow pack, use of fixed reference stakes, by changes in surface altitude measured by computerized remote sensing or by the geophysical location of buried markers. Variation in density through the snow pack must be assessed by trial pits. The winter balance for the whole glacier (B_w) or at selected points (b_w) is then the product of averaged snow pack density and volume across the area under study. Measurements are usually completed during late spring when the winter balance is at a maximum prior to the onset of summer melt conditions.

The winter mass balance is usually expressed as cubic metres of water equivalent (m^3/H_2O). Gauging the total volume of meltwater leaving the glacier in the summer and its subsequent subtraction from the net winter balance allows assessment of total net balance prior to new winter snowfalls. This method works with tolerable errors for temperate glaciers but on polar ice masses most melt waters in the accumulation area refreeze at depth in the snow pack so there is no net loss of mass. Where altitudinal changes in summer melt are of interest, a reference stake network can be employed to determine the summer balance (B_s; Fig. 4.1E).

When balance data (b_w/b_s) are plotted against altitudinal zones on a glacier (Fig. 4.1E) a clear picture of the changing net balance is seen and helps to define an <u>equilibrium line</u> where the net mass balance (b_n) = 0. The equilibrium line is of fundamental importance and divides a glacier into two unequal areas - an upper area of net mass gain (the accumulation basin) and a lower area of net mass loss (the ablation zone: Fig. 4.1G). Glacier erosion and deposition is

<u>Fig. 4.1</u> (Opposite) A) stress/strain curve for glacier ice. B) velocity profile with depth (h) from surface; note basal sliding (x) that is also evident in transverse velocity profiles (C). D) inclination of velocity vectors away from equilibrium line. E) winter (b_w) and summer balance (b_s) plotted against altitude (e) with net mass balance (b_n) for a typical valley glacier. E.L = equilibrium line where b_n = 0. The Activity Index is defined as the rate of mass increase (m) against rate of elevation increase (e) at the equilibrium line. F) Generalized mass balance curves for temperate (1) and polar (2) glaciers (after Andrews, 1975). G) Flow in a typical valley glacier. Note acceleration of flow towards and deceleration past equilibrium line. As a result of B and C above, sedimentary stratification in the accumulation basin is folded downvalley (e.g. Fig.4.2). Individual ice flow units (dotted lines) are related to sub-basins in the accumulation areas. Rock fall debris transported aong the margins of these flow units are exposed as medial moraines downvalley (solid lines).

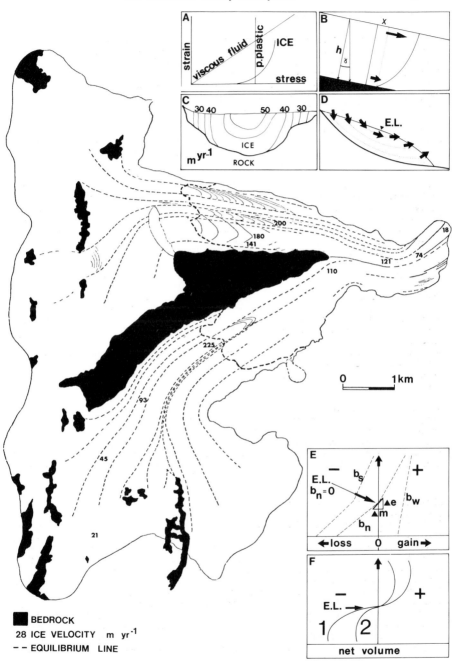

BEDROCK
28 ICE VELOCITY m yr^{-1}
-- EQUILIBRIUM LINE

accomplished as a result the throughflow of ice downglacier from one basin to another. Glaciers of high mass transfer, whose geological importance has been identified above, are characterized by a high gradient of net balance (b_n) at the equilibrium line (expressed as millimetres of water equivalent per metre change in elevation and referred to as an 'activity index'; Meier, 1960). Temperate glaciers with a high activity index can be distinguished from polar glaciers in arid high latitude areas with low mass transfers and correspondingly low activity indexes (Fig. 4.1F). In geological terms this means that temperate glaciers are more efficient agents of erosion compared with polar glaciers.

Valley glaciers are significant water resources and in many areas, glacier inventory and mass-balance studies are part of long term watershed management. The location of mining activity close to glacier margins also necessitates knowledge of likely future behaviour of ice masses (Fisher and Jones, 1971) and engineering construction, sometimes for hydro-electric power schemes using subglacial waters, has allowed many observations of processes associated with the ice/bed interface below valley glaciers (Mathews, 1964; Ostrem, 1975; Boulton et al, 1979; Wold and Ostrem, 1979; Vivian, 1980).

4.2.1 Ice Flow Along Glaciated Valleys

Ice flow in bedrock channels is strongly affected by frictional retardation against valleysides and highs on the channel floor (Fig. 4.1B,C). Ice flow occurs in response to gravity and occurs by internal strain deformation and sliding where the ice is pressure-melting at its base.

For glacier ice the rate of strain deformation is not linearly related to stress as it is for a viscous fluid. Instead stress and strain are related in the form of a power law (Fig. 4.1A) with a yield stress of about 1 bar (100 kN/m^2). According to Glen (1953, 1955) after whom this power relationship is named, strain rate, ξ, is equal to $\beta\sigma\eta$ where β and η are constants and σ is the value of applied shear stress. η commonly ranges in value from 2 to 4, its exact value depending on the orientation of crystallographic axes of ice crystals with respect to the direction of applied shear stress (e.g. ice flow direction) since ice crystals are strongly anistropic with regard to shear strength. The value of β is also dependent on ice temperature and varies over approximately two orders of magnitude over the range of englacial temperatures measured at polar and temperate ice masses (Sect. 2.2).

Shear stress (σ) is equivalent to Pgh sinα. Where P is equal to the density of ice (0.9 gm/cm), g is gravational acceleration (980 cm/sec/sec), h is equal to thickness and   is the slope of the ice surface. Vertical velocity profiles can be computed from this equation (Fig. 4.1B). Many studies demonstrate that surface velocities are proportional to the third power of surface slope and the fourth power of thickness (when η=3). Velocity rates are greatest at the surface even though values of shear stress are greatest at the bed. The above computation assumes a flat parallel sided slab of ice which is a reasonable approximation in an extensive lowland ice lobe but which is inappropriate to describe channeled flow where frictional retardation against the valleysides is significant (Fig. 4.2). Another complicating factor is that thickness is variable across a valley glacier and therefore the thickness factor (**h**) must be amended to refer to average thickness. This is most conveniently expressed as A/P where A = cross-sectional area of the glacier and P = ice-contact perimeter. The flow of ice can be modelled for different valley cross-sections but as a general rule, valley glaciers have a steeper surface slope, in response to frictional retardation compared to that given by equation 2.1 for ice sheet profiles. Such retardation can be clearly demonstrated by velocity measurements downglacier of a confluence between two major ice streams where it can be shown

Fig. 4.2. Frictional retardation of flow against valleysides; arcuate banding
(ogives) on the glacier surface results from higher centre line velocities.
Ogives are composed of annually formed couplets of dark (A) and light (B) ice
representing   one year's flow through icefalls upvalley.   Dark ice is
coarse-grained, contains pollen and windblown dust, having moved down the ice
falls in summer; light ice moved through under a snow cover in winter. Note the
rock fall debris (C), also deformed by flow, and the medial moraine (D).   The
arrowed boulder is the size of a two storey house (Fig. 4.6).   Photo:
T.D. Douglas, Austerdalsbreen, Norway.

that ice flow units increase their velocity as the frictional retardation of the
valleyside is left behind (Eyles and Rogerson, 1977).   Direct measurements of
basal shear stresses in subglacial cavities demonstrate that basal debris also
enhances frictional drag against the substrate with coefficients of friction
being dependent on debris concentrations (Boulton et al, 1979; Sect. 2.3.2).

The velocity fields along a glacier are systematically related to distance above
or below the equilibrium line since it is at that point that the greatest volume
of ice must be exported through the glacier cross-section (Fig. 4.1G).
Velocities of 200-800 m yr on temperate glaciers are not uncommon in the
vicinity of the equilibrium line with velocity declining both up and downglacier.
As ice accelerates towards the equilibrium line it experiences extending flow
(surface strain rates are negative) whereas below the equilibrium line
strain-rates are compressive with highest positive rates occurring in the
terminal area (Fig. 3.2). It is in this area that glacier margins are subject to
enhanced ablation as a result of heat re-radiated from valley sides and a
strongly convex transverse cross profile results. This leads to radial ice flow,

crevassing, and a further reduction in longitudinal velocities. Strongly compressive strain rates develop as more active ice upvalley piles into ice that is decelerating which results in a thrusted and folded glacier snout area (Fig. 4.3). By way of contrast, in the accumulation area snow drifting against the valley sides generates a concave transverse profile and resultant converging velocity vectors whereby flow rates are enhanced.

The long profile to the valley floor will also dictate local patterns of longitudinal compressive and extending flow; in overdeepended lows ice flow will be compressive whereas extending flow occurs as ice moves over substantial bedrock highs. However, on most valley glaciers any simple relationship between ice thickness, shear stress and ice velocity is complicated by seasonal changes in the area of the glacier base actually in contact with the bed. Ice/bed decoupling is common and is controlled by ice velocity and bed relief. Cavitation (decoupling) occurs in the lee of bedrock highs (Fig. 4.3) with the size of subglacial cavities exhibiting seasonal variation. In summer, rainfall and snow melt raise subglacial water pressures, the number of subglacial stream channels increase, and basal sliding velocities are enhanced resulting in cavity extension (Figs. 2.13). In winter, ice velocities decrease and cavities are reduced in size though 'regressive cavitation' sometimes occurs as basal ice 'buckles' up from points of contact with the bed (Vivian, 1980).

Periodic changes occur in the volume of snow and ice generated in the accumulation zone in response to long or short-term climatic changes. Fluctuations in mass are propagated downglacier as kinematic waves which are bulges of increased ice thickness with velocities of about four times normal ice velocity. The size of the bulge can be shown theoretically to increase downvalley in response to compressive flow but in reality waves subside by diffusion and can only be identified by zones of enhanced flow moving downvalley (Paterson, 1981). In addition, rockfalls from the valleysides may blanket the ablation zone, reduce ice melt and contribute to a more positive mass balance (Bull and Marangunic, 1968). Glacier ice may survive below a thick cover of debris as a rock glacier subject to independent downslope movement (Whalley 1974a for review). By way of contrast, thinner debris covers increase absorption of solar radiation by lowering surface albedos and are used to increase meltwater run off from glaciated basins for hydro-electric power production or to control potentially dangerous glacier termini (Eyles and Rogerson, 1977a).

4.3 Debris Transport By Valley Glaciers

Windblown dust and bedrock debris from exposed valleysides falling onto the glacier surface aove the equilibrium line, is buried within the snow column and transported through the glaciated basin. Several transport routeways can be identified within a typical temperate valley glacier (Figs. 1.7, 4.3; Sect. 2.4). Debris may be transported downglacier either as part of a thin, debris-rich basal traction layer in abrasive contact with the bed, or within the suspension zone close to the bed, but not in contact with it (Fig. 2.5). At the contact of converging flow units within a composite valley glacier the entire basal debris zone may be folded upwards by the intense compressive strain generated along the glacier contacts to be expressed as a medial moraine (Figs. 1.6, 4.3).

By way of contrast, debris falling onto the glacier surface below the equilibrium line is not ingested within the glacier as winter snows do not survive the following summer. Debris is transported along the sides of the glacier as a supraglacial lateral moraine or along the coalescent margins of two flow units as a supraglacial medial moraine (Fig. 1.6). With the melt of surface ice, further debris will be added from basal and englacial debris sources (Fig. 4.3).

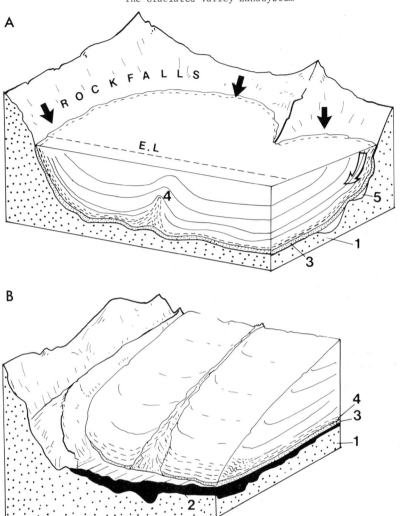

<u>Fig. 4.3</u>. Debris transport through a wet-based valley glacier.   E.L.;
equilibrium line.   A) accumulation zone;   B) ablation zone. 1, bedrock; 2,
lodgement till; 3, basal traction layer in contact with 1 and 2;   4, basal
debris zone folded upwards by lateral compression at the margins of iceflow units
and exposed in the ablation zone as a medial moraine; 5, basal cavity
(Fig. 2.13).   Folding and thrusting are common in the terminal zone where radial
flow results in a reduction in longitudinal flow rates and an increase in
compressive strain.

Multiple confluences will generate multiple supraglacial medial moraines (Fig. 4.1).

The term ´ablation till´ is widely used to refer to debris deposited from the glacier surface and associated with a hummocky surface topography along both glaciated valleys (Fig. 1.6) and ice sheet margins in lowland areas (Fig. 1.5). Thus the term is used in contradictory fashion, to refer both to material which on valley glaciers is largely coarse-grained and at the same time to material which is fine-grained since it is derived from basal debris (Figs. 1.7, 1.8, 3.1). The term that has been used to refer to the rubbly till transported on the surface of valley glaciers is supraglacial morainic till (Fig. 1.7; Boulton, 1976; Eyles, 1979) but the term coarse-grained supraglacial diamicts will be used below in recognition of the extreme variability of sediment type and grain size. In many cases such supraglacial diamicts are simply valleyside-derived rockfall debris transported in ´conveyor-belt´ fashion by the glacier without any glacial influence per se.

4.3.1 Basal Deposition By Valley Glaciers (see Sect. 2.5 et seq)

4.4 Supraglacial Deposition By Valley Glaciers

As discussed in Chapter 3, it is convenient to divide glacial landforms into two broad categories that describe the plan orientation of the individual landform with respect to the direction of glacier flow. The term ´controlled´ recognises that there is a close relationship between landform orientation (i.e. its long-axis) and iceflow direction. Controlled landforms result in a lineated topography where the constructional elements in the topography appear to be consistently aligned either parallel or transverse to iceflow direction. Conversely, uncontrolled landforms occur sporadically over the area (Sect. 3.3).

Fig. 1.6 shows the typical distribution of coarse-grained supraglacial diamicts on the surface of a valley glacier and as deposited landforms along the valley. Note that there are two principal locales of deposition, as medial moraines aong the valley floor and as lateral moraine ridges and kame terraces along the valleyside (Fig. 4.4).

4.4.1 Landforms Along The Valley Floor

Along the valley floor supraglacial diamicts are deposited as medial moraines in two states (Fig. 1.6);

(i)   a thick hummocky non-compact cover that is ice-cored during deposition, and

(ii) as a dispersed bouldery cover that is not ice-cored at the time of deposition.

(i)   Severe compression occurs in the terminal zone of glaciers and supra-
      glacial diamict thicknesses increase as a result of the enhanced
      concentrations of englacial debris. This cover insulates underlying ice
      from further melt. As the glacier recedes upvalley stagnant diamict-
      covered ice-cores are left along the valley floor with a hummocky high-
      relief surface (Fig. 1.6, 3.2). Subsurface melt of ice-cores produces
      surface collapse (kettle holes) comparable to that produced by solution
      in limestone terrains hence the term ´glacier karst´ (Clayton, 1964).
      Following several cycles of topographic inversion and final ice melt that
      may take several hundred years, an irregular topography consisting of
      steep-sided bouldery mounds, ridges and dead ice hollows is produced

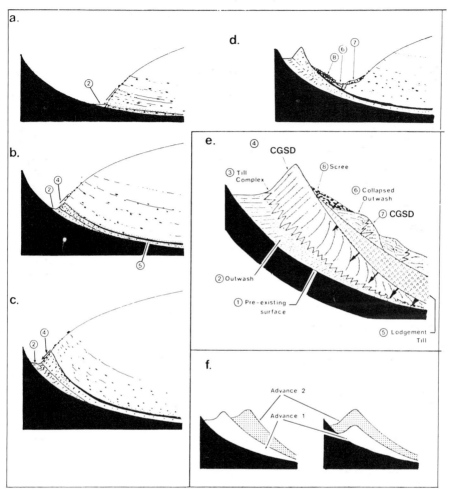

Fig. 4.4. Ground conditions and soils associated with the formation of lateral moraine ridges (from Boulton and Eyles, 1979). Key to numbers in e. C.G.S.D.; coarse-grained supraglacial diamict. The latter is consolidated in the moraine ridge and has a good clast fabric and outward dipping structure aiding high slope stability on the inner face (Whalley, 1975). The same diamicts when associated with kame terraces are non-compact and loose. (f) lateral moraine ridge construction by ice advances of different magnitude.

(Fig. 4.5B). Hummocky moraine, disintegration moraine, dead-ice moraine and stagnation topography are terms that have been used to describe such terrain overlying lodgement till at depth. The genesis of this topography is directly comparable to that associated with melt-out and flowed tills (Fig. 3.2). The underlying sediment types are different

however (Fig. 1.8 B,D) being much more rubbly in the case of valley
glacial deposits.

(ii) Where the diamict cover on the glacier surface is thinner, diamict is
shed directly from the icefront as it recedes.

Hummocky topography composed of coarse grained supraglacial diamict is frequently
present as one or more 'controlled' medial moraines along the valley floor
producing a longitudinally lineated surface. In large basins of complex geology,
each medial moraine may be composed of distinct bedrock lithologies reflecting
the origins of the moraines in different parts of the accumulation basin - a
feature used both for mineral prospecting in areas of Quaternary valley
glaciation (Evenson et al, 1979) and as an aid to geological mapping in remote
ice-covered parts of modern glacierized basins.

Non-compact supraglacial diamicts are easily bulldozed by the icefront during
winter readvances into large ice-cored push moraine ridges producing a diamict
surface that is transversely lineated. With the melt of ice-cores this
controlled lineation is partially destroyed.

A stagnant glacier terminus, insulated under a thick cover of supraglacial
diamict (Fig. 1.6) acts as a dam to more active ice upvalley and compressive
strains are correspondingly enhanced. Englacial debris is exposed by ablation at
the margins of ice flowing into the dam and very slowly the zone of dead ice is
propagated upvalley; a process already referred to as 'incremental stagnation'
(Sect. 3.4). In this way, active glaciers may deposit 'stagnation' topography.
The zone of stagnant ice may be periodically re-activated, bulldozed or
overridden as kinematic waves move downglacier in response to changes in glacier
mass balance (Sect. 4.2.1) as is typical for surging glaciers (e.g. Wright,
1980).

4.4.2 Landforms Along The Valley Margins

Formation of lateral moraine ridges results from the dumping of diamicts with
allied outwash deposition (Figs. 1.6, 4.4) in the trough between the glacier and
valleysides. Quantitatively, large volumes of supraglacial diamict may be
deposited because of the dynamics of glacier flow. In the ablation zone there is
a net radial component of flow away from the centre line (Sect. 4.2.1), the
effect of which is to ensure that larger volumes of diamict are transported in a
latero-frontal position. With destruction and burial of landforms by meltstreams
along the valley floor these lateral landforms dominate many glaciated valleys.

The formation of substantial moraine ridges is an obstacle to glacier activity
and may confine subsequent moraine ridge building to the inside of the outermost
ridge; a more powerful readvance leaves an overlapping moraine ridge (Fig. 4.4f)
that may bury organic debris and soils within the ridge stratigraphy
(Rothlisberger and Schneebeli, 1979). These ridges may also dam up large lakes
in tributary side valleys. During ridge formation, diamict is end-dumped as a
scree from the ice margin and as a result the long axis of boulders and stones
show an outward dip or imbrication parallel to a crude internal bedding. Coupled
with a higher degree of compaction and some cementation, this contributes to a
high angle of stability (70°) of the ice-proximal slopes of the ridge. The
achievement of such high stable slope angles has been of considerable
geotechnical interest particularly in view of using such material for embankment
rockfill dams (Zeller and Zeindler, 1957; Whalley, 1975) and it has been said
that glaciers appear to be better embankment builders than man. However large
lateral moraine ridges are also subject to the post-construction settlement
exhibited by large earth and rockfill dams as a result of the heterogeneity of

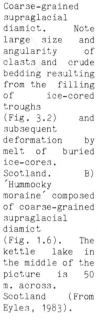

Fig. 4.5   A) Coarse-grained supraglacial diamict. Note large size and angularity of clasts and crude bedding resulting from the filling of ice-cored troughs (Fig. 3.2) and subsequent deformation by melt of buried ice-cores. Scotland.   B) ´Hummocky moraine´ composed of coarse-grained supraglacial diamict (Fig. 1.6). The kettle lake in the middle of the picture is 50 m. across. Scotland (From Eyles, 1983).

soils included within the ridge.

As the glacier continues to thin, meltwater streams occupying the trench between the glacier and moraine ridge, deposit thick sequences of sands and gravels (Ch. 7). These are left as terraces (kame terraces) as the glacier margin thins. The internal stratigraphy of these terraces is often multilayered and shows complexes of glaciofluvial sands and gravels, laminated clays (from the formation of ice marginal lakes), lenses of resedimented coarse supraglacial diamicts and valleyside fan and rockfall debris. In many ways the sedimentary sequences that accumulate in the sedimentary trough between the valleyside and the glacier margin (Figs. 1.6, 1.8D) are similar to those that are deposited in ice-cored troughs along the glacier terminus. Kame terraces frequently show internal deformation in the form of basinal saggings marking the collapse of strata deposited over melting ice-cores. Fine-grained strata are faulted but faults generally do not occur in gravels as the coarse particles move relative to each other and the shear-zone 'dilates'. Kame terrace formation continues at lower elevations as the glacier thins often resulting in terrace 'flights' (Fig. 4.8).

4.5 Geotechnical Properties Of Supraglacial Diamicts On Valley Glaciers

Coarse-grained supraglacial diamicts have not experienced comminution at the glacier base and as a result of washing on the ice surface, periglacial weathering and inclusion of silt and clay lenses during deposition , has a variable grading with a silt/clay fraction (mud) less than 15% (Fig. 8.1). It has a predominance of angular, oxidised clasts that lack striations, flat-iron shaping and preferred orientation typical of clasts in lodgement till and is best considered as a granular, effectively cohesionless soil (12 kN/m^2) whose shear strength (200-600 kN/m^2) is derived almost solely from particle strength and state of packing (Zeller and Zeindler, 1957; Marsal, 1969; MacArthur, 1969). Supraglacial diamicts are normally consolidated though overconsolidation occurs in the cores of lateral moraine ridges. Nests of boulders and mud lenses incorporated during deposition, result in a variable density (bulk; 1200-2200 kg/m^3, dry; 1050-2300 kg/m^3), high porosity (20-50%) and variable compressibility. With a low moisture content (6-14%) and reduced plastic and liquid limits (Fig. 8.2) supraglacial diamicts are sensitive to moisture changes which can be a problem in construction since these diamicts are found in wet upland areas of high intensity rainfall where dry spells are unreliable. Variable mud content causes problems with compaction for use as core material in rockfill dams since moisture contents are above their Proctor optimum value (6-8%). Mud and sand lenses also result in differential compaction when used for embankments and general fill. Because of the low state of compaction, supraglacial diamict is difficult to sample by standard methods and laboratory testing of representative samples containing large clasts is a major problem (Penman, 1971).

In the absence of glacial comminution, weak lithologies and those found in the more distant parts of the glacierized basin are over-represented in supraglacial diamicts compared to that of lodgement till. Since a large component is derived by rockfall from the valleysides it is frequently very bouldery; house-size boulders being not uncommon (Figs. 4.2, 4.6).

4.6 Deglaciation Of Glaciated Valleys: The Significance or Slope And Valley Infill Processes

4.6.1 Soil Slopes

Deglaciation of valleys leaves oversteepened valleyside slopes, often containing high groundwater levels and buried glacier ice and as a result lateral moraine ridges and kame terraces are rapidly dissected (Fig. 1.6). Soils are transported downslope by debris flow over valleyside fans (Broscoe and Thompson, 1969; Jackson, 1979). The coalescence of fans from opposing valleysides results in a U-shaped cross-profile commonly misattributed to glacial erosion. These fans are products of an environment transitional from predominantly glacial to predominantly fluvial conditions hence the term paraglacial (Ryder, 1971; Church and Ryder, 1972; Jackson et al, 1982). 200 m of paraglacial valley infill is not uncommon in glaciated valleys. The modern river often flows in a deeply entrenched channel below well developed terrace surfaces recording postglacial fluvial dissection of paraglacial infill (Fig. 4.8).

Paraglacial aggradation and infilling of valley floors is characterized by its extreme rapidity and a good example of such activity at the present day are the 'aluviones' of the Peruvian Andes where glacier recession in high valleys has

Fig. 4.6. The Big Rock of Okotoks, Alberta; a large quartzite mass of some 118,000 tons originating in the Jasper area west of the continental divide. Dumped by rockfall from an ice-free peak (nunatak) onto the Cordilleran Ice Cap surface (Fig. 1.1) it was transported east to the foothills near Calgary as part of the Foothills Erratics Train (Stalker, 1976).

left unstable moraine-dammed lakes. Failure of the moraine dam as a result of piping, earthquakes or large ice falls directly into the lakes result in a sudden flood of mud and boulders downvalley at velocities between 30 and 360 kph. Aluviones are a particular problem in the fertile and heavily settled Callegjon de Huaylas' (LLiboutry et al, 1977); 15,000 persons were killed by an aluvione in May 1970. After the disaster of December 1941 when the city of Huaraz was buried the Comision de Control de Lagunas de la Cordillera Blanca began to monitor moraine dammed lakes; recent remedial engineering work has involved tunnelling and revetment construction to increase freeboard around the lake dams.

In many valleys stratified glaciofluvial soils (Chap. 7) and bedded slope deposits similar to solifluction debris (Sect. 5.4.2, Fig. 1.8E, Sect. 10.1.2) are the dominant engineering soils; in glaciated valleys of the Appalachian Plateau for example Coates (1974) has described 300 m of glaciofluvial valley fill. Crandell and Waldron (1956) describe a stratified volcanic mudflow (a 'lahar') sequence, the characteristics of which are of general application to valley infill sequences. This extensive mudflow in the Cascade Ranges of Washington has a gently sloping (5–8 m/km) surface, lacks any glacial landforms and bedrock knobs protrude through it. The flow contains organic debris and is stratified with distinct upward fining textural gradations that typically arise as a result of the different particle support mechanisms that operate during flows (Johnson, 1970; Middleton and Hampton, 1976; Lowe, 1979; Sect. 3.8). The depth of the flow sequence shows an inverse relationships with valley width being up to 100 m thick in constricted valley sections. It may well be that present day aluviones of the Andean mountain chain and volcanic mudflows (lahars) of volcanically active areas provide good models for the rapid catastrophic paraglacial erosion of glacial landforms that would have occurred during deglaciation of many mountain valleys (Fig. 4.7). It is a very common observation that in many valleys very little in situ glacial material survives;

valley infill sequences and glaciofluvial/lacustrine sediments predominate.

At the present time, continuing postglacial valleyside slope activity is a hazard
to the utilisation of many mountain valleys particularly in North America where
development, deforestation and construction has been rapid and where there is a
lack of long term observations of slope activity. Conversely, in the European
Alps recurrent slope activity has been recorded historically and a variety of
preventive measures adopted ranging from remedial engineering works to
valleyslope zoning and avoidance of unstable areas. Movement of unconsolidated
soils downslope by debris flows appears to be the major problem in many areas
(Eisbacher, 1982). These flows are varyingly triggered by high rainfall, delayed
snow melt, deforestation, undercutting of slopes during construction activity and
flood water issuing from glaciers (Jackson, 1979). Flows generally emanate from
'reservoirs' at elevation, discharging through narrow gullies before widening
into a fan on the lower valleyside. The fan commonly grades into braided river
systems along the valley floor (Ch. 7). Debris diversion dams are employed to
contain debris flows moving over valleyside fans but the large volume of typical
flows (>10,000 m^3) makes such dam construction unsatisfactory without constantly
adding to the dam height or carrying out preventive measures (reforestation) in
the flow reservoirs.

In Europe, the systematic mapping of hazardous slopes in mountain areas commonly
recognises a ranking of slopes based on their activity within the past 150 years;
debris flows of a greater recurrence interval are difficult to identify in
practice (Eisbacher, 1982). Many maps combine the hazards due to debris flow,
rock slope failures and snow avalanches. Land speculation is avoided by the
legislated inability to insure construction on areas of marginal safety.

4.6.2 Rock Slopes

Rock slopes are commonly left in an oversteepened state by glacier retreat. A
complexly fissured and jointed rockhead is exhibited along many glaciated valleys
and results from rebound following glacial loading, glacitectonic stresses
imposed by basal ice and periglacial weathering. These effects are superimposed
on older tectonic lineaments and discontinuities along which glaciated valleys
are selectively eroded and therefore intimately associated.

A well defined system of joints, often defining sheet like rock slabs, is usually
found parallel to the valleysides and results from stress release into the valley
and relaxation of high horizontal stresses (Sect. 5.3). These have been referred
to as 'valley joints' (Bjerrum and Jorstad, 1968). Valley joints are limited to
the more competent lithological units in sedimentary successions (Fig. 5.7) and
are most characteristic of coarse-grained igneous and metamorphic rocks (Price
and Knill, 1967). A critical role in the release of sheeted rock slabs by
rockfall or slides is the development of high cleft water levels within the
rockmass which act to reduce effective stresses along joint surfaces (Peckover
and Kerr, 1976; Table 5.1). The presence of a glacier along the valley floor
also raises water levels in the adjacent valleysides and, following glacier
retreat and removal of lateral buttressing, rock slopes fail catastrophically.
Thus large rockfalls and slides are exceedingly common following glacier retreat
(Whalley, 1974; Cruden, 1976; Mollard, 1977). Piteau (1976) has also
demonstrated the importance of lateral river undercutting in dictating the
location of major landslides in the Coast Range valleys of British Columbia.
Growth of alluvial fans on one side of the valley displaces the river into the
opposite valleyside and bedrock masses subsequently fail as a result of toe-slope
erosion.

Fig. 4.7 a) Paraglacial fan sediments, Bow Valley, Canadian Rockies, Alberta. Poorly sorted, stratified debris flows exposed within large valleyside terraces. Figures for scale. b) Paraglacial fan sediments in the Alpine foreland, France. Note development of 'hoodoos' or 'mademoiselle coiffee' (boulder-capped pillars) at A. A crude stratigraphy parallel with surface slope is commonly developed similar to solifluction sequences (Fig. 1.8E).

In many valleys, hummocky rockslide debris, often extending from one valleyside to another, has been confused with moraine ridges (Porter and Orombelli, 1981). Various transport mechanisms have been suggested for the larger and more extensive rockslides (Voight, 1978) and the lubricating role played by saturated glacial soils swept up by slides is receiving increasing attention. Rockfalls and slides are a continual hazard to transport routes and settlements using glacially-eroded corridor valleys through mountain massifs (Bjerrum and Jorstad, 1968; Porter and Orombelli, 1981; Clague, 1981). A variety of remedial engineering work includes simple periodic scaling, rockbolting and buttressing, drainage measures, construction of debris fall trenches, retarding devices and rocksheds at the foot of rock slopes, large scale rockslope modification and tunnelling (Attewell and Farmer, 1975; Fookes and Sweeney, 1976; Voight, 1978). Simple avoidance of the site, detailed mapping (Merla et al, 1976), rockslope instrumentation and monitoring can also be mentioned. Well-defined seasonal cycles of rockslope activity during periods of snow melt and intense freeze-thaw activity during the spring and late fall are characteristic of many glaciated valleys (Grove, 1972; Peckover and Kerr, 1976, Church et al, 1979). On a longer timescale, periods of climatic deterioration (such as the so-called 'Little Ice Age', or 'Neuzeitlich' (=new ice time), of the period 1300 to 1800 A.D. in Europe) are associated with very large valleyside failures (Whalley, 1974) and enhanced rockslope activity (Rapp, 1960; Grove, 1972).

Peckover and Kerr (1976) describe remedial work along rockfall and slide prone transportation routes in Western Canada. Many problems are created during initial routeway construction work by over heavy blasting, the use of standard cut sections for rock slopes regardless of local conditions and deforestation of slopes during construction. A close relationship between forest clearance and subsequent landsliding is seen in many areas (e.g. Wu et al, 1979). Many design formulations established for large open pit rock slopes do not work along narrow transport corridors in mountainous terrain that cross many different rock types. Increasing emphasis is being placed on smooth wall blasting, where rockmasses are pre-split by light charges with millisecond delays along closely spaced aligned drill holes, and improved drainage during spring snow melt and major rain storms. Each part of the transportation route will offer a different problem requiring considerable care in examination of the rock slope and subsequent choice in remedial work.

Several workers have stressed the importance to rockslope instability of the release of residual stresses set up within mountain zones during episodes of tectonic and igneous intrusive activity and when the mountain topography is established (Gerber and Scheidegger, 1973; Kohlbeck et al, 1979). Rockfalls and slides are normal processes of mountain degradation and it has been suggested that the role of glaciation in high mountain areas is to evacuate pre-prepared bedrock debris (Gerber and Scheidegger, 1969; Whalley, 1974).

On the basis of geological sequence, relief, seismicity and intensity of precipitation, Eisbacher (1979) identified distinct landslide 'behaviour belts' within the Canadian Cordillera and demonstrated the value of regional zoning of landslide types as a basis of planning and remedial work. Rockfalls are particularly common in areas of high granitic massifs and volcanic metamorphic complexes where a wide range of discontinuities are created by stress-release and shearing. On the other hand large complex rotational slumps involve sandstone-shale-conglomerate sequences, whilst gravitational spreading of metamorphic rock slopes testify to slow creep. Catastrophic rock avalanches are frequent where bedding planes in carbonate and quartzite formations dip toward valleys (Cruden, 1976; Voight, 1978; Eisbacher, 1979a; Fig. 4.6). Rockslope processes affecting mixed sedimentary rock successions are discussed further in Sects. 5.4.4, Table 5.1 and Ch. 14.

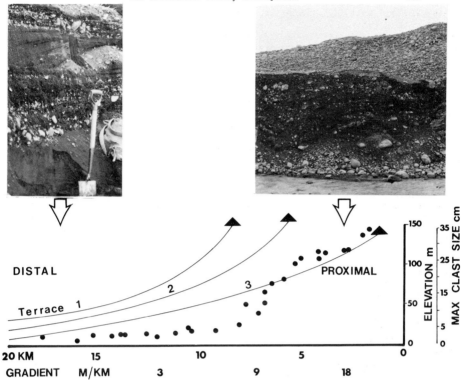

DISTAL

Terrace   1

2

3

PROXIMAL

ELEVATION m

MAX CLAST SIZE cm

20 KM   15   10   5   0

GRADIENT   M/KM   3   9   18

Fig. 4.8. Sediments and structures along a typical outwash fan and associated terraces. Note reduction in clast size and enhanced sorting from crudely bedded, matrix-supported (dirty) gravels, and cobbles to stratified sequences of clean gravels and cross-bedded sands. ▲▲ = ice-contact landforms such as moraine ridges, kame terraces (Fig. 1.6) or kettled outwash at the head of the outwash fan (Fig. 7.1B).

4.7 Glaciofluvial Erosion And Deposition

The predominant depositional process in many glaciated valleys has been the activity of high energy braided meltstreams. These streams are often responsible for a substantial debris discharge from the basin and are discussed in detail in Chapter 7. Another ubiquitous feature of valley glaciation is disruption and blocking of drainage and the formation of ice-dammed lakes. Many thick sediment sequences occupying buried valleys are composed of lacustrine clays and silts (e.g. Shaw, 1977b; 1979, Chs., 6, 14).

Channelled subglacial meltwaters with high velocities and abrasive loads are capable of eroding steep-sided flat-floored channels in bedrock and many of the so-called 'preglacial' valleys now buried by drift (e.g. Fig. 2.8) have been substantially modified by subglacial meltwaters (Woodland, 1970). Narrow gorge-like rock-cut or drift-cut valleys now dry or occupied by small 'misfit' streams, are ubiquitous in upland areas of moderate relief and were formerly considered to be the result of the overflow of ice-dammed lakes (Sugden and John, 1976). These channels bear no relationship to local topography often being cut across cols and spurs and are best explained as being cut by subglacial and ice marginal streams. Their engineering significance is that the side slopes are oversteepened and slope stability problems are common during construction (Glossop, 1968; Early and Skempton, 1972; Peckover and Kerr, 1976) due to the reactivation of old slope failures during excavations.

The glaciofluvial system is a dynamic one and meltwater discharges are characterised by violent summer floods superimposed on well-defined seasonal and diurnal discharge cycles posing costly river entrainment problems to engineers in modern glaciated areas. In general, fluvial processes in the glacial environment are similar to those of arid environments characterised by episodic, 'flashy' discharges, sparse vegetation cover and substantial fluxes of debris.

Thus significant and expensive problem in construction and maintenance of road and rail networks close to modern ice margins is episodic, frequently periodic, flooding, following the failure of ice-dammed subglacial or sidevalley lakes (Gilbert, 1971; Jackson, 1979). Perhaps the most notorious glacial floods are the Icelandic 'jokulhlaups' ('floods from the mountain') issuing from the Skeidarajokull ice lobe of the Vatnajokull Ice Cap (Nye, 1976). The scale of Pleistocene glacier ponding and flood discharges is even greater (Clague, 1975; Embleton and King, 1978; Sissons, 1981); the southward advance of the Cordilleran ice sheet (Fig. 1.1) along north-trending valleys in Washington for example, impounded several large lakes. The biggest, Lake Missoula, had a volume of 2130 Km^3. Its catastrophic failure has been graphically reconstructed by Bretz (1969).

Low sinuosity braided meltstreams migrate back and forth across the valley floor eroding primary landforms and sediments. As the process continues and downcutting occurs, older abandoned outwash fans are left as terraces on the lower valleysides, extending downvalley with remarkable persistence (sometimes referred to as 'valley train' after the characteristic slope of a bridal train). The older the terrace generally the steeper its surface gradient. Terraces are commonly 'paired' i.e. they are present at similar elevations on both sides of the valley marking the higher elevation at which the river formerly flowed prior to downcutting (Clayton, 1977). Terraces merge downvalley and may pass ultimately into fine grained marine sequences and raised beaches allowing important correlations between episodes of moraine ridge building (and therefore glacial activity) and sea-level change.

Fig. 4.8 shows the relationship between moraine ridges, outwash terraces and shorelines in southeast Scotland where an impressive databank of borehole logs, surveyed shorelines and large-scale topographic maps has been used to document land/sea level changes over the past 15,000 years along the Firth of Forth. Old raised marine shorelines dip to the southeast away from the former ice sheet centre in the Scottish Highlands to the northwest (e.g. Fig. 5.2). The oldest shorelines are found on the outer coast in East Fife. Each shoreline slopes less steeply than its older neighbour owing to declining rates of isostatic recovery as the ice sheet receded. Gradients range from 1.26 to 0.60 m/km and shorelines terminate upvalley in moraine ridges recording ice frontal positions. The most marked shoreline of this older series is the Main Perth shoreline. Subsequently, as isostatic recovery continued, estuarine muds were exposed, planed off by wave abrasion and buried by an extensive gravel outwash surface (the Main Late glacial shoreline). A substantial readvance of ice downvalley as far as Menteith is associated with a sequence of buried shorelines, (9, 10, 11; Fig. 4.9 A) inundated by a brief postglacial marine transgression when sea-level rose 7 m in

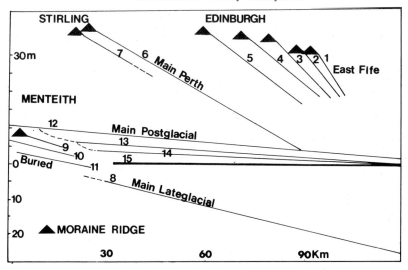

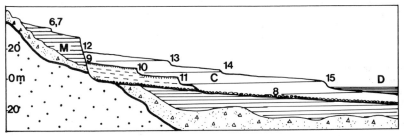

Fig. 4.9. A) Land/sea-level changes in southeast Scotland (After Sissons, 1976). Tilted shorelines (1-14) that slope west-east along the Firth of Forth result from greater isostatic crustal recovery towards the Scottish ice cap centre (approx. that shown in Fig. 5.2). See text for details. B) Diagrammatic cross-section at right angles to (A) showing ground conditions associated with shorelines 6-14. Dotted area is bedrock overlain by lodgement till. M) marine sediments cut by the abrasion platform and gravels of shoreline 8. Dashed horizontal lines represent beach sediments with peat on the beach surface, overlain by thick estuarine clays ('carse'; c). The modern shoreline (15) has extensive mudflats (D).

a thousand years at Menteith (Sissons, 1976). These shorelines are buried below marine muds ('carse') and peat. The Main Postglacial shoreline was abandoned about 6,500 y.b.p. with intermittent falls in sea-level thereafter recorded by postglacial beaches.

Perhaps the most famous examples of outwash terraces are those of the Alpine foreland in Bavaria, termed 'Deckenschotter', that can be traced upvalley into large terminal moraines marking former ice front positions (cf. Fig. 4.9). These outwash plain remnants were the subject of Penck and Bruckners' (1909) geological classic ´Die Alpen im Eiszeitalter´ whereby terrace horizons of the Gunz, Mindel, Riss and Wurm Rivers were ascribed to four major Quaternary glaciations - a model that subsequently has been very influential in Quaternary stratigraphy (Bowen 1978; Sect. 9.1).

In general, complex bedrock surfaces and old glacial sediments may be concealed by terrace gravels (Fig. 5.1, Sect. 5.2.4). Information as to the geometry of terraced sand and gravel units, their grading and engineering performance as aggregate can be found in varied reports of mineral assessment units operated by national, provincial and state geological surveys. Bowen (1978) provides a useful review of terrace stratigraphies in North America and Europe.

CHAPTER 5

Landforms and Sediments Resulting From Former Periglacial Climates

N. Eyles and M. A. Paul

INTRODUCTION

Large tracts of the mid-latitudes south of the limit of glaciation (Fig. 1.1) are recognized as containing relict landscapes and soils resulting from former cold, periglacial climates. The purpose of this chapter is to provide a check list of sediments (engineering soils) and ground conditions of particular significance to applied studies. The general distribution of many of these features in mid-latitude areas such as Britain and North American can be depicted and their typical geometries identified at the scale of individual site investigations.

The term periglacial is used elsewhere in this chapter as an umbrella-term ´for a wide variety of non-glacial processes and features of cold climates characterized by intense frost action regardless of age or proximity to glaciers´ (Washburn, 1980). This is an important statement for it should be noted that the meaning of periglacial, as used elsewhere in the literature, is ambiguous. In some cases it is reserved solely for those conditions where permanently frozen ground (permafrost) is present. The term is also employed to describe a fringing zone around modern or Pleistocene ice caps. Similarly, reference to a ´periglacial climate´ is unsatisfactory unless a particular climate is specified for there is a wide range of periglacial conditions with latitude and altitude. Periglacial environments range between the extremes of dry continental interiors with excessive seasonal extremes of temperature, to mountain zones in the mid and low latitudes where diurnal temperature changes are more marked and where mechanical break-up of rock masses by freeze-thaw cycles is at an optimum.

It must also be pointed out that many soils and landforms identified as ´periglacial´ form under conditions other than those of cold climates per se. Most of the features relating to modification of slopes by mass-wasting for example (Fig. 5.1) continue at present in the mid-latitudes albeit at a much slower rate. The character of fluvial and aeolian activity in many periglacial zones is similar to that occuring in present day semi-arid zones. The fundamental characteristics of landscape and soil evolution under a periglacial regime are in situ disturbance of underlying substrates (soil or rock) by frost and ground ice formation, production of weathered debris, its transport downslope by a variety of processes, valley floor reworking by high energy braided streams and wind activity. The existence of former cold climates has important engineering implications (Higginbottom and Fookes, 1971). For example, site investigations below slopes formed on clay bedrock often reveal the presence of solifluction sheets overlying relict shear surfaces on which the weathered soil has been reduced to its residual strength (Weeks, 1969). In other cases, deep excavations in valley floors, where the strata are composed of alternating hard and soft bedrock lithologies, have exposed disruption of the softer strata. Such ´valley bulging´ is associated with fracturing and displacement of the harder lithologies at their outcrops on the valley sides. Another example of the

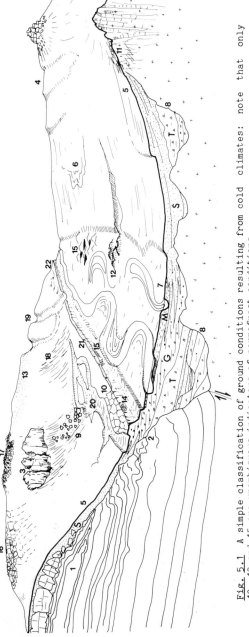

Fig. 5.1 A simple classification of ground conditions resulting from cold climates: note that only 10, 12 and 15 are unambiguous indicators of permafrost conditions.

A) Deep-Seated Disturbances

1. Flat-bedded sedimentaries cambered into valley; harder beds are fissured allowing individual blocks to founder and move downslope as complex slides or rafts riding on sheared-out soft lithologies. Distinctive fissure fills ('gulls') may be encountered upslope (Fig. 5.8).

2. Strata bulged and displaced upwards along the valley floor and cut across by undisturbed terraced gravels as part of a complex fill of a buried valley.

B) Slope Modification

3. Valleyside tors and cambered 'edges' exposed by slope-stripping and found in association with scree (clitter). Rockfalls common.

4. Summit tors on crystalline bedrock surrounded by blockfield of frost-heaved bedrock. Rock-cut nivation terraces cut by freeze-thaw and slopewash on the margins of snow patches (niveofluvial erosion).

5. Solifluction fans composed of downslope thickening, crudely-bedded sheets (s) of rubble ('head') turning downslope into slopewash on the lower slope. On carbonate terrains periglacial head may be cemented (e.g. coombe rock on chalk).

6. Mudflow underlain by low-angled shear planes - common on mudstone and shale bedrock.

7. Solifluction mantle interbedded with gravels deposited by periglacial braided streams and overlain by peats and silts (M) comprising the flood plain of the modern meandering stream.

8. Tills (T) lying in pockets in rockhead, below solifluction debris.

9. Stratified screes (grezes litees) and blockstreams.

C) Near Surface Disturbances

10. Fossil ice-wedge casts and polygonal patterned ground.   11. Frost-shattered bedrock.

12. Doughnut-like degraded circular rampart of former pingo; may also affect bedrock surface.

13. Clay-with-flints and other remanie soil mantles (plateau 'taele' gravels) resulting from mechanical churning of pedological soils and bedrock by frost heaving. Rockhead underlying such mantles is often highly irregular (cold water karst).

14. Soil involutions and contortions resulting from high pore pressures accompanying seasonal freezing of thaw-soaked soil (active layer; Fig. 5.4).

15. Faulted soil zones caused by ground contraction during freezing and/or collapse into subsurface voids following the melt of buried ground ice. The latter may also generate a pitted ground surface (thermokarst).

16. Patterned ground; polygonal nets and stripes.

17. Blockfield.

D) Fluvial
18. Alluvial fans.   19. Dry valleys.

20. Terraced gravels (G) deposited by braided flood-prone periglacial streams. The bedrock surface below these gravels may be diversified by scour-hollows.

E) Aeolian
21. Loess   22. Cover sand and dunes.

importance of former periglacial climates is provided by wind blown silt (loess).
This is perhaps the most characteristic periglacial soil and has an open
structure that collapses to a higher bulk density upon wetting. Such metastable
behaviour can result in severe construction problems.

Finally, it should be noted that it is not possible to link periglacial landforms
and sediments as a 'periglacial landsystem' in the same way as can be done for
glacial sediments. Fossil periglacial landforms in the mid-latitudes are most
commonly of an erosive not depositional origin (Fig. 5.1) and the underlying
periglacial sediment thickness may be thin or non-existent. Periglacial
landforms and sediments commonly form distinctive unconformities in many
stratigraphic sequences marking the cold climate weathering and erosion of a
landscape prior to renewed glacial or non glacial deposition.

5.1 The Distribution of Fossil Periglacial Landforms and Sediments

The generalized distribution of fossil periglacial phenomena in Britain, Ireland
and North America is shown in Figs. 5.2 and 5.3. Within the limits of the
periglacial zone which extended up to several hundred kilometres south of the ice
sheet limits, the type and density of forms varies widely; the concept of
extensive, maturely-developed periglacial landscapes simply does not apply. In
this regard, Fig. 5.1 depicts idealized landscape and ground conditions.

In North America, the most clearly developed forms are found along the
Appalachian chain, the Atlantic coast and within the Western Cordillera but apart
from the great expanse of loess and cover sands in the mid-west, periglacial
features are not well known in the continental interior (Smith, 1962; French,
1976). This is possibly the result of periglacial conditions that were too arid,
the rapid northward migration of displaced vegetation zones and the formation of
large postglacial lakes (Fig. 2.21).

Conversely, in Britain and Ireland, a more humid maritime position coupled with
extreme cold, results in an apparently greater density of periglacial forms, in
particular those relating to slope modification (Kerney et al, 1964; Chandler,
1972a; Sissons, 1979). In part this is a result of the smaller area and longer
observational history compared to North America but nonetheless eastwards, on the
more arid continental European mainland, a decreased maritime influence is
reflected in the predominance of periglacial wind transport and deposition
(Fig. 5.12).

An important consideration that has already been voiced in the introduction to
this chapter is that there are few features listed in Fig. 5.1 that are uniquely
periglacial. Of the many forms listed in that figure, slope processes are
probably of the greatest importance to the engineer but while periglacial
conditions saw an acceleration of rock and soil slope processes in many areas the
predominant control on rock slope stability is that of underlying bedrock
lithology. The importance of rock successions that encompass alternations of
competent and incompetent units that weather to engineering soils in dictating
slope activity can be stressed (Section 5.4.4). In this regard continued
research into the history of slope activity and its relationship with climatic
change is important.

5.5.1 Modern Day Periglacial Conditions

Modern day analogues for former mid-latitude cold climates probably do not exist;
the changing configuration of extensive postglacial seas and lakes as the ice
sheets withdrew (Figs. 2.21, 6.3) and the high sun angles found in mid-latitudes
are not a feature of modern high latitude periglacial zones. Detailed treatments

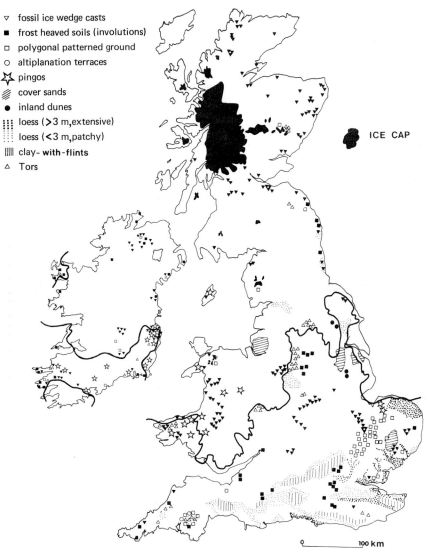

▽ fossil ice wedge casts
■ frost heaved soils (involutions)
□ polygonal patterned ground
○ altiplanation terraces
✕ pingos
▨ cover sands
● inland dunes
⦀ loess (>3 m,extensive)
⦙ loess (<3 m,patchy)
‖‖ clay- with-flints
△ Tors

ICE CAP

0 _____ 100 km

Fig. 5.2 Fossil periglacial features in Britain and Ireland. Note the overall increase in density outside the limits of the last glaciation (heavy line) and greater extent of aeolian cover in the east away from maritime influences (Fig. 5.12). Scale in km. The last phase of intense periglacial activity occurred about 11, 200 - 10,000 y.b.p., when an ice cap (the Loch Lomond Ice Cap; black area) reformed in Scotland (Sissons, 1979; Lowe et al, 1980). The effects of solifluction are ubiquitous.

of modern cold climates and geomorphological processes are found in French (1976) and Washburn (1980) and only general features of relevance to engineering investigations will be reviewed here.

The range of mean annual air temperatures in which periglacial climates are found at the present day is from -15 to +2°C compared to that in Britain, at present, of 10°C. Permanently frozen ground (permafrost) to depths of 1 km or more, underlie circumpolar areas where the mean annual air temperature is -7°C or less. Equatorward, in areas where annual air temperature is greater than -7°C, such continuous permafrost breaks down into discontinuous permafrost where the distribution of frozen ground is dependent upon snow and peat covers, water bodies, topography and aspect (Fig. 5.4). Permafrost may extend to depths of 200 m or more in deltaic lowlands with a deep sedimentary infill and submarine permafrost is present in shallow marine zones. The equatorward limit of discontinuous permafrost is about the -1°C isotherm.

Construction and design problems related to highways, hot oil pipelines, and sewage schemes in permafrost terrain are reviewed by Andersland and Anderson (1978) and Johnson (1981). 50% of the land area of Canada, 49% of the U.S.S.R. and 80% of Alaska are underlain by some form of permanently frozen ground. Recent interest in offshore submarine permafrost in the Beaufort Sea in Arctic Canada is a result of drilling for oil and gas, and construction of artificial islands (Vigdorchik, 1980a). Irrespective of soil or rock type, texture, water content or type of ice present, permafrost is defined as rock or soil material that has remained frozen for two or more years. The forms of sub-surface ice range from dispersed crystals to segregated masses such as lenses, wedges and pingo ice. Ice lenses are dominantly horizontal layers of ice ranging in thickness from millimetres to tens of metres and may extend over areas of several km. Ice wedges are vertical v-shaped ice lenses whilst pingo ice masses are plano-convex in section underlying a circular mound at surface often up to 50 m high. Each summer, permafrozen ground thaws to a depth averaging about 1 m resulting in a water soaked 'active' layer. The active layer is essentially a shallow zone of seasonal freezing and thawing overlying the permafrost table (Fig. 5.4). Below the permafrost table unfrozen ground (taliks) may exist.

Interest in the engineering behaviour of soils under modern periglacial conditions has promoted a number of soil classifications that recognize frost-resistant and frost-susceptible soils. The latter are soils in which segregated ice masses form during freezing leading to soil 'expansion and heaving at the ground surface. These soils are typically silts and silty-clays which are soils in which developing ice lenses are able to abstract water from the surrounding soil by virtue of enhanced capillarity. Thus classifications for the evaluation of the frost-susceptibility of soils emphasize the importance of silt content (Johnson, 1981). Better drained soils of low capillary rise such as sands, gravel, and tills are generally frost-resistant.

Soils with a high silt content are also thaw-susceptible in that they experience excessive porewater pressures and liquefaction following the melt of segregated ice. Similarly, well-drained thaw resistant soils are recognized.

5.2 Classification of Fossil Periglacial Phenomena

A simple classification of fossil periglacial phenomena commonly encountered in mid-latitude areas (Fig. 5.1) distinguishes four major groupings; 1) in situ near-surface and deep-seated disturbance of rock and soil masses; 2) slope modification processes involving downslope transport and deposition in topographic lows; 3) fluvial; and, 4) aeolian activity. It can be appreciated from Fig. 5.1 that the principal role of periglacial activity is to infill

v fossil ice wedge casts

x frost heaved soils (involutions)

o mounds of unknown origin, considered to result from ground ice

□ polygonal patterned ground

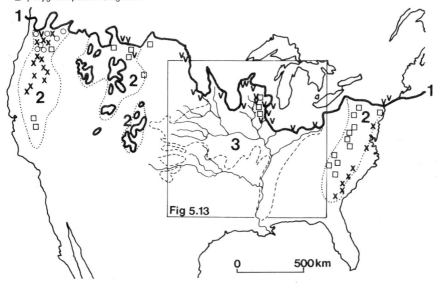

1   Wisconsin ice margin

2   Areas of frost-disturbed bedrock — solifluction mantles common

3   Extent of loess cover

<u>Fig. 5.3</u> Fossil periglacial features adjacent to the limit of the last (Wisconsin) glaciation in North America. The more southerly penetration of ice (38° lat.) compared to Europe (47° lat.) results in a more restricted periglacial zone. Few periglacial features developed on terrain exposed by glacier retreat as a result of large postglacial lakes (Figs. 2.21), high snowfall, steep climatic gradients south of the icefront and rapid colonization by vegetation. The inset shows area covered by Fig. 5.13.

pre-existing topography leading to an overall reduction in relief, with minor accentuation of relief in the higher areas. The grouping of periglacial phenomena established above will be followed below.

5.2.1 <u>Near Surface In Situ Disturbances: General Comments</u>

A complex range of weathered engineering soils and rocks is displayed at many sites and results from weathering processes that may operate singly or in combination and at different rates, under varying climatic conditions. Thus dessication cracking, as a result of lowered groundwater tables, soil contraction and cracking as a result of deep ground freezing, rebound and release of overburden or lateral earth pressures not only disturb the physical soil fabric but accelerate chemical weathering by promoting the activity of soil bacteria. In the long term, physical disruption, decalcification, acidification and mineral weathering, lead to the development of intensely weathered pedogenic soils. Pedogenic soils are commonly less than 2 m in depth and are usually divisible into three zones (A, B, and C), with fines and organic matter lost from the leached A horizon accumulating in the B horizon. The C horizon refers to parent material. This is a highly simplified picture of pedogenic soils (Curtis et al, 1976) and it is important to note that rock and soil masses are often weathered up to many tens of metres below the pedogenic soil. Igneous rocks, for example, may be deeply rotted as a consequence of Tertiary tropical weathering, ground water movement along fractures and hydrothermal alteration. The varying strength characteristics of weathered rocks and engineering soils are of crucial significance to the engineer in terms of excavation cost and foundation design; weathered rocks may in some cases be more effectively handled as soils (Sect. 9.5).

The subdivision of rock and soil masses into distinct weathering zones is used to describe physical field characteristics, to form a framework for laboratory analytical test data, and ensures that data from one site is compared to the same soil type elsewhere. Weathered materials are zoned according to the degree of alteration from the parent unweathered state. Zonation (commonly I, unweathered, to V, completely weathered) is determined on the basis of fissure intensity and fabric, extent of oxidation of matrix, fissure fillings, leaching of primary carbonate, extent of acidification and simple field strength and permeability properties. A working ad hoc weathering scheme can be quickly set up by engineering geologists working at any one site; several schemes are available for the more commonly encountered rock and soil types, e.g. igneous and metamorphic rocks (Knill and Jones, 1965; Deere and Patton, 1971; Dearman, 1974; Anon, 1977), chalk (Ward and others, 1968), mudrocks (Chandler, 1969, 1972b; Russell and Parker, 1979; Cripps and Taylor, 1981), and lodgement tills (Eyles and Sladen, 1981; Sect. 8.2.4). A major problem in engineering site investigations is assessment of the subsurface distribution of weathering grades from small diameter broadly-spaced drill holes. Trenching and test-pitting is cheap but may be prohibited by high water tables; large diameter auger holes permitting direct visual examination represents the best approach. In weathered strata in situ field testing is preferable to the laboratory testing of disturbed specimens.

5.2.2 Near Surface In Situ Disturbances

The periglacial environment is characterized uniquely by the high degree of in situ disturbance of soil and rock by frost penetration and the formation of ground icings. The disturbances that result are highly variable ranging from narrow but deep fossil ice wedge casts where the soil mass either side of the wedge is little affected (Fig. 5.7), to involutions, contortions (Fig. 5.6) and landscapes of large hummocky forms resulting from the growth and melt of extensive buried ground ice (Fig. 5.1). The engineering significance of these disturbances is that there are very rapid changes in soil type and properties with depth across small areas such as the average construction site. Near surface groundwater flow paths are highly disturbed and considerable nuisance value is created by water, following periglacial structures, emptying into construction sites (Morgan, 1971). The same soils that are prone to periglacial disturbance, i.e. silts, are also likely to prove problematical with regards to plant movement during wet weather.

The principle forms of in situ disturbance of rock and soil fabrics by periglacial processes relate to mechanical breakage and disruption by frost (congelifraction) and the churning and mixing of the weathered rock or soil mass (congeliturbation).

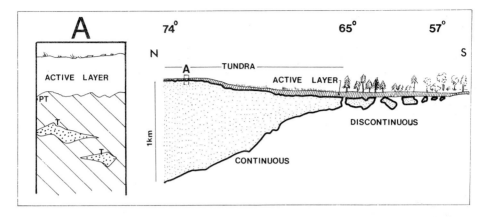

Fig. 5.4 Approximate distribution of permafrost along a north-south transect (110°W. long) in northwest Canada at the present day. Continuous permafrost north of the tree-line contrasts with discontinuous permafrost to the south. Submarine permafrost extends onto the continental shelf to the north (Vigdorchik, 1980,a). Sporadic permafrost occurs in peat bogs south of the limit of discontinuous permafrost. The surface active layer of seasonal freezing and thawing is generally between 1 to 3 m deep. Unfrozen taliks (T) exist within the permafrost body below the permafrost table (PT). Seasonal thaw does not penetrate below the permafrost table. Compare with Fig. 2.2.

5.2.3 Frost-Shatter

Mechanical disruption, caused by deep frost penetration and the freezing expansion of interstitial waters, is so widely reported that it should be anticipated at all engineering sites (Higginbottom and Fookes, 1971). Its character varies from intense fissuring to bodily heaving of rockhead. Frost action is not confined to any particular range of rock types but well-jointed rock masses are particularly susceptible. Frost disturbance down to 30 m is reported; chalk, mudrocks and shales may be completely remoulded to a structureless paste by repeated freeze-thaw cycles. Ward et al (1968) show a tenfold difference in the modulus of deformation for the end members of a five part weathering scheme for chalk; values which are probably not untypical of many fine-grained rock types that have been severely disturbed by ground icings and freeze-thaw. The failure to differentiate frost-shattered material from in situ bedrock during the site-investigation stage is frequently apparent once excavation proceeds (Section 5.4.3).

5.2.4 Ground Icings

Massive ground ice bodies, tens of metres thick, form in present day Arctic coastal lowlands that are floored by highly permeable gravels and sands from which water can be drawn for continued growth of subsurface ice.

Distinct conical mounds of buried ice up to 40 m high and 600 m in diameter, are termed 'pingos'. These may develop as 'open system' forms where groundwaters, moving downslope within unfrozen intrapermafrost taliks (Fig. 5.4), rise to the surface, freeze and generate updomed sediments (MacKay, 1978). Open system

pingos form in alluvial valleys in areas of discontinuous permafrost where
artesian water circulates; the largest known concentration of active forms is
found in unglaciated areas of Alaska and the Yukon (MacKay, 1979; Hamilton and
Obi, 1982). They are extremely rare in formerly glaciated areas (Flemal, 1976)
though Watson (1971) argued for open system formation for fossil pingos in Wales.
'Closed system' pingos on the other hand, form in areas of continuous permafrost
by the freezing of shallow lake bodies producing a closed pocket of unfrozen soil
and water at depth that is enclosed by frozen ground. High porewater pressures
generated in the unfrozen zone leads to intrusion of water into the soils above
and updoming of the soil mass.

In Britain, ill-drained peat-filled depressions up to 10 m deep and a hummocky
surface topography, probably resulting from thermal degradation (thermokarst) of
pingo forms, are reported from chalk surfaces in East Anglia (Sparks et al, 1972)
and from fluvial terraces cut across clay rocks in central London (Hutchinson,
1980). In the latter example, broad areas of terraced gravels flanking the River
Thames conceal a complex hummocky bedrock surface. Bedrock hummocks are
associated with local overthrusting and crumpling formed as a result of expansion
and subsequent decay of ground ice prior to burial by stream gravels (Berry,
1979). More fossil ground ice forms are likely to be identified in other valley
settings in former periglacial zones. The degradation of buried ground ice in
soils frequently results in foundering of overlying and adjacent strata into the
newly created subsurface voids with high angle, often reverse, faults either side
of large downward displaced blocks (Fig. 13.3). Normal faults record lateral
extension of the soil mass into subsurface voids (Shotton, 1965).

5.2.5 Cryoturbation

Soil churning, or cryoturbation, occurs at the end of the melt-season when high
pore pressures in the active layer are set up as soils freeze from the surface
down to the permafrost table (Fig. 5.4). Fine-grained silt-rich soils are
particularly susceptible; freezing of porewaters results in 9% expansion of the
soil framework which is expressed as heave at the surface. This is compounded by
the disruption generated by the growth of ice lenses parallel to the descending
freezing-front. The upward 'heave' or 'upfreezing' of stones and boulders
contributes to soil churning. Upfreezing of clasts is the result of ice
formation around the clast (reflecting their greater thermal conductivity) and
the extrusion of the clast up through the soil mass; other processes also
operate (French, 1976). The heave of smaller particles results from the upward
movement of soil water towards the descending freezing plane. A hummocky
surface, referred to as patterned ground (Fig. 5.5), develops by continued soil
churning and is sometimes underlain by soils showing 'cryoturbation' structures
of complexly involuted flame and festoon structures (Fig. 5.6). The depth of
cryoturbation is a guide to the depth of summer thaw to be permafrost table at
time of formation. The effects of frost churning are particularly clear where
the affected zone originally consisted of horizontal layers of different soil
types; deformation by churning is then easily discerned. In other cases, the
occurence of vertical heaved stones is particularly diagnostic.

Patterned Ground

A hummocky patterned ground surface showing closely grouped circular forms is a
result of continued soil churning and the heaving of material to the surface
(Fig. 5.5). Soil churning on a shallow slope results in extension of circular
surface patterns into soil stripes as churned material creeps downslope
(Fig. 5.1). Other larger polygonal patterns in soil and bedrock are the result
of soil cracking during intense cold; the freezing of water penetrating down
cracks results in an ice wedge which may expand, by repeated seasonal cracking

Fig. 5.5 Periglacial patterned ground composed of sorted circles. Results from continued frost churning (congeliturbation) at the circle centres and radial movement of clasts. Spitsbergen.

Fig. 5.6 Periglacial involutions; deformation by high porewater pressures set up during freezing of the active layer. An upper sand layer penetrates a lower gravel bed. North Wales.

and filling, to dimensions of 10m deep and 3 m in width. Their slow thaw at the end of periglacial conditions, and infilling with soil leaves a fossil ice wedge cast (Fig. 5.7). Sand wedges, where the contraction crack was filled by windblown sand, are also common (Black 1976). It should be stressed that wedge-shaped soil infillings and soil 'tongues' develop by a variety of processes not simply by thermal contraction (Walters, 1978). Fossil patterned ground is easily recognized from air photographs by the response of vegetation during dry spells when differences in soil moisture exist across the surface (Shotton, 1960; Morgan, 1972).

5.2.6 Dissolution of Carbonate Bedrock

On soluble carbonate-rich rocks, frost churned soil may disguise a highly irregular rockhead showing large vertical pipes and deep, enclosed, steep-sided hollows resulting from solution. An illustrative example is that of the Cretaceous chalk dipslopes in southern and eastern England and the Champagne country of northern France where a local relief of 10 m is not unusual. The deeply indented surface is overlain by 'clay with flints' (Hodgson et al, 1974) a ferruginous, flint-rich, calcareous sandy clay resulting from solution of the chalk surface and congeliturbation of residual masses of Tertiary sandstones. A windblown silt (loess) component is also recognized. An alternative suggestion that clay with flint deposits are residual decalcified till masses from an early as yet unrecognized glaciation is discussed by Shephard-Thorn (1975). The 'plateau drifts', mapped last century by the Geological Survey throughout southern England, are also in situ frost-churned and weathered relict soil masses. Deeply dissolved limestone surfaces ('karst') are frequently encountered below thick drift covers. The collapse of soil into subsurface voids is often expressed on the drift surface (Fig. 13.5) and can be recognized from air photographs and mapped in detail using simple geophysical apparatus; the approach is similar to that used for the location of shallow subsurface mine workings (Dearman et al, 1977a).

5.3 Deep-Seated In Situ Disturbances

In areas where hard and soft bedrock lithologies alternate, it is common for the harder rocks to form cappings to hills and escarpments. This situation occurs widely along the Jurassic and Cretaceous outcrops of southern and midland England, on Carboniferous strata in south Wales and northern England and on Cretaceous and Tertiary mudstones and sandstones of Alberta, Saskatchewan, Manitoba, eastern Montana and the Dakotas. In these areas lithological successions of clay, shale, mudstone, limestone and sandstone are found in several combinations. If the lower slopes, cut into soft lithologies, suffer periglacial disturbance, the cap rock will start to move downslope. This phenomenon is termed cambering, and is frequently encountered on the margins of escarpments overlooking clay lowlands (Straw, 1966). Cambered blocks are usually a few tens of metres across, and dip inwards toward the valley at 30° or more. This is usually out of character with regional structures, and often provides the first evidence for the existence of cambering.

Valley bulging is a deep-seated structural disturbance usually associated with cambering (Fig. 5.8). Deep excavations in the floors of valleys overlooked by cambered strata often reveal contortions in the soft layers in which there appears to have been upduming as a result of horizontal shortening. These structures were first noticed in quarry workings (Hollingworth et al, 1944) and excavations for dam cut-off trenches and appear to be quite widely distributed. Their engineering significance lies in the fact that the distortion produces polished shear surfaces within and adjacent to the bulge, which often dip at steep angles. The adverse influence of such features on the stability of deep

Fig. 5.7 Ice wedge cast in sands and silts. Fast Anglia, England. Note normal
faults resulting from collapse of strata as ice wedge thawed.

foundations or excavations for dams and other structures does not need to be
emphasized (Walters, 1971; Fig. 14.5).

Significant engineering problems are associated with the dislocations between
cambered blocks (Fig. 5.8). Open tension cracks, termed ´gulls´, may be several
metres wide. These are not always open to the surface and so may form
‚substantial voids within the cambered formation. Gulls are commonly filled with
till or washed material containing fossil fauna and floral remains with a
significant windblown silt (loess) component (Shephard - Thorn, 1975). Gulls are
a frequent nuisance in foundation design and are associated with   rapid
unpredictable subsidence particularly when filled with metastable loess (Worssam,
1981; Sect. 5.6). Gulls result from extensional movement as cambered cap rocks
move downslope (Hedges, 1972). They are frequently invisible on air photographs

but a series of deep (3 m) trenches at right angles to the slope contours is
usually effective in locating them during site investigations. Geophysical
exploration is made difficult given the widespread distribution of open joints
and frost-shattered rock to irregular depth. Gulls are typically arranged en
echelon and terminate at major cross-cutting joints. Hawkins and Privett (1981)
describe a ´swarm´ of gulls that completely penetrate a cambered limestock
caprock. Downslope riding of the limestone over underlying clay can be
approximated to a wedge failure and the factor of safety calculated and plotted
against a range of residual shear strengths for various water heads within the
gulls. In complex cases (e.g. Horswill and Horton, 1976), the cambered formation
may be disrupted by systems of normal faults which serve to lower the cap rock
down the valley slope (Fig. 5.8). Although most prominent downslope from the
slope crest, cambering and related phenomena have been traced up to 1km back from
the edge of caprock escarpments.

There is little agreement on the origin of valley bulges. It is generally
considered that they are genetically related to cambering, in view of their
spatial association and the fact that the shortening required for the bulge is
complementary to the extension required for the cambering, at least in a
qualitative sense. They appear to be restricted to areas that have suffered
glacial or periglacial activity but it has been suggested (Peterson, 1958;
Ferguson, 1967; Matheson and Thompson, 1973) that stress relief following valley
incision may have caused the updoming and associated lateral movements. Raised
valley rims indicate updoming along the valley floor (Fig. 5.8). Vaughan (1976),
has shown that rebound after valley excavation could cause a bulge-type structure
which initially would be smaller than the observed features but might grow as
result of repeated freeze-thaw activity over long periods. In many shales,
rebound occurs in two stages; an initial period of rapid elastic rebound
followed by a slower rate of expansion in which the shale is disturbed by
increasing water content and disruption to the clay fabric. Deep ground freezing
may be more disruptive in its effects given the development of such higher
moisture contents.

Matheson and Thompson (1973) observed that along valleys cut in Cretaceous marine
mudstone and sandstone sequences, total rebound was of the order of 2 - 10% of
valley depth. Rebound rates of 1 cm yr are reported by Fleming et al (1970).
Because of the presence of unrelieved stresses in nearly all rock types
(Nichols,1980), local rebound phenomena, separate from those of regional
glacio-isostatic rebound (Chapter 6), should be more common in glaciated terrains
characterised by rapid valley erosion, than they actually appear to be from the
published data. The fact that rockhead is in most cases thickly veneered by
sediment and is seldom directly observed may be significant. Thus it is probably
not a coincidence that local bedrock structures caused by the release of
overburden pressure are well known on the glacially-scoured Palaeozoic limestone
plains abutting the Canadian Shield in eastern Ontario and adjacent New York
State where there is no sediment cover. Here ´pop-up´ structures involve in situ
decoupling and arching of the uppermost 1 - 3 m of limestone (and thus are
strictly shallow features) and appear to be postglacial. An elongate
drumlin-like form has been employed to indicate the direction of maximum
horizontal stress remaining in the ground. Release of overburden pressure
(´pop-up´) is rapid, similar to that experienced with buckles seen in quarry
floors (Coates, 1964; Lo, 1978). Nichols (1980) has identified high values of
horizontal stress (2 - 10 MPa) in the area of shield terrains and shield margins.

In summary, the relationship between deep-seated ground structures and former
periglacial climates is difficult to specify in the light of strong evidence for
the importance of rebound processes.

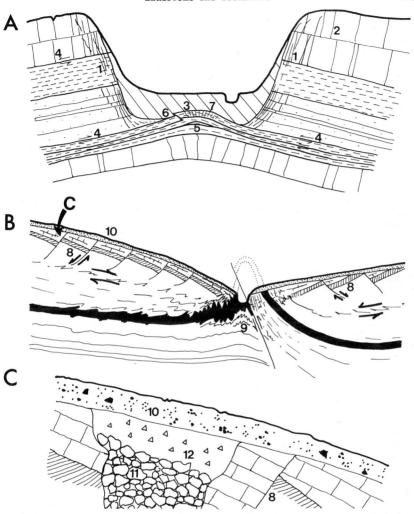

<u>Fig. 5.8</u> A) Schematic section through a valley cut in mixed lithologies showing
raised valley rims.
1) Stress relief joints   2) Master joint due to tensile fracture
3) Tension cracks   4) Bedding plane shears
5) Open voids   6) Overthrust fault
7) Buckled beds (bulged)
        B) Deformation in soft rocks below cambered cap rocks.
8) Normal faults in cap rock and rotation of fault-bounded blocks
9) Deep-seated compression and shearing accompanying valey bulging
    (Fig. 14.5).
10) Solifluction sheet
        C) Detail in B; a ´gull´ formed in the caprock.
11) Rubble with voids   12) Till.

5.4 Periglacial Slope Modification

The periglacial environment is one in which slope processes are particularly active due to the ground disturbance that occurs as a result of repeated freeze thaw cycles, and also because of the availability of spring meltwater from ice and snow. Slope processes are encouraged by the poor drainage of areas underlain by permafrost, which restricts the movement of soil water to the active layer at the ground surface. The depth and stability of the active layer commonly controls slope development under periglacial conditions.

5.4.1 Solifluction

Solifluction is perhaps the most widely known process of periglacial mass movement and is the flowage of soil and rock debris as a water soaked mass. Some authors prefer gelifluction to describe the process when it occurs as a result of ice melt in thaw susceptible soils and rocks. The terms include both true flows, in which movement occurs throughout the depth of the moving mass, and slides in which most movement occurs across a basal shear plane (Chandler, 1970; Chandler, 1972a; Harris, 1977, 1981, 1981a). Similar remarks appertain to flowed tills (Sect. 3.8).

Solifluction occurs as a result of the excess pore water pressures that are generated during thaw. The basic theory has been presented in a series of papers by Morgenstern and co-workers describing the thaw of frozen ground with regard to the stability of dams constructed on thawing ground, and thaw 'bulbs' developed around hot oil pipelines (Morgenstern and Nixon, 1971; Nixon and Morgenstern, 1973).

When frozen soil thaws, there is a transference of load from the interstitial ice to the soil skeleton. This raises the effective stress, and so the soil consolidates. The transferred load is taken initially by the soil water (as in the usual consolidation model) and so an excess pore pressure is created whose subsequent decay controls consolidation ('thaw-consolidation'). The magnitude of the excess pore pressure is determined by the balance between the rate of thaw and the rate of consolidation which can be expressed as the thaw-consolidation ratio (R). R is given by the expression:

$$R = \frac{\kappa}{2\sqrt{C_v}} \tag{5.1}$$

where κ is the thaw constant that describes the rate of progress of the thaw plane into the soil ($\kappa\sqrt{time}$) and C_v is the coefficient of consolidation. If $R=0$, dissipation is instantaneous whereas if $R=\infty$, no dissipation occurs. The maximum value of the excess pore pressure (U) at depth as a fraction of the effective stress ($\sigma_{v'}$) that acts vertically is given by

$$U / \sigma_{v'} = \left[\frac{1}{1 + \dfrac{1}{2R^2}}\right] \tag{5.2}$$

If the thawing soil lies on a slope, the increase in pore pressure leads to a decrease in soil strength and thus a reduction of the angle of ultimate stability against landsliding. McRoberts and Morgenstern (1974) have shown that under conditions of thaw consolidation the angle of ultimate stability (ψ) is given by the expression:

$$\tan = \frac{\lambda'}{\lambda}\left[1 - \frac{1}{1 + \dfrac{1}{2R^2}}\right]\tan \phi' \tag{5.3}$$

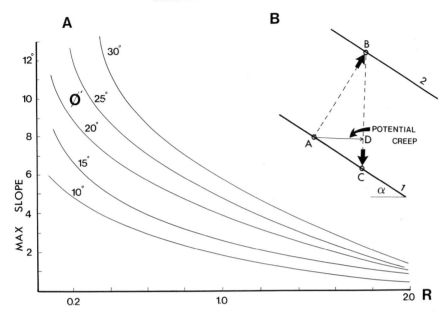

Fig. 5.9 A) Maximum stable slope under solifluction conditions for different values of thaw consolidation ratio (**R**) and angle of internal friction $\emptyset'$. B) Downslope movement of a particle during one freeze-thaw cycle. A particle A, with the ground surface at level (1) will be raised to B by frost heave which lifts the ground surface to position (2). During the subsequent thaw, it will settle to point C. The horizontal distance AD is the potential creep for a single freeze-thaw cycle. Since AD = ABsin α, the potential creep is greater on steeper slopes or for larger heaves. In practice, the movement can be greater or less than AD due to interference with adjacent particles.

for a cohesionless soil of unit weight λ and submerged unit weight λ', and friction angle $\emptyset'$. Insertion of typical values into this equation shows that movement can occur on clay slopes as shallow as 1° - 2° (Sect. 5.4.3). Typical solifluction rates range from 1 to 6 cm/yr.

In addition to an effective stress model, it is also possible to analyze the mechanism of solifluction using total stresses. Hutchinson (1974) showed, in the case of solifluction mantles on clay slopes, that there is an inverse relationship between the thickness of the solifluction layer and the slope angle and concluded that the mantle moved into position by undrained failure at constant shear strength. Values of undrained strength are probably related to moisture content and data suggest that the soil was soliflucted into position at a liquidity index (Sect. 8.2.3) of between 0.6 to 1.0 (moisture content=50 - 75% approximately). These values indicate substantial water contents in the newly thawed soil, which agree with the high pore pressures predicted by the effective stress theory of thaw consolidation.

During flow episodes related to spring snow melt, weathered rock particles and matrix are transported more rapidly as a watery slurry or debris flow. The transport mechanism involves both true suspension in which the particles are supported by fluid turbulence, and interparticle interference whereby particles are supported by momentum transfers due to collisions. The character of the deposits so formed can reflect either support mechanisms: thus sedimentary structures can occur in proximity to massive matrix-supported units in which larger clasts ´float´ (Sect. 3.8). The actual flow mechanics are complex, but it appears that such debris flows behave as pseudo-plastic liquids which give rise to lobate flow morphologies (Hutchinson, 1970).

Slow creep may occur at other times of the year when soil or rock particles are displaced from their rest positions by the growth of ice crystals such as needle ice (´pipkraker´; Washburn, 1980). This displacement occurs in a direction normal to the ground surface (Fig. 5.9). When the ice melts, the particle settles vertically, and so moves downslope by a small distance. Repeated cycles of growth and melting of crystals lead to overall movement of particles downslope by an irregular, ratchet-like process (French, 1976; Fig. 5.10). Field measurements of present day creep are rare but movement rates of about 1 - 4 cm/year are reported. Since the amount of movement clearly depends on the slope angle, creep is most significant on steeper slopes. It should be emphasized, of course, that creep is not exclusive to periglacial conditions; wetting and drying cycles also produce similar movements.

5.4.2 Solifluction Sequences

The character of solifluction debris varies greatly according to the material from which it is derived. South of the limit of glaciation solifluction debris consists of frost shattered bedrock of local origin whereas to the north, glacial soils may be moved downslope to form valley infill sequences that can be very difficult to distinguish from in situ glacial soils (Fig. 1.8E).

In Britain, solifluction debris is ubiquitous and it is very rare for landforms and rock slopes not to show downslope thickening accumulations (Fig. 5.11). North of the limit of glaciation, solifluction debris is composed of disaggregated tills, outwash, scree, slipped and fallen bedrock masses and therefore a confusing range of stone shapes and lithologies. The long axis of stones and boulders lie downslope parallel with the surface slope (Fig. 5.11) and a crude internal bedding results from the successive accumulation of downslope thickening sheets of debris. Angular clasts predominate as a result of freeze-thaw weathering of exposed bedrock but the odd striated and shaped boulder will be encountered having been eroded from till. Softer bedrock lithologies are disturbed by solifluction and creep processes (Fig. 5.11). The packing of solifluction debris is normally poor with numerous voids between the larger angular clasts. Fines washed out of the debris accumulate downslope as sandy slopewash which may be well stratified forming ´grezes litees´ (Sect. 5.5). Solifluction sequences commonly show interbedded gravels and sands that record periods of enhanced snow melt or summer storms and increased surface run off. Distinct upwards fining or coarsening of grain size (Fig. 1.8E) occurs in response to cyclic solifluction and slope wash events or to exhaustion of upslope supply of coarse debris. Solifluction sheets frequently block drainage along valley floors resulting in the formation of shallow lakes and bogs. Peats are buried episodically by further solifluction debris. The final valley floor sequence shows alternating peat horizons, clays and silts, gravels and solifluction rubble. The peats and clays are particularly troublesome due to their high compressibility.

Dines et al (1940), Mottershead (1971) and Benedict (1976) give excellent reviews of the character of solifluction sequences and of the diverse terms that have

Fig. 5.10 The effect of long-term creep processes (Fig. 5.9) in the periglacial environment. House posts have been rotated downslope by freeze-thaw creep cycles since 1943 (Jahn, 1976); Longyearbyen, Spitsbergen.

Fig. 5.11 Deformed bedrock rotated downslope by periglacial creep and overlain by a solifluction mantle 1.5 m thick. Note coarse grading and downslope orientation of clast long axes. Dartmoor, England.

been used to describe them. Dines et al (1940) show that several episodes of solifluction are recorded in southern unglaciated Britain. Dissected terraced remnants of once extensive sheets contrast with younger slope sequences at lower elevations (see also Crampton and Taylor, 1967). In Britain, periglacial slope deposits are well exposed along the coast (Dines et al, 1940; Watson and Watson, 1967). In many situations the modern cliff truncates old wave-cut platforms buried by beach deposits and solifluction debris derived from a buried cliffline inland. Thick slope deposits are bedded recording the interplay of slope wash and solifluction as the degrading cliff was progressively entombed under periglacial conditions accompanying lowered glacial sea-levels (Fig. 4.9). A good example is the buried chalk cliffline of Yorkshire and Lincolnshire which dates from the previous interglacial (Fig. 9.3) when sea-level was some 15 m higher than that present (Fig. 15.1). The volume of solifluction debris on wave cut platforms has enabled estimation of cliffline recession rates under periglacial conditions (Williams, 1968).

Solifluction debris is varyingly referred to as 'head' (because such debris always caps or 'heads' the local stratigraphy), morainic or upland drift, or guck (glacial muck). It is frequently confused with in situ till and many 'local' or mountain tills, with clasts derived only from the underlying bedrock, are in fact thick accumulations of solifluction debris. By way of contrast solifluction is also an extremely effective way of stripping slopes in mountain and upland areas and the absence of any glacial soil cover in many upland areas frequently records the efficiency of solifluction rather than ice-free conditions during glaciation.

The distribution of solifluction debris is markedly asymmetric along many valleys with varying degrees of bedrock exposure from one valleyside to another. Inactive north facing rock slopes, where the depth of thaw was restricted, contrast with thickly obscured thaw-prone south facing slopes. In some other contrasted cases, south facing slopes dried out more quickly and thus become inactive. Active solifluction fans may push the valley floor stream over to the other valley side which will be undercut and stripped of any soil cover.

5.4.3 Geotechnical Aspects Of Solifluction Deposits

The geotechnical properties of solifluction material are difficult to specify largely because of variability in parent material. Generally it is of higher density, increased strength, lower compressibility and permeability and is more cohesive compared with coarse-grained supraglacial diamicts (Sect. 4.5) and may rarely, overlap with the range of properties exhibited by lodgement till (Chapter 8). Limited over-consolidation is achieved by freeze-drying in the periglacial environment similar to flowed tills (Sect. 3.8). Solification debris is distinguished from lodgement till primarily by its downslope thickening stratigraphy, absence of glacial landforms and well-developed stone fabric parallel to surface slope. It cannot be distinguished from in situ tills on the basis of textural analysis alone no matter how sophisticated (Falconer, 1972; Baker, 1976; Fig. 1.8). The presence of silt 'caps' on clasts, formed by the washing of fines down through the thawed solifluction mantle (Harris, 1981) and vertical 'heaved stones' are useful diagnostic characteristics of solifluction debris. Carter and Hart (1977) demonstrate the value of micropalaeontology in discriminating between in situ frost shattered rock masses and solifluction debris. The latter commonly contains transported fossil fauna and other exotics. Its most troublesome characteristic is its high lateral permeability such that ingress of water into excavations through solifluction debris is a common nuisance (Sect. 13.3). Its variable matrix content is a problem when, in upland valleys suitable for dam and reservoir construction, it is the only locally available material for clay cores of dams. Locating a source of matrix-rich material is often a headache for the site engineer.

Table 5.1 Processes acting to decrease rock slope stability. (After Johnson and Wathall, 1979, with additions).

A. **Factors leading to an increase in shear stress**

1. Removal of lateral or underlying support
   undercutting by streams or removal during construction activity
   weathering of weaker strata at the toe of a slope or at the base of an acquifer
   washing out of granular material by seepage erosion and piping

2. Increased internal pressure
   build up of pore-water pressure
   cyclic loading and unloading of slope forming materials

3. Rebound from glacier loading or valley cutting, stress release into valley

B. **Factors leading to a decrease in shearing resistance**

1. Materials
   beds which decrease in shear strength if water content increases (shales, mud-stones)
   low internal cohesion of rocks involved in shearing in bedrock: faults, bedding planes, valley bulging and cambering

2. Weathering changes
   weathering reduces effective cohesion and to a lesser extent the angle of shearing resistance; freeze thaw cycles result in frost shatter.
   expansion of clay minerals in mudstones and shales disrupt fabric. Chemothermal weathering in metamorphic and igneous rocks. Loss of cement and dissolution in sedimentary rocks

3. Pore-water pressure increase
   high ground water table levels in the past resulting from permafrost, the mantling of slopes with superficial debris or to increased precipitation due to changes in climate
   formation of "cleft" pore-water pressures during glaciations
   high rainfall events
   frost-bursting

In Britain, on clay, shale or mudstone strata outcropping in broad strike valleys at the base of harder escarpment forming units, many low angle slopes ($3 - 4^{\circ}$) lie at angles about half that of ultimate stability against landsliding under present climatic conditions. At the present day, conditions of least stability for other slopes occur when the groundwater table coincides with the ground surface (which is a common winter condition in many areas in Britain). Under these conditions the angle of ultimate stability on normal slopes is about one-half the angle of friction which for a clay soil at its residual strength is about $7^{\circ} - 8^{\circ}$.

Site investigations in slopes on clay bedrock with anomalous low slope angles ($3^{\circ} - 4^{\circ}$) have revealed the presence of fossil slip surfaces at shallow (2 or 3 m) depths. The soil on these surfaces has been reduced to its residual strength, and the surfaces themselves are polished and slickensided. They frequently lie within a shallow, highly disturbed zone which is interpreted as a solifluction mantle whose lower limit is thought to lie at the level of the former permafrost table (Chandler, 1970, 1970a, 1977a). Their presence has obvious implications for the stability of engineering works on these slopes and there have been cases

of foundation movement as a result of their disturbance during construction. The
best known cases involve failure of the earthworks associated with major road
schemes (Weeks, 1969).

The initiation of slope movements on such low-angled slopes requires, by
definition, conditions different from those obtaining at the present day. The
generally accepted view is that periglacial thaw-consolidation (Sect. 5.4.1)
raised pore pressures high enough for failure to occur. Skempton and Weeks
(1976) have presented a study of the solifluction mantle that overlies the Weald
Clay at the foot of the Lower Greensand escarpment in southeast England. These
deposits appear to have been derived at various times during the last and
penultimate glaciations and have moved down slopes of as little as 2°. They are
in general about 2 m thick and are bounded by slickensided shear surfaces that
extend into the brecciated upper surface of the underlying clay. The suggestion
that they moved under thaw conditions is supported by the estimate of an R value
of between 0.98 and 1.63 (the uncertainty being due to lack of knowledge of the
thaw rate). Such values would lead to movement on slopes greater than about
1.5^o, assuming a residual value of $\emptyset'$ around 15° (Fig. 5.9). It is clear that
excess pressures due to thaw consolidation can account for the movement on the
slopes in question. In general, thaw consolidation may provide a possible
explanation for movements on many low-angled slopes on clayey bedrock.

In the case of deposits derived from clayey bedrock, soliflucted debris consists
of bedrock (lithorelics) set in a softened remoulded matrix. Such a fabric has
been noted in solifluction deposits derived from overconsolidated marine clays
such as Lias clay (Chandler, 1972b) and the Keuper Marl (Hutchinson et al, 1973).
It is generally similar to the fabric associated with mud slides occuring at the
present day along eroding coasts where cliffs are cut in the same clays
(Hutchinson, 1970). On the chalk terrain of southern England 'coombe-rock'
refers to cemented chalk solifluction rubble infilling dry valleys (coombes); a
high silt component sometimes results in metastable characteristics (Fookes and
Best, 1969).

The major engineering problems related to solifluction sequences concern the
reactivation of slope failures by excavations in lower slope zones. In many
cases the existence of solifluction fans and fossil slip surfaces at depth is not
indicated by site investigation and costly highway re-routing has been necessary
(Symons and Booth, 1971).

5.4.4 Rock Slope Processes

Just as the role of periglacial processes is difficult to isolate with regard to
deep-seated bedrock structures (Section 5.3) so similar uncertainty surrounds the
influence of periglaciation on more shallow rock slope behaviour. Certainly the
importance of freeze-thaw cycles and cleft water in joint systems can be
demonstrated by well developed seasonal patterns of rockfall frequency in modern
cold climates (Sect. 4.6.2) but many workers in periglaciated landscapes of the
mid-latitudes have demonstrated the greater importance of geological structure
and sequence in determining valleyside slope stability. To take a celebrated
example, Black (1969) has argued that in southwest Wisconsin, periglacial
climates failed to produce any distinct slope forms. Current processes account
for most observed slope features. This is significant because the area is that
of the classic unglaciated 'driftless area' (Fig. 9.2) where periglacial
landscapes might be expected to be the most maturely developed in North America.
The area is one of a maturely-dissected cuesta topography underlain by sandstone
and dolomite successions that were clearly not susceptible to periglacial
disturbance. Varying bedrock susceptibility of different rock types to
periglacial processes is well known (Smith, 1962; Martini, 1967; Williams,

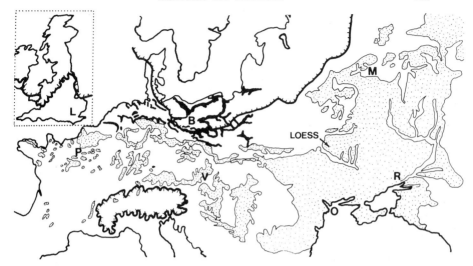

<u>Fig. 5.12</u> Loess cover in Europe related to limits of the last glaciation and source areas represented by major meltwater channels (black). Increasing cover to the east is related to continentality. After Flint (1971). 1 cm = 230 km; L, London;, P, Paris, B, Berlin, V, Vienna; M, Moscow, R. Rostov, O, Odessa.

1968).

With regard to the rapid lithological alternations seen in sedimentary lowlands it is not uncommon for slopes to be more rapidly degraded under the present postglacial climatic conditions. In many valleys for example, well defined periglacial phenomena such as solifluction fans and fossil cemented screes, etc. are widespread (Fig. 5.1) but the structures of dominant engineering significance are cambered valleyside 'edges', bulged valley bottoms and large deep-seated rotational failures in the harder, more competent, rock units moving on complex coalesced basal slip planes through weathered incompetent shales or mudstones (Knox, 1927; Brunsden and Jones, 1972; Ingold, 1975; Varnes, 1975; Skempton and Hutchinson, 1976; Mollard, 1977; Johnson and Wathall, 1979). Fine-grained lithologies in sedimentary bedrock successions are particularly prone to such movements (Scott and Brooker, 1968; Anderson and Richards, 1981) and maps showing the distribution of bedrock lithologies (e.g. Fig. 9.7) can be considered as depicting susceptibility to slope instabilities and the location of potentially disturbed ground.

The importance of movements along the contacts of incompetent/competent lithologies is largely the result of high porewater pressures that build up at these permeability contrasts, combined with mechanical, chemical and bio-chemical weathering (Table 5.1). Postglacial sea level rise has also contributed to elevated porewater pressures in coastal successions (Hutchinson, 1975, <u>et al</u>, 1980). Again, it should be noted that none of the processes enumerated in Table 5.1 are uniquely periglacial.

5.5 Fluvial Erosion and Deposition Under Periglacial Conditions

The term niveo-fluvial describes erosion and deposition by seasonal spring meltwaters derived from snow patches and ground icings. The development of deep ground-freezing results in surface water flow on rock types that do not usually allow surface drainage (e.g. chalk, limestones). The gently folded Mesozoic strata of southern and eastern England and northern France, that include clays, sands and soft chalk lithologies, are particudlarly susceptible to freeze-thaw cycles and periglacial meltwaters were able to rapidly evacuate thaw-saturated surface debris. The rapidity of such erosion on chalk terrains has been demonstrated by Kerney et al (1964); large cirque-like forms in the chalk escarpment in southeast England were cut by several hundred years seasonal snow melt. Saturated chalk debris moved downslope and accumulated as large fans at the escarpment foot. Successive periods of fan formation are recorded by buried pedogenic soils containing organic matter that can be radiocarbon dated thereby allowing the slope history to be put into a context of climatic change.

An increased density of surface drainage under periglacial conditions is demonstrated in many areas by dense networks of dry valleys (referred to as dells, bournes, coombes, vallons en berceau; French, 1976). These valleys may flood during high rainfall drawing the comment that they may also result from short-lived spring flow. This potential for flooding should always be considered when siting structures in dry valleys; sites of spring emergence and boggy terrain should be looked for in the walk-over phase of any site-investigation. Dry valleys are associated at depth with deeply frost-shattered rock and on chalk terrains may be filled with re-cemented chalk solifluction rubble or ´coombe-rock´ (Sect. 5.4.3).

The repeated washing downslope of fines on thawing slopes results in smooth ungullied slopes underlain by thick, bedded scree-like accumulations for which the term grezes litees is used (French, 1976). They are identified by their slope foot position, downslope thickening and planar internal bedding. An upper limiting slope angle of 24° is common (Boardman, 1978).

Nivation is an important periglacial process on many slopes and refers to the combined effects of freeze-thaw weathering and slope transport at the margins of snow patches lying in hillside hollows (Fig. 5.1). Frost wedging and disintegration at the head of the snow patch extends the hollow into the hillside. Weathered material may then either slide over the snow patch surface or be moved downslope by both solifluction and slope wash. The terrace, or tread, on which the snow patch sits and which is progressively widened by headward erosion, is referred to as an altiplanation or cryoplanation terrace (Reger and Pewe, 1976); slope angles of between 2 and 8° are common (Fig. 5.1). In southwest England (Fig. 5.2) they are well developed on metamorphic terrain surrounding large granite intrusions (e.g. Dartmoor). Typical backscarp heights lie between 2 - 12 m, terrace width is commonly 10 - 100 m. On the granite outcrops, summit tors are the exhumed remains of unweathered bedrock deeply penetrated by Tertiary tropical weathering and stripped by periglacial processes. Summit blockfields of frost-heaved bedrock masses commonly surround the tors (Linton, 1955) and may extend downslope as blockstreams (Smith, 1953; French, 1976; Washburn, 1980). Valleyside tors form when ´edges´ of harder lithologies project from valleyside slopes cut in sedimentary strata of different resistance for weathering (Linton, 1964). These edges are frequently ´cambered´ into the valley (Fig. 5.1).

Worsley (1977) in a review of former periglacial climates in Britain, highlighted the importance of valleyfloor fluvial processes and showed that sequences deposited by seasonal multi-channeled braided streams, migrating back and forth

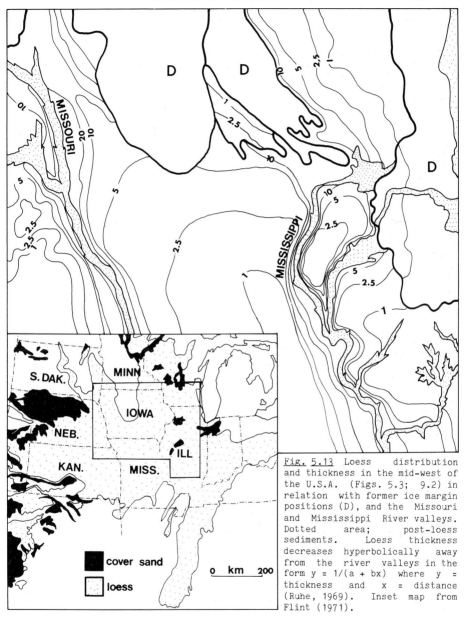

Fig. 5.13 Loess distribution and thickness in the mid-west of the U.S.A. (Figs. 5.3; 9.2) in relation with former ice margin positions (D), and the Missouri and Mississippi River valleys. Dotted area; post-loess sediments. Loess thickness decreases hyperbolically away from the river valleys in the form $y = 1/(a + bx)$ where y = thickness and x = distance (Ruhe, 1969). Inset map from Flint (1971).

across the valley floor, often account for the bulk of periglacial deposits in
many valleys. The form and internal composition of these fluvial sequences is
discussed in Chapter 7. Valleys throughout the glaciated mid-latitudes often
show upper terraces composed of coarse outwash (Figs. 4.9, 5.1) on whose surface
multiple braided stream channels can be still identified. These contrast with
the modern meandering river imprisoned in a single channel by vegetated cohesive
soils. These rivers transport dissolved and suspended load in the absence of
coarse non-cohesive debris produced under modern climates. The ground conditions
along many valleys show a modern flood plain merging downvalley into an estuary
underlain by buried drift-filled valleys that are infilled with glacial and
marine soils (Figs. 4.9).

5.6 Aeolian Deposition

Wind scouring of unvegetated periglacial surfaces and inactive outwash plains has
left a veneer of wind blown silts (loess) and sands (cover sands) as the most
widespread of periglacial sediments (Figs. 5.2, 5.12, 5.13, 5.16) which in some
areas attain substantial thicknesses.

Periglacial cover sands are not of great engineering significance though they may
be of local importance for glass and ceramic manufacture (e.g. the Shirdley Hill
Sand in northwest England and feldspar-rich sands in Illinois (Williman and Frye,
1970; Tooley, 1977). Uses for aggregate and for local ground water supplies are
also recorded (Ryckborst and Leusink, 1980). Fields of sand dunes, with sand
depths of up to 8 m, lie downwind of former Pleistocene lake basins, coasts,
valley infills and outwash plains indicating that the dunes are composed of
redeposited glaciofluvial sands (Maarleveld, 1960; Catt, 1977). Entrenchment by
early postglacial streams into thick valley fill sequences is important as it
exposes sediments susceptible to wind erosion (deflation). However, many
'periglacial' dune systems are periodically active under present day conditions
following loss of vegetation cover by fire or dessication as groundwater tables
are lowered by man.

The most distinctive windblown sediment is loess (German; losen, to dissolve), a
calcareous silt with a well defined and restricted grading, restricted
plasticity, high porosity (>40%) and low moisture content (c. 15%). It is
classified as a collapsing soil given its metastable behaviour of collapsing to
higher densities upon wetting (Fig. 5.14). Under low moisture contents
fine-grained soils will not compact to their optimum density; a restatement of
the classic relationship between density and moisture contents established by the
Proctor test. Loess is deposited in arid periglacial environments at moisture
contents well below those needed to achieve optimum density.

The in situ density of loess is perhaps its most useful index property since it
gives an indication of the ultimate settlement and shear strength when
subsequently consolidated by wetting (Lutton 1969). Extreme post-wetting
settlements are common at bulk densities lower than 1,100 kg/m^3 whilst optimum
Proctor densities of 1,500 kg/m^3 are not uncommon where natural moisture contents
rise above 20%.

Many foundation problems with loess have concerned uncontrolled and unforeseen
wetting of foundations during storm ponding and, in the case of housing
subdivisions, by watering of lawns, poor drainage and leakage from public
utilities. Settlement of large areas of newly irrigated farmland and reservoir
floors provide classic illustrations of metastable behaviour (Holtz and Gibbs,
1951; Clevenger, 1958; Sheeler, 1968). In some cases the pre-consolidation of
loess by wetting and allowing the soil mass to consolidate under its own weight
has proved attractive in canal and earth dam construction. In the mid-west of

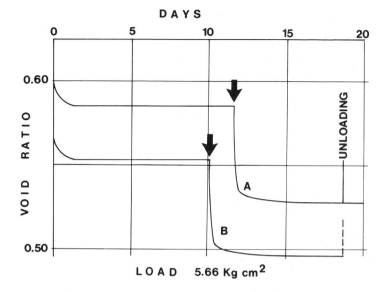

Fig. 5.14 Structural collapse of loess in the oedometer after wetting (arrow). Collapse consists of an initial rapid compression followed by slower time-dependent compression as samples approach saturation. Co-efficients of subsidence (R) for the samples are 3.52 (A) and 3.44 (B). After Feda (1966).

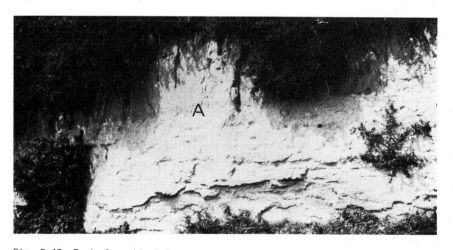

Fig. 5.15 Typical vertical loess bluff about 6 m high with characteristic prismatic jointing. A; base of postglacial pedogenic soil (Fookes and Best, 1969). Loess has been mapped in the past as 'brickearth' referring to its use for brickmaking. Kent, England.

the U.S.A., consolidation is commonly accelerated by surcharging, i.e. constructing a thick temporary embankment over the site. Plate-load tests show bearing strengths greater than 6 kg/cm^2 when dry (15% moisture) and as low as 0.3 kg/cm^3 when saturated at moisture contents greater than 24%. Feda (1966) defined the coefficient of subsidence (R) as;

$$R = \frac{hp - hw}{hp} \qquad\qquad (5.4)$$

where hp = sample thickness in the oedometer after consolidation at pressure, p, before wetting and hw = height of the sample after wetting. R values of 3 - 4 are common (Fig. 5.14). Handy (1973) identified the significance of the relationship between saturation moisture content and liquid limit. Loess will collapse when the liquidity index (Sect. 8.2.3) exceed 1 (i.e. moisture content = liquid limit). Structural collapse is not uniquely a loess problem and is shared by wind blown subsident sands and certain residual soils on granites. Analogous behaviour is also exhibited by sensitive soft marine ´quickclays´ (Chapter 6).

Site investigations in areas of loess cover emphasize determination of depth to competent subloess soil or rock for cast-in place pier foundations and likely future changes in moisture contents due to varying land-use (Scheeler, 1968). The use of compaction by saturation (pre-flooding) or by underground explosion is receiving increasing attention.

A distinctive feature of loess is its ability to stand with vertical faces (Fig. 5.15). Faces of 90 m high are reported in China (Yong-Yan and Zong-ho, 1980). Such structural stability results from calcium carbonate cementation and in some cases from cement-lined root holes inherited from previous vegetation. Cliff failure occurs by undercutting, either by streams or by slowly increasing moisture contents on the slope foot (Lutton, 1969, 1971).

Extensive loess belts are principally associated with dry continental interiors such as the Great Plains of the U.S., (loess cover=800,000 km^2) the Ukraine of the U.S.S.R. (500,000 km^2) and Northwest China (630,000 km^2). Moister maritime areas such as Britain have only scattered occurrences of loess (Catt, 1977; Figs. 5.2, 5.12) dating from the end of the last glaciation (Wintle, 1981; Burrin, 1981) and deposited by winds emanating from intense anticyclonic conditions centred over Scandinavia (Lill and Smalley, 1978). Mapping and subsurface investigation demonstrates that much loess material has been reworked by solifluction and slopewash. Loess sheets commonly drape pre-existing topography and may obscure suites of river terraces, against which the loess is banked, having been dropped by wind in their lee. A precise hyperbolic decrease in loess thickness over several hundred kilometres away from source areas has frequently been identified (Fig. 5.15). In central Europe, sequences of thick loess units with intercalated soils and fossil faunas have yielded an exceptional record of both climatic and meteorological change (Schultz and Frye, 1965; Kukla, 1977; Bowen, 1978; Veklich, 1979; Sect. 9.1).

Details of the extensive Chinese loess belts between latitudes 47 and 33°N (Fig. 5.16) have only recently been made available (Zin and Liang, 1982). These soils are most extensively developed in the middle basin of the Yellow River (Huang He) and adjacent provinces. Thick Early Quaternary loesses (Lishian or ´old loess´) are compact and stable as a result of natural _in situ_ collapse following groundwater changes, while overlying layers of younger loess (Malanian) dating from the last glaciation are metastable and liable to collapse. Total thicknesses of loess reach 300 m (Fig. 5.16) with younger loesses making up between 10 to 30% of local thicknesses. There are systematic regional changes in clay fraction from northwest to southeast with a corresponding decrease in

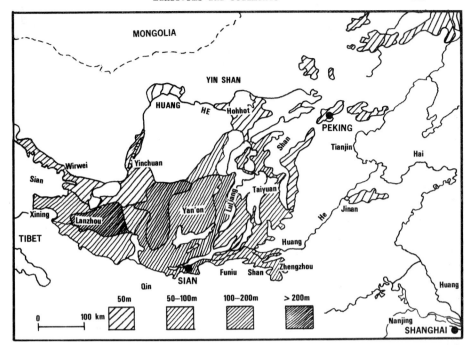

Fig. 5.16 Distribution and thickness of loess in China (After Lin and Liang, 1983).

collapsibility and thickness. This decrease in collapsibility with distance from source areas is typical and allows engineering geological zonation of loess belts. Thus Lutenegger (1983) pointed out that the liquidity index of Chinese loess soils decreases systematically in a southeast direction in parallel with a decrease in the potential for collapse.

CHAPTER 6

Glaciolacustrine and Glaciomarine Clay Deposition; a North American Perspective

R. M. Quigley

INTRODUCTION

The object of this chapter is to discuss clays in general terms, highlighting those depositional processes and concepts considered to be of direct relevance to geotechnical engineering particularly with regard to problematical 'soft' clays. To achieve this objective, the chapter is divided into three major subject areas as follows:

(1) The three-dimensional distribution of soft clays in North America with special reference to Wisconsin deglaciation and glacio-isostatic rebound. (2) The sedimentology of soft clays as it influences their appearance, stratigraphy, and fabric. (3) The geochemistry and mineralogy of soft clays as they influence their geotechnical behaviour.

The references cited are selective and of recent vintage. An attempt has been made to stress concepts of general application to glaciated terrains elsewhere.

6.1 Distribution Of Soft Clays In North America

Soft clay soils are largely the products of sedimentation in proglacial and postglacial water bodies that existed during retreat of the Laurentide ice sheet between 18000 and 6000 y.b.p. Generally speaking, the oldest sediments exist in the south and the youngest in the north. In arctic Canada and some more southerly mountainous regions, proglacial lakes still exist in contact with active glaciers.

The enormous area ultimately covered by both freshwater lakes and inland marine seas is illustrated in Fig. 2.21. Laminated clays were deposited in landlocked freshwater lakes in deep crustal depressions at the ice front, whereas in coastal regions, saline seawater entered the depressed areas and flocculated marine clays were deposited. The evolution of this proglacial lacustrine system between 18000 and 6000 y.b.p. is related to climatic factors, which controlled the rate of the Wisconsin withdrawal, combined with glacio-isostatic rebound and eustatic rises in sea level. A brief review of these controls is vital to understanding the distribution and stratigraphy of soft soils.

6.1.1 Ice Retreat, Isostatic Rebound, And Eustatic Sea Levels

Glacio-isostasy refers to altitudinal changes in the earth's crust as a result of loading and unloading by large thick ice sheets. Glacio-eustasy refers to changes in sea-level caused by the growth and decay of the same ice masses as they alternately lock-up and release large volumes of the earth's water.

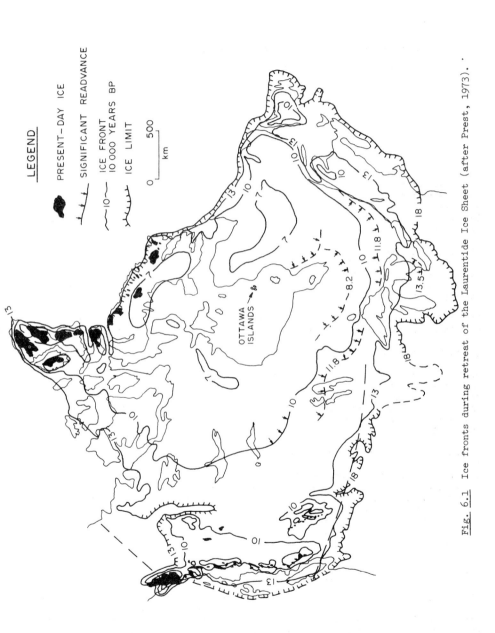

Fig. 6.1 Ice fronts during retreat of the Laurentide Ice Sheet (after Prest, 1973).

Common estimates of sea-level lowering range from 90 m to 160 m (see review by Budd, 1979). Recent work clearly shows however that there can be no simple relationship between sea-level and global ice volumes. A number of variables work against the employment of any standard eustatic sea-level history (or curve; e.g. Fig. 6.5) that can be widely employed, outside the immediate sedimentary basin. Such variables include sediment, water and ice sheet loading, lithospheric plate motions (plate tectonics), deep seated transfer of mantle material, changes in ocean basin volume and changes in the shape of the ocean surface (geoid). The geoid is a gravitational equipotential surface and has been shown to have high and low points which deviate by several tens of metres from the general equipotential surface. Local histories of sea-level change are created as the irregular surface of the geoid migrates or changes in amplitude (see Morner, 1979, Newman et al, 1979, and Peltier, 1980 for ´state of the art´ reviews).

In North America, the Laurentide ice sheet is estimated to have been up to 5000 m thick at the time of maximum advance 18000-20000 years y.b.p. (Fig. 9.9). At this time, eustatic ocean water levels were at least 120 m below present-day levels and in addition the earth´s crust in the Hudson Bay area may have been isostatically depressed 1000 m by the weight of the ice (Andrews 1972).

The ice retreated in stages, often with large or small proglacial lakes at the ice front. The retreat was erratic and interrupted by both stoppages and readvances of major ice lobes. A schematic presentation of the late Wisconsin glacial retreat is shown in Fig. 6.1. Major ice- front positions are shown as solid lines for 18, 13, 10, and 7000 y.b.p. These particular positions represent stages where extensive lakes or seas existed in the depressed basins in front of the continental ice mass. For example, the Champlain Sea existed between 12500 and 10000 BP in the St. Lawrence and Ottawa lowlands (Fig. 6.3B) until crustal rebound "expelled" the sea. Major readvances of ice at 13.5, 11.8, and 8200 y.b.p. overran extensive deposits of soft clay. Examples are the major readvance out of Hudson Bay that overran up to 250 km of Barlow-Ojibway clays at 8200 y.b.p. (Fig. 6.3D), and the readvance over Lake Agassiz sediments about 11800 y.b.p. (Fig. 6.3B). The importance of low profile ice margins buoyed up by high porewater pressures in glaciolacustrine clays has been discussed with regard to rapid surge-like movements of ice sheet margins by Boulton and Jones (1979).

As the ice sheet thinned and retreated northward the earth´s crust rebounded isostatically as is shown in Fig. 6.4 for Hudson Bay. Total rebound since retreat started 18000 y.b.p., is estimated to be close to 800 m of which 200 m has occurred since final deglaciation at 8000 y.b.p.

Maximum postglacial rebound in Canada is shown in Fig. 6.2. The uplift contours indicate three major centres of glaciation: Bathurst Inlet, Ellesmere Island and eastern Hudson Bay which has rebounded more than 250 m since deglaciation compared with 200 m for Ottawa and 150 m for Cape Cod, the approximate southern limit of Wisconsin glaciation. Inadequate uplift data exist on the rebound chronology of western Canada.

The rebound map is a significant geotechnical aid since, for the arctic, it can be used directly to predict the maximum elevation at which one might find postglacial marine sediments including soft clays. Corrections for the different rate at which sea-level itself has risen (eustatic changes) are not required since the sea had risen to within 10 m of its present level by the time the lower arctic was deglaciated about 7000 y.b.p.

Eustatic sea level changes were a dramatic feature of coastal and some interior

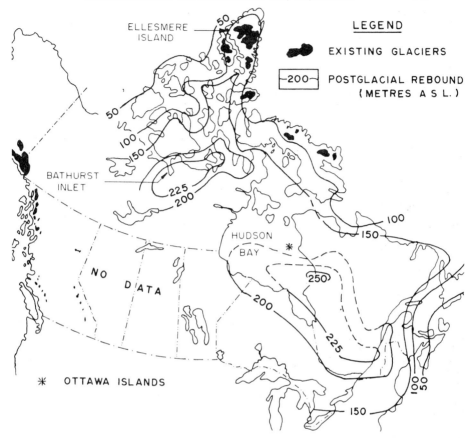

Fig. 6.2 Maximum postglacial rebound of Eastern Canada: insufficient data are available for western Canada (After Andrews, 1972).

regions between 18000 and 7000 y.b.p. Two generalized sea level curves, are shown at the top of Fig. 6.5. The curves show a long irregular 120 m rise in sea level from approximately 20000 y.b.p. to the present. Extensive ^{14}C dating has now established a fairly clear chronology of glacial retreat and its interruptions.   Dated readvances in the Great Lakes area summarized by Dreimanis (1977a,b) are shown as arrows on Fig. 6.5. Although changes are continually being made to the eustatic sea level curve the essential features of the curve and the geotechnical interpretations remain valid.

A recent eustatic sea level curve for Hudson Bay is also shown in Fig. 6.5.   It shows that the Tyrrell Sea (Fig. 6.3D) had essentially reached present levels by 6000 y.b.p. It also shows two abrupt drops in sea level about 8200 and 7500 y.b.p., the former correlating with the major Cochrane glacial readvance from James Bay.

6.2 Major Proglacial Freshwater Lakes And Marine Seas

The proglacial and postglacial lakes and seas that developed and disappeared as
the late Wisconsin ice sheet retreated were related to three factors: (1) the
volume of water supplied by the melting ice, (2) the rate of glacio-isostatic
rebound of the depressed trough within which the bodies of water collected south
of the ice front, and (3) the rapid opening and (or) damming of drainage outlets
that caused huge and abrupt changes in lake levels.

In this section, the spatial and temporal history of the proglacial lake systems
is briefly reviewed using, as examples, the five ice-front conditions shown in
Fig. 6.1, namely, 13000, 11800, 10000, 8200 and 7000 y.b.p.   The general
mineralogical trends of proglacial lake sediments are also briefly presented in
this section.

13000 y.b.p. Ice front. Many of the major soft clay deposits in the northern
United States were deposited prior to 13000 y.b.p. (Fig. 6.3a) such as the
Boston freshwater and marine clays, Connecticut Valley clays in glacial Lake
Hitchcock, Lake Chicago clays, and early Lake Agassiz clays in North
Dakota-Minnesota. In Canada, the main early freshwater lakes were in the Lake
Erie basin in southern Ontario and probably in southern Alberta and Saskatchewan
between the Cordilleran and Laurentide ice sheets.   Marine submergence of the
coastal regions of the Canadian maritime provinces probably correlates with this
episode.

The Port Huron readvance about 13500 y.b.p.   (Figs. 6.1, 6.5) was a major
geologic complication in southwestern Ontario. The ice sheet was partially
floating and overrode previous soft clay deposits with little or no
consolidation.   Most of the soft clay deposits in the Sarnia area are considered
to be glaciolacustrine diamicts that are essentially normally consolidated at
depth (Quigley and Ogunbadejo 1976; Sect. 2.9).

The composition of the Lake Erie basin clays is dominated by glacial erosion of
Paleozoic shales and carbonates, which have produced clays rich in illite,
chlorite, calcite, and dolomite.   The chlorites are an unstable ferruginous
variety of Devonian origin that weathers by oxidation to smectite, producing
surface soils of considerably greater activity than unweathered clays below 2 or
3 m depth (Sect. 8.2.4).

11800 y.b.p. Ice front. Major water bodies that existed between 13000 and 10000
y.b.p. were the Champlain Sea, Lake Algonquin, and Lake Agassiz as shown in
Fig. 6.3a. The 11800 year y.b.p. ice front is significant because it represents
the Greatlakean advance (Dreimanis 1977a,b), which overrode soft clay sediments
in all three of the above-named water bodies.

Champlain Sea. The Champlain Sea occupied the St. Lawrence lowlands from
approximately 12500 to 10000 y.b.p. (Cronin 1977). The sea is believed to have
invaded the lowlands from the Gulf of St. Lawrence when the ice dam melted in the
vicinity of Quebec City. It mixed with freshwater lakes in the valley and marine
clays were subsequently deposited over freshwater varved clays in many areas.

The ice front is known to have formed the northern boundary of the sea for many
hundreds of years, oscillating back and forth in some areas forming clay-rich
moraine ridges such as the St. Narcisse moraine in Quebec which may be a
grounding line of a floating ice margin. Only in ice distal regions of the
ancient seas does one find thick, massive clays. In proximal areas the clays are
extensively interlayered with sands and gravels of deltaic or turbidite origin
(Donovan, 1978) or even interlayered with glacial till associated with glacial

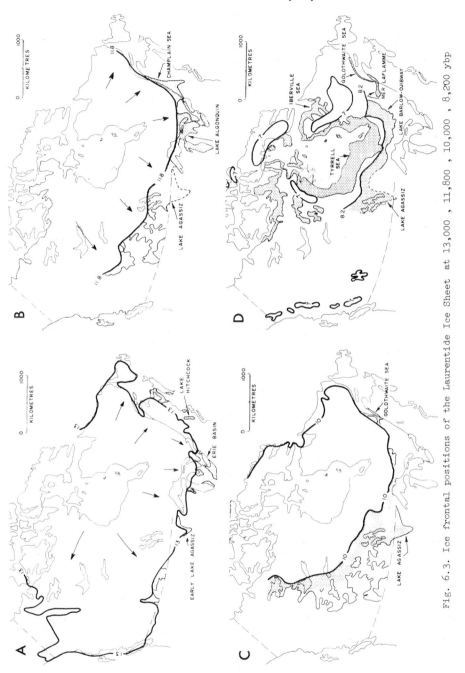

Fig. 6.3. Ice frontal positions of the Laurentide Ice Sheet at 13,000 , 11,800 , 10,000 , 8,200 ybp

pulses. Some clays are so laden with ice-rafted stones and boulders as to be till-like (Section 2.9.).

There were two major sources of sediment for the Champlain Sea; from the igneous rocks of the Canadian shield and from sedimentary and metamorphic rocks of the Appalachians. The composition of clay from the Champlain Sea is dominated by shield-derived primary minerals such as quartz, feldspar, amphibole, mica, and chlorite, usually with small amounts of smectite of controversial origin (Soderman and Quigley 1965; Gillott 1971; Bentley and Smalley 1978b). Another important component of these clays, which is receiving increasing attention, is glacially ground amorphous material (Hendershot and Carson 1978; Yong et al. 1979). Carbonates are present in certain Leda clays where the influence of Paleozoic or older limestones was felt. Marine shells also contribute significantly at a level of about 1-4% carbonate.

Lake Algonquin. The early phases of Lake Algonquin existed south of the 11800 y.b.p. ice front in the Georgian Bay, Lake Simcoe, and western Superior areas. Lake Algonquin probably connected with Lake Agassiz as a lake phase as well as a fluvial outlet as shown in Fig. 6.3. All clays in these regions are of freshwater origin and are characterized by a wide variety of varved clays. Two major drainage events affected the levels of Lake Algonquin: an outlet at Kirkfield, which corresponds with low lake levels dated 11000 y.b.p., and the major drainage event through North Bay about 10400 y.b.p. (Karrow et al. 1975). This latter event supplied enough freshwater to the Champlain Sea to greatly reduce its salinity.

Source rocks for sediments deposited in Lake Algonquin are complex. In northern regions, the rock flour was essentially of Precambrian igneous origin and like most Leda clays, most varved clays consist of quartz, feldspar, amphibole, chlorite and mica. South of the Hudson Bay lowlands carbonates and smectites of Paleozoic and Mesozoic origin occur in the sediments. In southerly regions, Paleozoic bedrock supplied carbonate, illite, chlorite, and some smectite.

Lake Agassiz. The geology and mineralogy of Lake Agassiz clays have been extensively reviewed by Elson (1967) and more recently by Baracos (1977). The lake was of enormous size, extending from the Manitoba escarpment (Fig. 6.3) easterly at least to Lake Nipigon and southerly from near Hudson Bay 350 km into the United States. Early Lake Agassiz, which drained southward into the Mississippi, was already in existence at 13500 y.b.p. and its northerly late phases lasted to 8000 y.b.p. when it drained northward into Hudson Bay.

During intermediate phases, the lake drained eastward into Lake Superior resulting in large fluctuations in lake level as the Superior outlets were opened and closed by ice retreats and advances. These events (at least five) were very significant since they introduced drying intervals (crusts) on exposed soft clays and erosional surfaces within the clay deposits.

Teller (1976) described three major sedimentary units that apparently relate to three major phases of the lake. Early Lake Agassiz sediments in the U.S.A. were overriden by ice about 13000 y.b.p. (Port Huron advance), after which the lake expanded rapidly northward and extensive, deep-water unit 1 clays were deposited. Continued ice retreat northward then opened lower drainage outlets to the east, lake levels fell, and unit 1 sediments were subjected to a long period of erosion and desiccation. Probably at least two major ice advances occurred, corresponding to the Greatlakean and Algonquin substages; unit 2 sediments, which are rich in ice-rafted detritus, represent a major readvance of the ice and renewed deep-water conditions (Fig. 6.5).

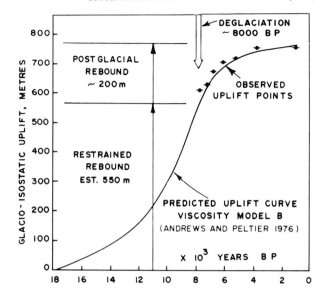

<u>Fig.</u> <u>6.4</u> Observed and predicted glacio-isostatic rebound of southern Hudson Bay,

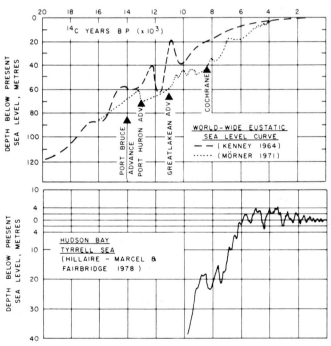

<u>Fig.</u> <u>6.5</u>. Eustatic sea-level curves and related glacial advances interrupting the retreat of Wisconsin ice.

The siltier surficial unit 3 sediments represent the final shallow-water phase of the lake just prior to drainage into Hudson Bay about 8000 y.b.p. All units of central and southern Lake Agassiz are dominated by montmorillonite from Cretaceous age shales. Some of the sediment supplied to Lake Agassiz may have come from the west since early proglacial lakes in Alberta and Saskatchewan are believed to have drained easterly around the ice front.

The varved clays of easterly Lake Agassiz in western Ontario probably were derived essentially from rock flour off the Canadian shield immediately to the north, supplemented by carbonate and shale from the Hudson Bay lowlands.

10000 y.b.p. Ice front. This ice-front position (Fig. 6.3c) is significant in that it represents the end of the Champlain Sea and the start of Mer La Flamme in the St. Jean region of Quebec. The extensive varved clay deposits in the Northwest Territories and the northern Prairies also correlate with this ice-front position.

^{14}C dates on Mer La Flamme sediments indicate that it existed from about 10400 to about 8000 y.b.p. Gadd (1975) speculates that the La Flamme sediments reflect a small source area of ultramafic composition.

8200 y.b.p. Ice front. Lake Barlow-Ojibway south of James Bay existed as a variably sized proglacial lake from about 10000 to less than 8000 y.b.p. (Vincent and Hardy, 1977). During its later stages it is presumed to have been interconnected with shrunken Lake Agassiz (Fig. 6.3d).

The Cochrane readvance of glacial ice out of James Bay (Fig. 6.5) occurred about 8200 y.b.p. overriding soft clays for about 250 km. The same clay deposit is normally consolidated south of the ice limit and overconsolidated north of it. The 8200 y.b.p. position shown in Fig. 6.3 represents approximately the limit of the readvance. Final retreat of the ice was apparently very rapid because Cochrane till rarely has soft clays overlying it.

Mineralogical studies of Barlow-Ojibway sediments again show a dominance of shield source rock flour. A variable is carbonate content, which varies from zero to approximately 13%.

7000 y.b.p. Ice Front. Final phases of soft clay deposition in Canada occurred in the Tyrrell and Iberville Seas surrounding present Hudson Bay (Fig. 6.3). Clay deposition was extensive in the southern regions. Although the sea level has been more or less constant since 6000 y.b.p. the land has risen 200 m since deglaciation and marine deposits are thus exposed well above sea level. Ice retreat within Hudson Bay was extremely rapid and invasion by seawater was apparently via Hudson Strait in a manner quite similar to that of the Champlain Sea (Andrews and Peltier 1976; Hillaire-Marcel and Fairbridge 1978).

Detailed mapping. Recent mapping by St. Onge (1972) has shown that glacial Lake Edmonton was in reality a complex sequence of many lakes decreasing in age by about 2000 years from the phase 1 ice front to the phase 6 ice front as shown in Fig. 6.6a. The step-by-step evolution illustrated in the figure seems to be typical of most of the "short-life" lakes found on the western Prairies.

Christiansen (1979) has presented a similar province-wide chronology of lacustrine events in Saskatchewan (Fig. 6.6b). From an engineering point of view, Fig. 6.6 illustrates the typical complexity of soft clay distribution in western Canada. The lakes were small, shortlived and restricted to the depressed area in front of the continental ice sheet.

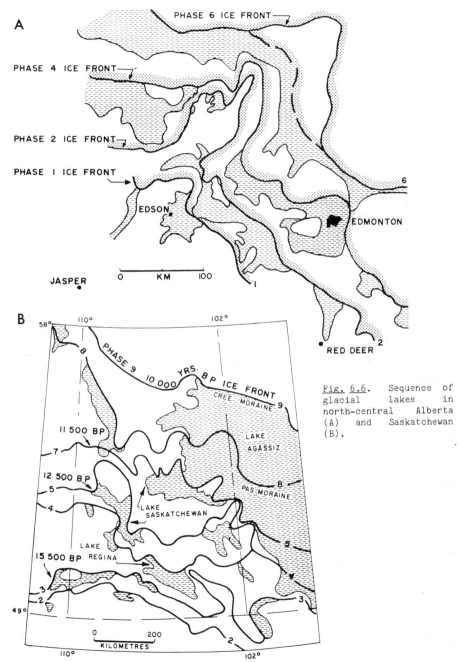

Fig. 6.6. Sequence of glacial lakes in north-central Alberta (A) and Saskatchewan (B).

6.3 Sedimentology

The main types of soft soil that may be encountered in geotechnical practice are
waterlaid tills, and mudflows (coarse-grained material in a fine clayey matrix),
laminated (varved) clays and marine clays.

Waterlaid Tills. Although still poorly understood, waterlaid tills form an
important part of many soft clay sections. Recent data demonstrates that the
term 'till' is probably inappropriate (Sect. 2.9). These deposits include pebbly
lake clays deposited directly at or below a floating ice margin into minimal
grain-size sorting. Some crude stratification may be visible, the soil will be
till-like yet be soft and of lacustrine origin.

Such deposits up to 30m thick may or may not overlie lodgement till and contain
interbeds of laminated silts and clays (Fig. 2.20d). Variable consolidation is
achieved by dewatering as groundwater levels are lowered by lake drainage,
removal of overlying soils and other processes (Sect. 2.9.2). Subaqueous mud
flows result from the slumping of pebbly lake clays and are intimately associated
with density underflows as the slump accelerates (Morgenstern, 1967) resulting in
laminated silty clays. In the writer's experience variably consolidated stony
lake clays, slump deposits and laminated clays are a very common feature of
proximal glaciolacustrine sediments. At Sarnia, Ontario massive non-stratified
stoney clays are considered to be waterlain below floating ice (Quigley and
Ogunbadejo 1976). The importance of these soils in the Great Lakes basins and
elsewhere (Fig. 2.21) can be stressed.

6.3.1 Varved clays in proglacial lakes.

Varved clays, are layered sediments deposited in glacial freshwater lakes. A
single varve, representing 1 year of deposition, consists of a couplet of summer
silt and winter clay; this time framework is difficult to demonstrate however.

The role of turbidity currents in the formation of varved clays is generally
accepted as having been the most important depositional mechanism in proglacial
lakes fed by sediment-laden streams (Jopling and McDonald, 1975; Ashley, 1975;
Shaw, 1977b).   Kenney (1976) reviewed the sedimentological processes of varved
clay formation including the three major types of lake current, namely,
heavy-density turbidity currents, interflows, and low-density overflows. If the
sediment load of a cold inlet stream is about 1 g/L or more, a heavy-density flow
will occur in a cold (0-6°C) proglacial lake (Fig. 6.7A). This is because water
is close to its maximum density over this temperature range (Fig. 6.9) and
density differences are controlled by suspended sediment loads. (See Matter and
Tucker, 1978 and Schluchter, 1979, for reviews).

A bottom turbidity current may flow for miles, even along a fairly flat lake
bottom, depositing a graded bed of sand and silt. The currents may be continuous
over several days or just short pulses associated with storms, extra warm,
high-melt runoff events or ice berg gouging an slumping on delta front
(Fig. 2.20). The multilaminated nature of the summer layer of varves is
attributed directly to such intermittent bottom flows, especially in the proximal
region of proglacial lakes. Although varved clays thin with distance from their
source, they must be considered as remarkably continuous, a feature created by
the long travel distance of heavy-density bottom currents. During the winter,
the inlet streams were deficient in sediment and were thus of lower density than
the sediment-laden, thermally stratified proglacial lakes. Overflows were thus
characteristic of winter conditions as shown in Fig. 6.7B and the slow settling
of clay particles to form the winter layer was relatively undisturbed.

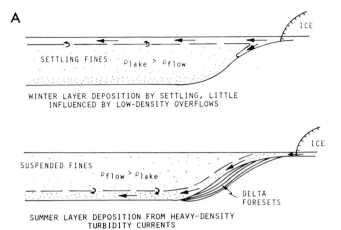

A

SETTLING FINES $\rho_{lake} > \rho_{flow}$

WINTER LAYER DEPOSITION BY SETTLING, LITTLE
INFLUENCED BY LOW-DENSITY OVERFLOWS

ICE

SUSPENDED FINES $\rho_{flow} > \rho_{lake}$

DELTA
FORESETS

SUMMER LAYER DEPOSITION FROM HEAVY-DENSITY
TURBIDITY CURRENTS

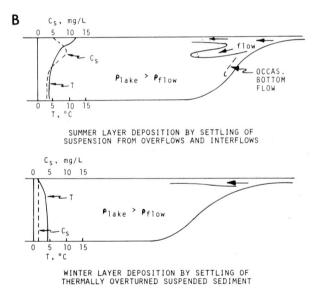

B

C_s, mg/L

flow

C_s

$\rho_{lake} > \rho_{flow}$

OCCAS.
BOTTOM
FLOW

T

T, °C

SUMMER LAYER DEPOSITION BY SETTLING OF
SUSPENSION FROM OVERFLOWS AND INTERFLOWS

C_s, mg/L

T

$\rho_{lake} > \rho_{flow}$

C_s

T, °C

WINTER LAYER DEPOSITION BY SETTLING OF
THERMALLY OVERTURNED SUSPENDED SEDIMENT

Fig. 6.7. Deposition of laminated (varved) clays by A) turbidity currents (summer layers) and settling (winter layers) in cold proglacial lakes.

B) settling from suspension from sediment-deficient overflows and interflows. Cs = concentration of suspended solids. T = temperature, C.

Continuous deposition by turbidity currents occurs where large amounts of sediment are delivered to the lake year round.

Present-day lacustrine systems, similar to the above are described by Gilbert (1975) for Lillooet Lake, B.C. and Gustavson (1975) for proglacial Malaspina Lake in Alaska. Lacustrine processes during deglaciation of valleys are discussed by Shaw (1977b), Shaw and Archer (1979).

It is very unfortunate from a sedimentological viewpoint that engineers describe any rhythmically laminated fine-grained sediment as 'varved'. There is increasing recognition that many sequences previously described as varves are multiple turbidite sequences of graded silt to clay units (Fig. 6.8) without any obvious seasonal control on sedimentation. The formation of varved silty-clays requires the cessation of melt runoff into the lake during winter to create a closed lake system in which precipitation of clay particles can take place. In many cases where large ice lobes or glaciers sit or float in lakes, there is year round delivery of sediments and turbidite activity occurs almost continuously resulting in graded laminae that are not true varves. These turbidity currents deposit single or multiple graded (fining up) laminae (Fig. 6.8) and clay laminae may be thin or non-existent as a result of infrequent quiet water conditions in proximal areas. Consistent clay layer thickness and sharp textural division between silt and clay components are the principal diagnostic criteria for varve recognition (Ashley, 1975).

6.3.2 Varved And Laminated Clays In Postglacial Lakes.

Current regimes and sedimentary conditions were different in postglacial freshwater lakes no longer in contact with the ice front. Suspended sediment loads were greatly reduced (probably less than 0.1 g/L) and under such conditions, streams enter lakes as overflows and interflows (Fig. 6.7B). Heavy-density bottom flows would only occur during spring floods or as products of slumping.

Present-day sedimentation in Hector Lake, Alberta (Smith, 1978) probably reflects conditions in many postglacial lakes. During the summer, warm stream water with its suspended sediment enters the lake as overflows and interflows, overriding heavier cold water (Fig. 6.9). Even during reduced winter stream flow, the very cold ($\sim$0°C) stream water overflows heavier 4°C water in the lake. Deposition of sediments is largely by year-round settling with silt and fine sand settling out during the summer and clay during the winter. Smith (1978) reports classical varve couplets in Hector Lake near the inlet delta where there is adequate sediment to produce them but the varves rapidly decrease in thickness with distance, becoming thin laminae or even massive clays only 2 km from the inlet. This rapid attenuation is largely related to the inefficiency of overflows and interflows as carriers of sediment along with the small sediment loads. (See also Gilbert and Shaw, 1981).

6.3.3 Varved Clay Structure.

A sketch showing the structure of a typical distal varved clay is shown in Fig. 6.10. Variations in water content, Atterberg limits and grain size are plotted through two complete varves.

The grain-size curve shows a silt-rich summer layer (80% > 2um), a distinct upwards fining through the transition zone and a clay-rich (80% < 2 um) winter layer. Coarse summer layers of distal varves will show little lamination since the turbid bottom flows will already have lost most of their coarser fraction during proximal sedimentation.

The transition and fine layers represent settlement of suspended solids during the autumn and winter. The very high moisture content (75%) of the clay layer indicates an open flocculated structure as shown by Quigley and Ogunbadejo (1972) using X-ray methods. It seems highly probable that flocculation is related to sediment density and interparticle collisions to produce flocculation. The graded bedding shown by Kenney (1976) indicates that flocculation does not occur prior to settling. In a proximal varve the coarse summer layer is much thicker (up to 1 m), multilayered (Fig. 6.10) and may even be ripple marked, indicating high velocity bottom flows near the sediment source as with any thick turbidite unit (Leeder, 1982). The transition zone and winter layer are likely to contain silt laminae representing autumn or winter thaw periods, since they are close to the glacial meltwater source. Foundation designs on glaciolacustrine clay substrates are discussed in Sect. 12.2 and 13.4.

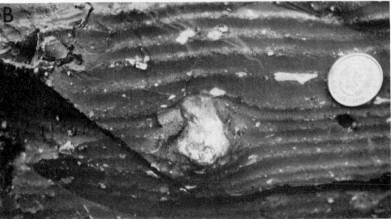

Fig. 6.8 A) ´Varved´ silty clays which on closer inspection are multiple graded turbidite units. B) Multiple turbidite units consisting of graded beds from silt (light) clay (darker). Note ice-rafted debris. Photographs courtesy of P. Fralick.

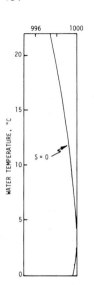
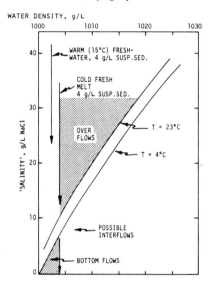

Fig. 6.9.   Water
density
vs. temperature   and
salinity.

6.4 Clay Deposition In Inland Marine Seas.

The salt dissolved in seawater (primarily NaCl) gives it a very high density (1020 g/L at a salinity of 35%) as shown in Fig. 6.9. Since suspended sediment loads in meltwater streams were normally 1 or 2 g/L and rarely greater than 4 g/L, then virtually all freshwater, sediment-laden, surface streams would have entered the sea as overflows. The density relationships in Fig. 6.9 indicate that cold subglacial streams should actually rise to the surface after entering the sea at the ice front. Bottom turbidity currents appear to be a remote possibility even under brackish water conditions, unless they develop from an accelerating submarine slump already containing saline pore fluid.

The early phases of the Champlain Sea were extremely deep, as illustrated in Fig. 6.11. The lower parts of both the Champlain and Tyrrell Seas are inferred to have been near normal marine salinities of 35 g/L. The Champlain Sea probably remained saline until crustal rebound reduced its depth during late phases about 10000 y.b.p.

The characteristics of a typical freshwater overflow for present-day Howe Sound, British Columbia are shown in Fig. 6.12a. Mixing of freshwater and saltwater proceeds by diffusion and turbulence down to a depth of about 5m. This upper 5 m down to the halocline temperature maximum (htm) constitutes the surface overflow within which initial low-salt flocculation of clay particles commences. Below this zone, extensive biologic activity by plankton modifies the clay floccs by ingestion and subsequent excretion, a process generally known as pelletization.

An electron photomicrograph (Fig. 6.13a) shows a small inorganic edge-to-edge floccule taken from Howe Sound surface water at S = 3%. It therefore represents initial flocculation within the freshwater overflow. Figure 6.13b shows a larger, more face-to-face floccule taken from deeper water at S = 20%.

Zooplanktonic agglomeration and pelletization produce very complex, organic-rich

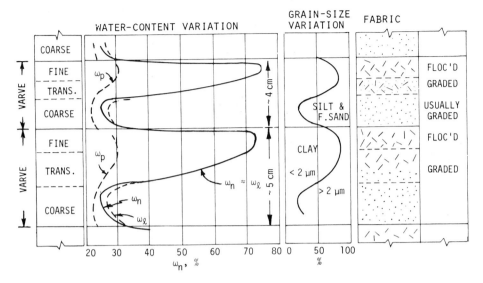

TYPICAL PROXIMAL VARVE

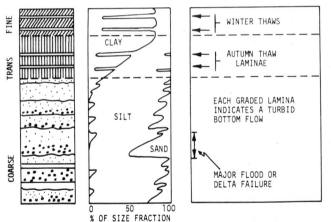

Fig. 6.10.   a) Typical varves from New Liskeard, Ontario (Chan and Kenney, 1973). b) Schematic representation of complex structure of a typical varve.

floceules and fecal pellets. Figure 6.13c shows the fabric of a water-borne fecal pellet from Howe Sound. Figure 6.13d shows, for comparison purposes, domains or peds in Leda clay. Perhaps many of the peds are fecal pellets.

Many clay flocoules contain significant amounts of organic matter along with siliceous and calcareous shell fragments. These organic-rich flocoules provide suitable location for *in situ* bacterial activity and probably explain, in part, the black mottling common in marine clays.

The salinity of the Champlain Sea has attracted considerable attention for both

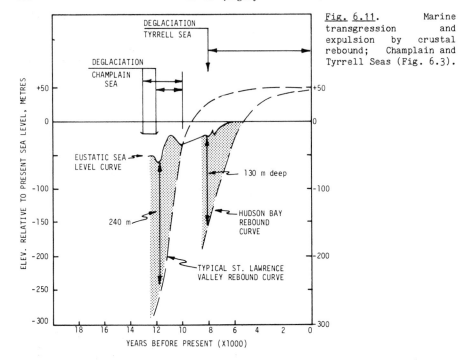

Fig. 6.11. Marine transgression and expulsion by crustal rebound; Champlain and Tyrrell Seas (Fig. 6.3).

geological and geotechnical reasons and has been studied using fossil indicators (Cronin 1977) and boron analyses (Gillott 1970). From a geotechnical point of view, the major consideration is flocculation of the input sediments to form a soil having a high void ratio.

In summary marine clays of proglacial origin are generally deep-water deposits representing a variably cemented accumulation of inorganic floccules, organic agglomerates and fecal pellets which will retain their structure *in situ* until destroyed by consolidation.

6.5 Geochemistry

An attempt will be made in this section to isolate those geochemical and mineralogical aspects of soils that control sensitivity. Sensitivity is the major consideration for all soft clay deposits and refers to their ability to be transformed spontaneously from solids to liquids by shock or vibration. It is this aspect of soft clay deposition both in Canada and Scandinavia that has attracted most interest in view of the environmental hazards such 'quickclays' present to the human settlement of many areas (Tavenas et al., 1971; Penner and Burn 1977; Fransham and Gadd, 1977; Brand and Brenner, 1981). Skempton and Northey (1952) identified a 'quick' soil as one having a sensitivity of greater than 16 whilst the Swedish Geotechnical Institute identify the sensitivity of a quickclay as being greater than 50. Sensitivity values greater than 500 are reported for certain Leda clays of the St. Lawrence Valley (Penner, 1965).

The factors playing dominant roles in establishing sensitivity are summarized in

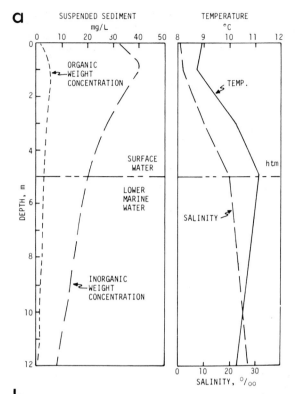

a

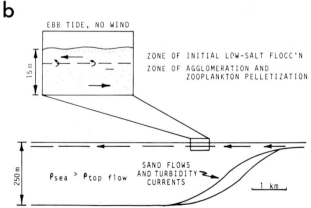

b

SUMMER AND WINTER DEPOSITION BY SETTLING
OF FLOCCULATED SEDIMENTS

<u>Fig. 6.12</u> a) Vertical variation in suspended sediment, salinity and temperature showing surface overflow at Howe Sound, B.C.

<u>Fig. 6.12</u> b) Deposition of marine clay by flocculation and settling from low density·overflows.

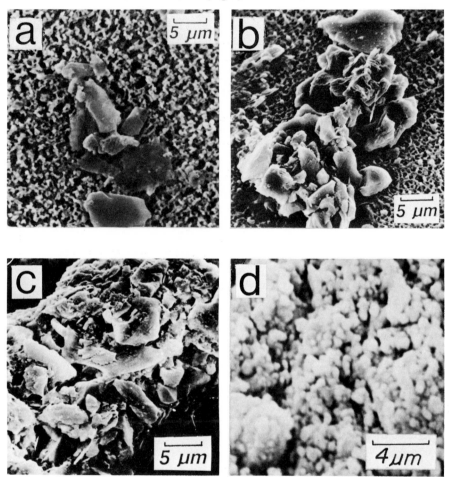

Scanning electron photomicrographs of marine clays: a) small floccule from water at S = 3%; b) larger floccule from water at S = 20%; c) clayey ped formed by mineral bearing fecal pellet from Howe Sound, B.C.; d) ped or domains of clay in Leda clay.

Fig. 6.14. The numerator lists factors producing a high undisturbed strength and the denominator lists factors creating a low remoulded strength. The nature of the overall sensitivity phenomenon will be discussed first followed by a more detailed look at each component of Fig. 6.14.

6.5.1 Factors Producing A High Sensitivity.

It is generally accepted that sensitive soils, be they clayey or silty, have acquired an open, flocculated structure during sedimentation and that this

	FACTORS PRODUCING HIGH UNDISTURBED STRENGTH AND HIGH SENSITIVITY

	1. DEPOSITIONAL FLOCCULATION
	    SALINE (LOW ZETA POTENTIAL)
	    HIGH SEDIMENT CONC.
	    DIVALENT CATION ADSORPTION
	
	2. SLOW INCREASE IN SEDIMENT LOAD
	
	3. CEMENTATION BONDS
	    CARBONATES & SESQUIOXIDES (AMORPHOUS)

$S_t = \dfrac{Su \ \ UNDIST}{Su \ \ REMOULD}$	**FACTORS PRODUCING LOW REMOULDED STRENGTH AND HIGH SENSITIVITY**

	1. HIGH WATER CONTENT ($\omega_n \geq \omega_\ell$)
	    LITTLE CONSOLIDATION OR DECR. IN $\omega_\ell >$ DECR. IN ω_n
	2. LOW SPECIFIC SURFACE OR SOIL GRAINS
	    HIGH SILT CONTENT OR HIGH ROCK FLOUR IN<2 μm FRACTION
	    HIGH PRIMARY MINERAL $\equiv$ LOW CLAY MINERAL CONTENT
	3. HIGH ZETA POTENTIAL
	    (EXPANDED DOUBLE LAYERS = HIGH INTERPARTICLE
	    REPULSION = DISPERSED OR PEPTIZED STATE)
	    LOW SALINITY BY LEACHING (<2 g/L)
	    ORGANIC DISPERSANTS (ANION ADSORPTION)
	    INORGANIC DISPERSANTS (ANION ADSORPTION)
	    HIGH MONOVALENT CATION ADSORPTION RELATIVE TO
	    DIVALENT CATIONS
	4. LOW AMORPHOUS CONTENT
	5. LOW SMECTITE CONTENT

Fig. 6.14. Geochemical and mineralogical controls on sensitivity of clay soils.

structure remained open even though subsequently buried by many metres of soil. A slow rate of deposition accompanied by simultaneous development of bonds is necessary for retention of this open fabric. Strength gain by cation exchange, increases in Van der Waals attractive forces and cementation by precipitation are all possible (6.14).

The factors listed in the denominator are more critical than those in the numerator since they create a very low remoulded shear strength. An in situ water content close to or greater than the liquid limit (high liquidity index; Sect. 8.2.3) is a characteristic of sensitive soils. This implies: (1) little compression of the soil due to cementation bonding, or (2) the liquid limit has decreased by chemical or biochemical change more than the in situ water content has decreased by consolidation.

Some of the most sensitive soils are characterized by a very low specific surface since they consist essentially of silts with little clay mineral matter; the liquidity index is close to unity, or higher, and cementation bonds are believed to be a significant factor in maintaining the natural water content above the liquid limit. For this type of quick soil, geochemical factors are probably less significant than for clayey soils.

In the case of clayey soils, geochemical factors control sensitivity by peptization (dispersion); the more dispersed the soil the lower the remoulded strength and the higher the sensitivity. A convenient way to express dispersion is by zeta or electrokinetic potential. The major factors causing dispersion in natural soils are low salinity, organic and inorganic dispersants, and monovalent cation adsorption.

Finally, low smectite and low amorphous contents (both active constituents of soil) are required to achieve very low remoulded strengths.

Most of the factors listed in Fig. 6.14 operate simultaneously in a soil deposit and affect both the numerator and denominator of the sensitivity equation. Therefore, although sensitivity is a relatively simple engineering concept, it is a very complex mineralogical and geochemical phenomenon (see Brand and Brenner, 1981).

6.5.2 Depositional Flocculation.

In saline environments, flocculation and sedimentation of peds has been discussed under "Sedimentology".

The mechanisms whereby a flocculated fabric is achieved in the clay layers of freshwater varves is probably related to particle concentration and collisions close to the sedimentary interface as discussed under "Sedimentology".

6.5.3 Cementation Bonding.

A slow rate of deposition accompanied by a simultaneous development of bonds is necessary for retention of the open fabric after burial (Quigley and Ogunbadejo 1972). In the case of near-surface sediments, postdepositional adsorption of divalent cations and increases in Van der Waals attractive forces may be adequate to resist compression. However, for sediments subjected to the high stresses of deep burial, cementation bonds are probably required, and of the possible cementing agents, calcium carbonate and amorphous sesquioxides (including silica) seem the most probable. The undisturbed strengths and preconsolidation pressures induced by such cementation may be significant (Kenney and Folkes, 1979).

The presence of amorphous materials as cementing agents in marine clay has been a matter of speculation for years. The amorphous constituents of soils consist of complex hydroxides of silica, alumina, and iron, produced by glacial grinding. Their importance has been emphasized by authors such as Yong et al. (1979), and Bentley and Smalley (1978a). In Fig. 6.15b, a relationship between sensitivity and total amorphous matter is shown for Gatineau and St. Alban Leda clays. The amorphous material coats particles and peds and is itself so chemically active that it behaves like a clay mineral thickener if present in large amounts. In this latter case, it would increase the remoulded shear strength and effectively reduce sensitivity.

Carbonate precipitation as a bonding agent (Townsend 1965) may be more significant than generally realized for both freshwater and marine clays. The St. Jean de Vianney soils in Quebec on which the notorious flow slides are developed (Tavenas et al, 1971) contain more carbonate than most marine clays, possibly related to the presence of Paleozoic limestones in the area. Whereas most marine clays contain 1-4% carbonate, probably of organic origin (shells, etc.), samples of the quick St. Jean soil contain 6-13% carbonate (Moum and Zimmie 1972. Certain quick clays in the Arnprior region northwest of Ottawa also contain up to 13% carbonates. Carbonate contents of 30% have been reported for quick Tyrrell Sea deposits at Port Rupert on James Bay.

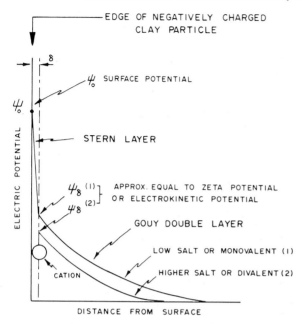

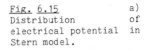

Fig. 6.15 a) Distribution of electrical potential in Stern model.

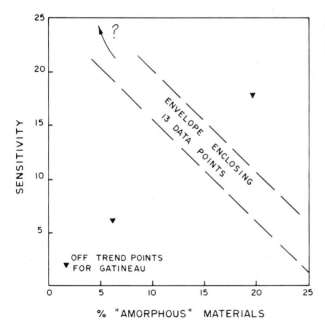

Fig. 6.15 b) Sensitivity vs. amorphous material content for Gatineau and St. Alban Leda clay. (After Yong et al., 1979).

6.5.4 High Zeta Or Electrokinetic Potential.

Any in situ chemical reaction that effects an expansion of the Gouy double layers
and increases the zeta potential (Fig. 6.15a) reduces the remoulded strength of a
soil. Such soils are dispersed or peptized relative to their original
flocculated state at deposition. Penner (1965) clearly illustrated the
correlation between electrokinetic potential and sensitivity for most Ottawa area
clays of low salinity (Fig. 6.16a).

There are many natural chemical and biochemical reactions that will peptize a
soil, the two most important probably being a decrease in salinity and organic
anion adsorption as discussed by Rosenquist (1966). Two other important factors
in natural soils are inorganic anion adsorption and monovalent cation adsorption.

The importance of a low salinity is well established (Torrance 1975, 1975a) and
need not be considered further here. The adsorbed cation regime may be
important, and certainly in Penner's (1965) paper, high adsorbed (Na + K)
relative to Ca + Mg) seemed to play a significant role in establishing a high
zeta potential and a high sensitivity (Fig. 6.16a).

Postdepositional adsorption of Ca^{2+} and Mg^{2+} and possibly K^+ at the expense of
Na^+ may desensitize a soil as shown by Moum et al. (1971) for the bottom of a
Norwegian quick clay. The remoulded strength is actually increased by cation
exchange so that an originally peptized quick clay is rendered nonquick. The
source of Mg^{2+} remains in contention but may be normal groundwater exchange or
possibly in situ degradation of soil minerals. This does not appear to have
happened at most Leda clay deposits where salt leaching has apparently dispersed
the soils and caused concentrations of landslides as described by La Rochelle et
al, (1970).

A plot from Yong et al. (1979) is also drawn on Fig. 6.16a to show the large
differences in relationships from site to site. The St. Alban clays are quite
high in "amorphous" matter (~12%) and this may in part explain the difference in
the two data sets.

The role of naturally occurring inorganic and organic dispersants remains a
complex area. The work of Soderblom (1966) illustrated the importance of complex
humic compounds in creating "quick" varved clays overlain by freshwater peat
deposits.

In the case of marine clays, the problem is one of "anaerobic transformation of
organic material sedimented together with clay particles" as quoted from
Rosenquist (1966). Recent work by Donovan and Lajoie (1979) emphasizes the
importance of considering sediments as anaerobic reducing systems. The Eh versus
pH diagram in Fig. 6.16b taken directly from Donovan and Lajoie, shows that Leda
clay samples from near Trois Rivieres, Quebec are in a highly reduced state with
FeS_2, Fe^{2+}, and maybe free sulphur as stable porewater species. Free iron is
probably produced from amorphous iron hydroxides, free sulphur is produced by
reduction of porewater sulphate and iron monosulphides and possibly pyrite
precipitate from solution. Zones of black mottling in Leda clay are probably a
complex mixture of organic matter and iron sulphides, which rapidly disapear on
exposure and oxidation.

Donovan and Lajoie (1979) show an important increase in bicarbonate content in
the aquifer groundwater accompanying the decrease in sulphate by in situ
reduction and suggest bacterial resiration as a source for the CO_2. The sulphide
bicarbonate equilibrium phenomenon is very complex but from a geotechnical point
of view, the significance is that both bicarbonate and sulphide are inorganic

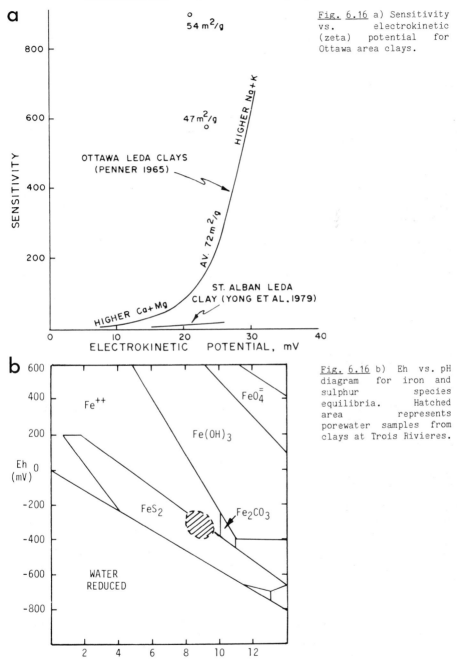

a

Fig. 6.16 a) Sensitivity vs. electrokinetic (zeta) potential for Ottawa area clays.

b

Fig. 6.16 b) Eh vs. pH diagram for iron and sulphur species equilibria. Hatched area represents porewater samples from clays at Trois Rivieres.

anions and if adsorbed they effectively expand the double layers, operate as peptizers, and produce very low remoulded shear strengths.

Laboratory measurement of bicarbonate and sulphide species is very difficult because of CO_2 loss and rapid sulphide oxidation upon exposure of soil samples to the atmosphere. Furthermore, all laboratory characterization of the physicochemical properties of remoulded soils is necessarily done on artificial material, fully oxidized compared with the highly reduced in situ state.

If humic dispersants are also produced by anaerobic bacterial activity related to digestion of organic matter in marine clays, then extremely low remoulded strengths and very high sensitivities may be obtained (Soderblom 1966) provided the salinity is low. One would expect such features to be relatively less important for glacial freshwater clay deposits where organic matter is usually present in much lesser amounts. The presence of iron sulphides in Leda clays is shown in Fig. 6.17 where numerous agglomerations of pyritic framboids are shown cementing silt grains together. The bacterium desulphovibrio desulfurican is thought to play a critical role in iron and sulphur reduction.

All factors that increase the zeta potential and disperse or peptize a soil will cause some reduction in undisturbed strength along with the large reductions in remoulded strength. A decreased resistance to erosion is a predictable result for marine soils rich in clay materials but might not be so easily predictable for those consisting essentially of rock flour, composed of primary mineral grains of clay size.

6.5.5 Low Smectite Content.

Because of their gel-like, thixotropic characteristics, smectites operate as thickeners maintaining a fairly high remoulded shear strength and a fairly low sensitivity. Nevertheless, sensitivity values as high as 11 or 12 are reported by Buck and Parry (1976) for Lake Bonneville clays at Salt Lake City, Utah. These clays contain abundant montomorillonite and were deposited in a flocculated state in saltwater of glacial Lake Bonneville. Buck and Parry concluded that peptizing phenomena controlled sensitivity, especially saltwater leaching. A high correlation was also obtained for organic contents greater than 32%. Although these data are probably not applicable to freshwater smectites of the western Prairies, they may well have some importance for marine clays of the west coast of British Columbia.

6.5.6 High Liquidity Index

In the case of soils consisting dominantly of rock flour composed of primary minerals rather than clay minerals, values of liquidity index of 2-4 are possible, and the primary explanation appears to be cementation bonding immediately following deposition. Exceptionally high sensitivities for these soils are attributed by many authors such as Cabrera and Smalley (1973) to be as much related to liquefaction as to electrokinetic phenomena.

In the case of clayey soils, peptizing reactions including leaching, ion exchange, and dispersant adsorption serve to decrease the liquid limit more than the natural water content is reduced by consolidation. Since the liquid limit is in fact a moisture content defining a remoulded shear strength, studies of its variation with chemistry should correlate with sensitivity (Fig. 6.14).

Hendershot and Carson (1978) illustrate a significant drop in liquid limit upon removal of amorphous material from a Gatineau area marine clay and speculate, like Yong et al. (1979), that amorphous material plays a significant role in establishing sensitivity. The conundrum appears to be that although amorphous

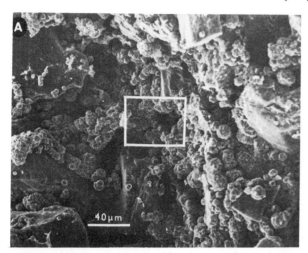

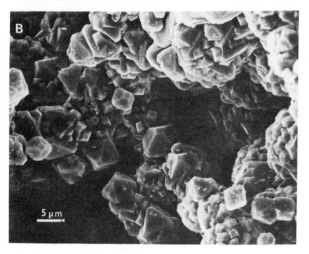

Fig. 6.17.   Pyritic
framboids   from   11   m
depth, Hawkesbury Leda
Clay.   A)   irregular
agglomerations   between
silt   particles.   B)
close up (from Haynes
and Quigley 1978.)

material influences sensitivity, it operates in a complex way on both the top and
bottom of the sensitivity equation.

6.5.7 High Primary Mineral Content

The failure mechanism of quickclay soils has been discussed by Smalley (1971) and
Moon (1972, 1974, 1978) in terms of a high content of rockflour in which primary
mineral grains of clay size predominate in contrast to true clay minerals. These
primary minerals are quartz and feldspar and produce ´inactive´ clay size
abrasion products (subglacial ´washout´: Sect. 2.4) that may be held together
simply by a carbonate cement. Rupture of this cement by sudden shock releases
the fine particles into the porewater and the clays then become ´quick´.

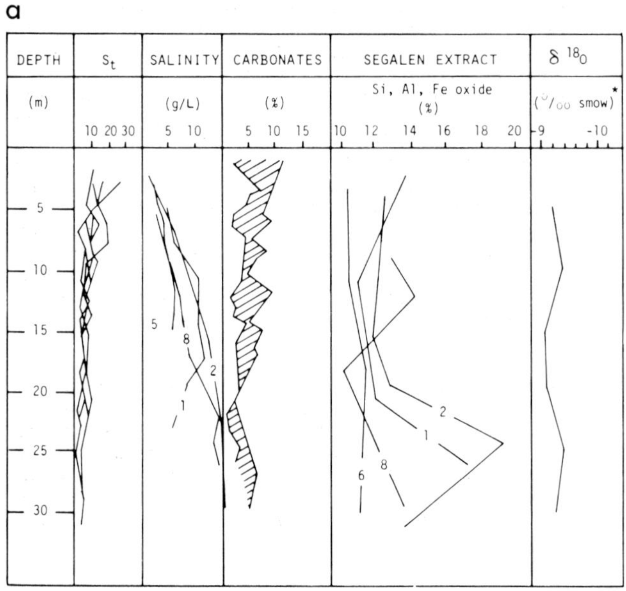

a

Fig. 6.18.   a)
Sensitivity, salinity,
carbonate. ´amorphous´
matter and ^{18}O vs. depth
in Hawkesbury Leda Clay.
SMOW = Standard Mean
Ocean Water.
b) sensitivity, salinity
and porewater cation
chemistry vs. depth,
Hawkesbury Leda Clay.

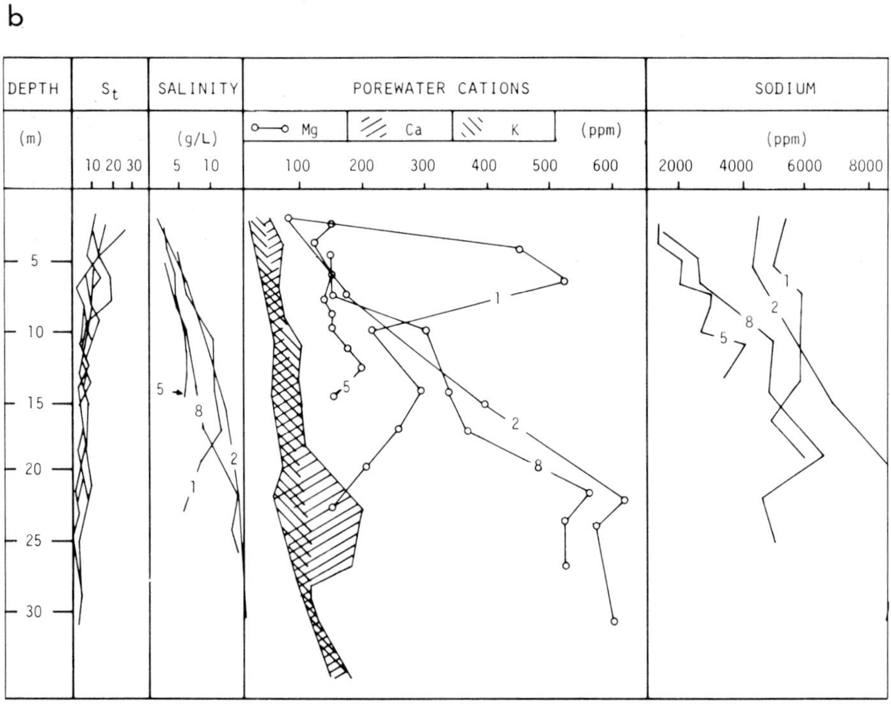

b

Geotechnical profiles. Major efforts to produce comprehensive geochemical and mineralogical profiles of marine clay deposits are becoming more common (Torrance 1975, 1975a; Haynes and Quigley 1976; Yong et al. 1979). Rarely, however, does a single laboratory have the equipment, manpower, or funds to assess all aspects of the complex sensitivity equation. The field sampling of sensitive clays also poses problems (Penner and Burn, 1977) and a sample quality better than class 2 or 3 (Table 11.1) is sometimes difficult to achieve (Graham, 1979). In Fig. 6.18, composite profiles from four boreholes in Leda clay near Hawkesbury, Ontario are presented as examples of the types of correlations being attempted. In Fig. 6.18a sensitivity appears to correlate directly with carbonate content, inversely with salinity, and not at all with "amorphous" matter. In Fig. 6.18b it is seen that $Na+$ dominates the pore water with secondary Mg^2+, Ca^2+, and $K+$, yielding salinities varying from about 2 to 4 g/L at surface to about 15 g/L at depth. At these relatively high salinity values and depressed double layers, the effects of other peptizing constituents in the soil such as organic dispersants and bicarbonate are probably neutralized.

Figure 6.18a also contains a column showing ^{18}O values for pore water squeezed from the Hawkesbury samples. The values are typical of either (1) present-day rainwater at this latitude, or (2) mixed glacial meltwater and glacial seawater. Since the pore water is still saline, it is inferred that the ^{18}O value probably indicates a brackish water environment of deposition. Interpretation of oxygen and hydrogen isotope work is likely in future years to yield much new data on the environments of marine and freshwater clay deposition (Desaulniers et al, 1981).

6.6 Summary

The importance of understanding the role of sedimentology in influencing soil composition, fabric, structure, and engineering behaviour cannot be over-emphasized.

The physiochemistry and mineralogy of soft clays have been introduced by discussing the many factors that influence the sensitivity equation. Emphasis is placed on amorphous matter, dispersion or peptization as related to electrokinetic potential extraction problems, and problems of correlating sensitivity with physicochemical and mineralogical data. Particular stress is placed on the very severe problem of duplicating the highly reducing, sulphide- and bicarbonate-enriched, anaerobic conditions that exist in the field compared with oxidized, aerobic laboratory conditions.

CHAPTER 7

Glaciofluvial Transport and Deposition

A. D. Miall

INTRODUCTION

Glaciofluvial outwash rivers are low-sinuosity, multiple-channel (braided) type and are part of distributary (fan) systems up to at least 100 km in length (Fig. 7.1). Grain sizes decrease from gravel-dominated near the ice front to sand or silt in distal reaches. Sedimentary processes are dominated by aggradation and lateral accretion of gravel longitudinal bars in proximal regions (Scott-type), changing to sandy linguoid bars downstream (Platte type). In upper to mid reaches a heterogenous assemblage of lithofacies reflects sedimentation in different topographic levels with contrasting energy levels (Donjek type). The deposits here may be strongly cyclic. Larger-scale cycles are developed by long-term glacial advance-retreat phases. Glaciofluvial deposits pass laterally into interfluve pond, aeolian or lacustrine sediments, and downstream into marginal marine environments (Figs. 4.8, 4.9).

7.1 The Glaciofluvial Environment

Valley glaciers or continental ice-sheet lobes which terminate on land release large quantities of meltwater that drain broad outwash plains. The rivers which cross these plains erode and rework previously deposited glacial sediments, redepositing them as a distinctive suite of glaciofluvial lithofacies.

This reworking process begins under the glacier itself. Vigorous sub-ice meltwater streams may deposit fluvial sands and gravels in channels cut into lodgement till. These may be overlain by and interbedded with younger tills as ice-flow and meltwater tunnel patterns change resulting in a complex local stratigraphy (Fig. 2.11). At the glacier terminus some of the sediment load may be dumped in kames which are isolated hummocky mounds deposited against the ice front. Well-defined linear ridges of glaciofluvial sediment, termed eskers (Fig. 2.9), are formed by deposition in meltwater tunnels, or by retreat of tunnel mouths during glacial retreat, or by the let-down of englacial and supraglacial stream debris. Details of the formation, stratigraphy, grain size characteristics and sedimentary structures of eskers have been described by Price (1973), Banerjee and McDonald (1975) and Saunderson (1975). Kame and esker deposits are frequently intimately associated with flowed tills of the distinctive supraglacial landsystem (Fig. 1.5) formed by supraglacial melt-out and downslope mass movement (Figs. 1.8B, 3.6, 3.7). These deposits commonly grade into fine-grained sediments of ice-marginal lakes (Gustavson et al, 1975; Shaw, 1975) particularly as melting of stagnant ice causes slumping, faulting and the creation of small lake basins.

Another common ice-contact landform composed of outwash is kame and kettle topography, also described as pitted or kettled outwash and kamiform topography. Rich (1943) and Price (1973) have drawn attention to the very rapid aggradation

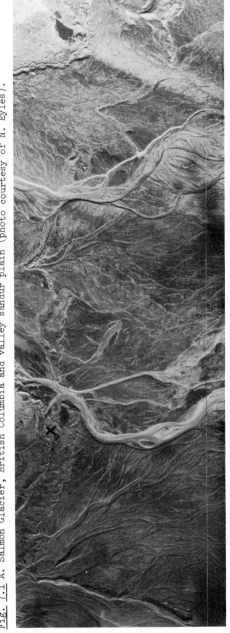

Fig. 7.1 A. Salmon Glacier, British Columbia and valley sandur plain (photo courtesy of N. Eyles).

B. Braids on an Icelandic sandur plain with moraine ridges at X (photo courtesy of R.J. Price).

Fig. 7.2
The formation of kettled outwash at an Icelandic ice front. Rapid aggradation of
outwash by meltstreams (X) has buried the gently sloping ice margin (foreground).
The surface of the outwash plain is now elevated above the ice margin in response
to glacier thinning. Kettle holes (Y) record the melt of the buried ice margin
and will ultimately form a hummocky pitted outwash surface. Photo courtesy of N.
Eyles.

of outwash fans at the icefront and the frequent burial of the thinning terminal
zone of the glacier under the fan. A pitted outwash surface develops when buried
ice subsequently melts (Fig. 7.2).

Beyond the immediate ice terminus lies the outwash plain, or sandur (plural:
sandar, Icelandic; Fig. 7.1, 7.3). Glaciofluvial facies are the dominant
deposits here, and the purpose of this chapter is to summarize the sedimentology
of these deposits. Modern sandar are well developed in Alaska, Arctic Canada and
Iceland and have been described by Fahnestock (1963), Williams and Rust (1969),
Rust (1972), Church (1972), Bluck (1974, 1979) Gustavson, (1974), Bothroyd and
Ashley (1975), and Boothroyd and Nummedal (1978). Most of the sedimentological
characteristics of the deposits have been described in general reviews of the
braided river depositional environment by Miall (1977, 1978), Rust (1978), and
Bluck (1979). The economic significance of braided river deposits in connection
with placer mineral occurrences has been most recently discussed by Pretorius
(1979) and Smith and Minter (1980).

7.2 Fluvial Morphology

Glacial outwash rivers are typically of multiple-channel (braided) type, and are
normally of low sinuosity. Near the glacier terminus there may be one or two
main channels but downstream these typically diverge into several or many
channels, separated by active and inactive bars and islands. Under appropriate
climatic conditions inactive channel margin areas may become vegetated (Alaska),
or they may be areas of wind deflation or aeolian dune migration (Iceland).

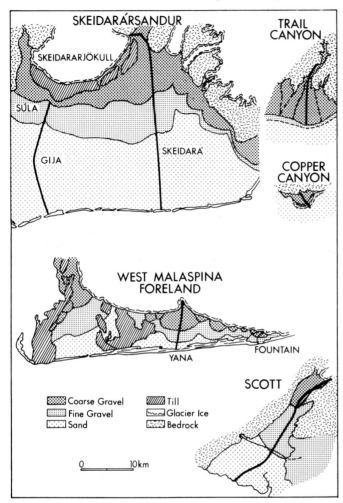

<u>Fig. 7.3</u> Selected outwash plains, showing scale and facies distribution. Skeidararsandur is in Iceland, the Malaspina and Scott systems are in southern Alaska. Also shown are the Trail Canyon and Copper Canyon fans, two small examples of alluvial fans in intermontane basins (Death Valley, California). Reproduced from Boothroyd and Nummedal (1978).

There are several reasons for the predominance of the multiple-channel river patterns of sandar. These are the presence of a large bedload, variable discharge and non-cohesive channel banks. Braided rivers typically are unable to transport more than a small fraction of the available bedload except during extreme floods (Ostrem, 1975). At other times gravel and sand are deposited as a variety of bedforms and bars. These further constrict and divert the flow, a

Fig. 7.4 Vertical airphoto of part of the Donjek River, Yukon territory, showing the four topographic levels defined by Williams and Rust (1969) (part of photo A15728-89, National Airphoto Library, Ottawa, Canada).

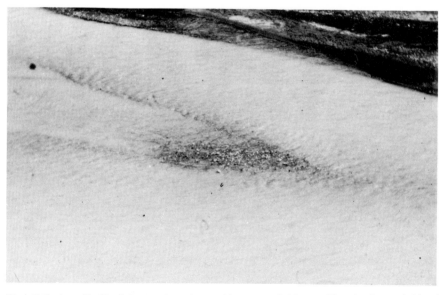

Fig. 7.5 Longitudinal bar undergoing active aggradation. Flow is from right to left. Width of field of view about 30 m. North Saskatchewan River, Alberta. Photo courtesy of D.G. Smith.

Fig. 7.6 Imbrication of gravel clasts. Turnabout River, Ellesmere Island, Arctic Canada.

process facilitated by the easily erodible banks. Glacial outwash streams are characterized by strongly seasonal discharge variations, so that for most of the year, active bar growth and channel migration are occurring. The processes have been described at length by Bluck (1974, 1979).

Stream gradients vary from 2 to 50 m/km, which though steep, is much less than the small alluvial fans of intermontane basins (Boothroyd and Nummedal, 1978). Modern sandar range in length up to about 100 km, between glacier terminus and the sea. Typically there is a downstream decrease in grain size from coarse gravels to sand (Figs. 4.8, 7.3). Maximum clast size may reach 50 cm near the ice margin.

Outwash plains may be divided into a series of topographic levels, reflecting stages of progressive channel abandonment due to lateral migration or vertical downcutting. These were first recognized in the Donjek River, Yukon (Fig. 7.4), and may be summarized as follows (after Williams and Rust, 1969).

Level 1: This is the level of the main channels, and is the principal sediment dispersal route, with little or no vegetation cover and bars exposed only during low water

Level 2: This level is active only during flood stages, with few active channels at other times and sparse vegetation cover.

Level 3: Here there is little continuous water movement except for low-energy flow during flood stages, with moderate vegetation cover in humid areas.

Fig. 7.7 Formation of sand lenses with dunes and ripples in abandoned   channels.
North Saskatchewan River, Alberta.   Photo courtesy of T. Jerzykiewicz.

Fig. 7.8 Avalanche slope at the termination of a sandy linguoid bar.   Bar is
capped by ripples. Athabasca River, Alberta.

Table 7.1: Lithofacies types (after Miall, 1978).

Facies Code	Lithofacies	Sedimentary structures	Interpretation
Gms	massive, matrix supported gravel	none	debris flow deposits
Gm	massive or crudely bedded gravel	horizontal bedding, imbrication	longitudinal bars, lag deposits, sieve deposits
Gt	gravel, stratified	trough crossbeds	minor channel fills
Gp	gravel, stratified	planar crossbeds	linguoid bars or del-taic growths from older bar remnants
St	sand, medium to v. coarse, may be pebbly	solitary (theta) or grouped (pi) trough crossbeds	dunes (lower flow regime)
Sp	sand, medium to v. coarse, may be pebbly	solitary (alpha) or grouped (omikron) planar crossbeds	linguoid, transverse bars, sand waves (lower flow regime)
Sr	sand, very fine to coarse	ripple marks of all types	ripples (lower flow regime)
Sh	sand, very fine to very coarse, may be pebbly	horizontal lamination, parting or streaming lineation	planar bed flow (l. and u. flow regime)
Sl	sand, fine	low angle (<10°) crossbeds	scour fills, crevasse splays, antidunes
Se	erosional scours with intraclasts	crude crossbedding	scour fills
Ss	sand, fine to coarse, may be pebbly	broad, shallow scours including eta cross-stratification	scour fills
Sse, She, Spe	sand	analogous to Ss, Sh, Sp	eolian deposits
Fl	sand, silt, mud	fine lamination, very small ripples	overbank or waning flood deposits
Fsc	silt, mud	laminated to massive	backswamp deposits
Fcf	mud	massive, with freshwater molluscs	backswamp pond deposits
Fm	mud, silt	massive, desiccation cracks	overbank or drape deposits
Fr	silt, mud	rootlets	seatearth
C	coal, carbona-ceous mud	plants, mud films	swamp deposits
P	carbonate	pedogenic features	soil

Level 4: This consists largely of dry islands and interfluves with either dense vegetation in humid areas or areas of aeolian deflation and dune migration.

Close to the ice terminus three or four of these levels may be recognized within the valleys of the main channels. Downstream, large areas of the outwash plain may be at a single topographic level, reflecting progressive abandonment of parts of the channel network as the distributaries migrate (avulse) across the plain. Shifts in channel position come about as aggradation and slope reduction occurs,

Table 7.2: Facies assemblages in outwash rivers. Based on Miall (1978) and Rust
(1978).

Name	Environmental setting	Main facies	Minor facies
Scott type	proximal gravel-dominated rivers	Gm	Gp, Gt, Sp St, Sr, Fl Fm
Donjek type	medial gravel-sand rivers	Gm, Gt, St	Gp, Sh, Sr, Sp, Fl, Fm
Platte type	distal sand-dominated rivers	St, Sp	Sh, Sr, Ss, Gm, Fl, Fm
Slims type	distal silt-dominated rivers	Fl, Fm	—

with consequent diversion of the runoff down steeper parts of the sandur. Bluck
(1979) showed that lateral shifts of braided channels may average rates of 100
m/year,   partly through avulsion or reoccupation of older courses, and partly
through lateral bar accretion and consequent bank erosion.   The existence of
these topographic levels has important implications for the preservation of
characteristic vertical profiles in the sediments, as discussed later.

7.3 Sedimentary Processes

Studies of sandar in Iceland and Alaska show that there is a progressive
downstream change in grain size, and in bar and bedform type (Boothroyd and
Ashley, 1975; Boothroyd and Nummedal, 1978), permitting a subdivision into
proximal and distal reaches.

7.3.1 Proximal Outwash Rivers

In the proximal reaches of active channels, longitudinal bars are the dominant
bar type (Fig. 7.5). These are composed of gravel and are diamond-or
lozenge-shaped in plan, elongated parallel or subparallel to flow direction.
They develop by gradual clast accretion over diffuse gravel sheets. Flow
commonly passes obliquely or diagonally over the bar, causing downcurrent
progradation to take place at an angle to the overall channel direction.   In this
way significant lateral accretion can occur, causing continual changes in channel
patterns.   Longitudinal bars are typically less than 1 m in height and several
hundred metres, or less, in length.

The details of internal structure and growth pattern depend largely on flow
stage, as documented in detail by Bluck (1974, 1979) and Hein and Walker (1977).
Under flood conditions the bars are completely submerged and are thought to
evolve as planar gravel sheets. Under less energetic flow conditions flow over
the bar may be shallow, flow separation occurs at the downstream margins and the
bar may develop steep downstream avalanche faces.   In the first case crude
horizontal stratification is the result, and clasts commonly show a
well-developed imbrication fabric (Fig. 7.6).   In the second case the gravels

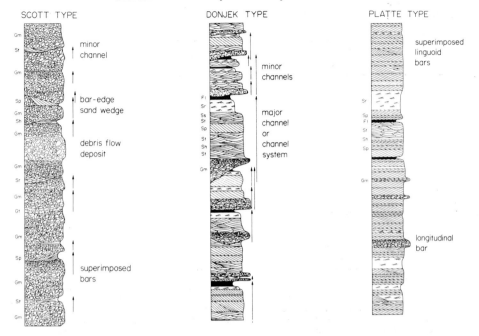

Fig. 7.9 Vertical profile models for the most common types of glaciofluvial sequences (from Miall, 1978).

develop crossbedding. Bluck (1974, 1979) described a range of complex bar responses to variations in grain size, bar and channel morphology and flood stage, but few of the structural details observed have been recorded in ancient rocks, suggesting that the uppermost exposed parts of bars that Bluck studied have a low preservation potential because of their susceptibility to scour during subsequent channel migration.

Even in actively aggrading rivers most of the larger exposed gravel bars may be bounded on at least one side by the cutbank of an active channel. The deposits that are eventually buried and preserved therefore consist of welded bar fragments, cut by remnants of channel scour surfaces. Crude horizontal stratification is by far the commonest internal structure with crossbedded gravels only becoming common in more distal reaches.

In proximal reaches of outwash rivers sand is deposited only during low-water periods, typically toward the end of the peak melt-season floods. Minor channels, particularly at higher topographic levels, may be progressively abandoned by falling water, at which time they are filled with sandy bedforms including dunes and climbing ripples (Fig. 7.7). Sheet run-off from exposed gravel bars builds small deltas or sand wedges toward active channels. Mud drapes may settle over bar and channel alike as flow dries up completely.

7.3.2 Medial And Distal Outwash Rivers

Downstream the dominantly gravelly bedload changes to pebbly sand and, eventually, to sand with little or no gravel. The distance over which this transition occurs varies from less than a kilometre to more than 10 km, depending on slope and discharge magnitude.

Active channels in sandy outwash plains are dominated by the migration of linguoid or lobate bars. These have gently dipping upstream surfaces which may be covered with dunes or ripples. Downstream they terminate at sinuous avalanche slopes (Fig. 7.8). The bars vary in width from a few metres to 150 m, and in length up to 300 m. Height ranges up to 2 m but 50 to 100 cm is more typical. These bedforms are similar to, but larger than sandwaves, and it is not clear whether they are in fact genetically different (Harms et al, 1982). In both cases the internal structure is planar crossbedding. Variations in the geometry of the foresets, and the presence or absence of of reactivation surfaces and backset ripples depend on flow stage variations, and have been described and illustrated by Collinson (1970).

The bars may be capped by planar bedded sand sheets, dunes or ripples. Large dunes may form in the deeper channels.

In elevated parts of the outwash complex, including topographic levels 3 and 4 (the overbank or floodplain part of the outwash plain) sand deposits may only be formed during short lived flood events. Ripples (including climbing ripples) are the typical internal structure. Laminated silts and muds with roots are the other most characteristic deposits, formed by sediment baffling and entrapment as overbank runoff filters into the vegetated areas of the plain.

In the most distal reaches grain size may decrease to silt grade. Channels and bars become of low relief and are poorly differentiated. Ripples are the dominant bedform.

Outwash rivers eventually empty into a lake or the sea (Figs. 2.20, 6.7). Commonly there are numerous active distributaries, and the term outwash fan or fan delta is used to describe this general morphology (Galloway, 1976).

7.4 Lithofacies Classification

The lithofacies of braided rivers have been extensively described and illustrated in the various publications referred to earlier. Miall (1977, 1978) erected a generalized scheme for lithofacies classification which is summarized in Table 7.1. There is not the space to discuss this scheme thoroughly, and the following notes are intended only as a brief clarification of the table.

The lithofacies have been subdivided into three main groups: gravel, sand, and fines. These are coded by the letters G, S and F. The lower case letters which follow are intended as mnemonics for distinctive internal textures or structures.

Some of the lithofacies shown are not common in sandur deposits. Gms is to be expected only immediately adjacent to the ice front where it represents reworked till. Eyles et al (1983) have proposed a revised method of classifying such deposits, using a coding system based on the letter D, for diamicts (Fig. 1.8; Sect. 1.4). Sse, She and Spe were defined by Boothroyd and Nummedal (1978) based on observations in Iceland. These earlier facies occur mainly in interfluve areas marginal to the active channels.

Fig. 7.10 Scott-type outwash sequence. Deposits are almost exclusively gravel including multistorey units of lithofacies Gm, formed by superimposition of longitudinal bars (upper half of photo), and a transverse bar deposit, lithofacies Gp, (lower half of photo). Exshaw, Alberta.

Fig. 7.11 Interbedded gravels and sand lenses, intermediate between Scott and Donjek types of braided profile. Exshaw, Alberta.

The recognition of the various lithofacies types is an essential prerequisite for interpreting Quaternary or older outwash deposits in terms of the sedimentary processes described in the preceding section.

7.5 Outwash Facies Models

Facies models for outwash rivers were proposed by Williams and Rust (1969). Rust (1972, 1978), Boothroyd and Ashley (1975) and Boothroyd and Nummedal (1978). These ideas were incorporated into generalized facies models for braided rivers by Miall (1977, 1978) and Rust (1978).

Miall (1978) recognized six principal facies assemblages in gravel- and sand-dominated braided rivers. Three of these are commonly developed in outwash rivers, and are listed in Table 7.2. The three assemblages can be summarized in characteristic vertical profiles as illustrated in Figure 7.9. An additional tentative model for silt-dominated streams was proposed by Rust (1978) based on the Slims River, Yukon (Table 7.2).

7.5.1 Scott Type

This facies assemblage is named after the Scott outwash fan, one of the Alaskan braided systems studied by Boothroyd and Ashley (1975). The assemblage typifies the proximal part of this river. The dominant sediments are multistorey gravel units built up by aggradation of longitudinal bars. Minor crossbedded gravels may also be present (Fig. 7.10). The gravel may show crude horizontal stratification, resulting from clast sorting over the bar. Each bar tends to preserve an upward decrease in clast grain size.

Interbedded with the gravels are thin lenses of sand, representing deposition in abandoned channels, or bar-edge sand wedges. Fining- upward gravel-sand cycles 1-2 m thick are commonly deposited during the waning stages of flood events. Silt and mud are rare in the Scott-type assemblage and sand-silt-mud together comprise less than 10% of the total vertical thickness.

Channel scour surfaces are common but are hard to identify as they commonly separate gravel units of virtually identical texture, grain size and fabric.

7.5.2 Donjek Type

The downstream transport of gravels in a fluvial system is limited by the frequency of highly competent run-off events, so that there is a rapid decrease of grain size downstream and in most sandar, within a few kilometres of the ice front a significant quantity of sand becomes interbedded with the gravels in the resulting deposits. Such is the case in the middle reaches of the Donjek River studied by Williams and Rust (1969), and Rust (1972), whose work provided the basis for the Donjek facies assemblage model of Miall (1977, 1978).

The assemblage is defined as containing between 10 and 90% gravel in cumulative thickness, and a wide variety of lithofacies types is present (Table 7.1). At this point on a sandur the differentiation of the channel-interfluve system into distinct topographic levels may be particularly marked, and this is reflected in the heterogenous nature of the lithofacies assemblage (Fig. 7.11). Bar gravels dominate the lower, most active channels, while sands and pebbly sands occur in the more elevated reaches, and significant accumulations of silt and mud may be present in long-abandoned areas of the outwash plain or in local interfluve ponds. The tendency of the channels to migrate laterally because of bar growth (Bluck, 1979), or to switch position by avulsion, leads to a process of channel aggradation and gradual or sudden abandonment. A fining-upward sequence several

Fig. 7.12 Planar crossbed set (lithofacies Sp), overlain by silt lens. Blackfalds, Alberta.

metres in thickness is a typical result. Thinner cyclic sequences may represent flood events, as in the Scott-type assemblage.

The strongly cyclic nature of the deposits of this assemblage is its most distinctive characteristic. Statistical studies of the cyclicity using Markov chain analysis have been reported by a few workers, including that by Corner (1975) on the terraces of the Tana River, Norway.

7.5.3 Platte Type

In distal reaches of sandar plains the run-off is distributed amongst numerous shallow distributaries. Topographic differentiation of the channels is less marked than in the Donjek-type environment, and average channel depth and slope smaller. Sand is the dominant deposit. The Platte River of Colorado and Nebraska is a non-glacial river which shows these characteristics (Smith, 1970, 1971, 1972; Blodgett and Stanley, 1980), and was used as the basis for erecting the Platte braided stream model by Miall (1977).

The dominance of sandy bedforms, mainly sand waves and linguoid bars plus, less commonly, dunes, leads to a deposit characterized by abundant crossbedding. Planar crossbedding is the dominant type (Fig. 7.12). Minor gravel lenses and overbank fines may be present. Cyclic sequences are rare.

7.5.4 Slims Type

In the most distal reaches of an outwash river, or where the sediment source is dominated by fine-grained wind blown material, such as loess (Sect. 5.6), the resulting deposits may consist mainly of silt. The Slims River, Yukon, is a good example, but has received little detailed study. A tentative model was suggested by Rust (1978) based on his own reconnaissance work and that of Fahnestock (1969).

The river is characterized by low relief bars and channels, in which the main sediment types are massive, laminated and ripple cross-laminated sandy silt. Coarser channel sediments are virtually absent. It is not known what minor lithofacies may be present or what the characteristic vertical profile is, because no Quaternary or ancient sandar deposits of this type have been described.

7.6 Outwash Depositional Systems

Few modern sandar have received the kind of large-scale sedimentological analysis required for an understanding of the regional environmental and facies variations encompassed by the term "depositional system" (Sect. 1.5). Until such work is carried out, stratigraphic and environmental interpretations of Quaternary and older periglacial deposits will be severely hampered.

Facies studies described in the previous section have documented proximal-distal changes in glaciofluvial deposits. Boothroyd and Nummedal (1978) and Edwards (1978) have briefly described lateral transitions into aeolian or lacustrine environments, and downstream passage into marine environments. The close association between major outwash river plains and aeolian activity is seen in many areas of Quaternary glaciation (Figs. 5.12, 5.13, Sect. 5.6). Edwards (1978) proposed a tentative stratigraphic model for a glacial advance-retreat cycle, in which it was suggested that the subaerial periglacial facies would be deposited mainly during the retreat phase. Glaciofluvial and lacustrine deposits overlie fully preserved or partially eroded till sequences and the fluvial deposits may show a gross upward fining in response to a decay or retreat of the glacial sediment source. Coarsening-upward glaciofluvial cycles may be formed in front of an advancing glacier, as exemplified by many Quaternary sequences where the successions are capped by till (Fig. 15.1, Fulton and Halstead, 1972; Miall, 1980).

Large-scale coarsening- and fining-upward cycles of this type are 10 m or more in thickness, and should be distinguishable from the within-channel cycles of Donjek-type sedimentary profiles by their greater thickness and lateral extent. Another type of large scale cycle that has not yet been identified in outwash deposits is that produced by lateral channel migration and switching on large alluvial fans. Miall (1980) discussed several examples of this from ancient non-glacial fan deposits, in which it was suggested that fining-upward cycles up to about 40 m thick were formed as a distributary system gradually combed across its own deposits, resulting in progressive channel abandonment and energy decrease. Patterns of topographic levels and vegetation growth on modern outwash fans suggest that similar processes might be occurring at present (e.g. see Boothroyd and Nummedal, 1978), and long-term lateral channel migration has been documented on a small outwash system in Iceland by Bluck (1979). However no large-scale stratigraphic studies of glaciofluvial deposits have been undertaken to investigate this possibility.

A few studies of ancient glacial deposits have demonstrated some of these large-scale facies relationships, and have documented the existence of coarsening- and fining-upward cycles such as those described here. The Dwyka Formation of the Karoo Basin, South Africa, is a good exmple (Visser, 1982).

However, only the most generalized stratigraphic information is available, and the kind of detailed depositional systems analysis proposed here has not been attempted.

CHAPTER 8

Geotechnical Properties of Lodgement Till — A Review

J. A. Sladen and W. Wrigley

8.1 INTRODUCTION

Lodgement till sequences, found in almost all glaciated lowland areas underlie such a large proportion of the industrialized world, particularly in Northern Europe and North America (Figs. 9.1, 9.2) that their geotechnical properties are of great interest to civil engineers and other earth scientists and therefore warrant detailed discussion. Whilst lodgement till is in many ways unique among the soils with which the geotechnical engineer is concerned, it lends itself to perhaps more consistent generalizations than any other such widespread sediment. This chapter demonstrates how simple engineering index properties of lodgement tills can be related to the geological processes which gave rise to them, and how these index properties can be used to predict more complex geotechnical properties such as shear strength, compressibility and in situ stress conditions. The geological processes responsible for deposition of lodgement till have been described in detail in Chapter 2. Geotechnical notation used in this chapter are listed in Appendix III.

The geotechnical properties of fluvial and lacustrine sediments which are commonly found in association with lodgement till (Fig. 1.4) are not discussed in this chapter although some of the effects which these associated sediments have on the geotechnical properties of lodgement till are briefly indicated. Many of the problems encountered during civil engineering construction in areas where lodgement till is the predominant soil type are related to the presence within till sequences of sand, gravel and clay deposits (Fig. 1.4, Sects. 13.4, 13.5).

8.2 Engineering Classification Of Lodgement Till

8.2.1 Particle Size Distribution.

Lodgement tills are invariably well-graded (unsorted), having been subject to sorting by neither wind nor water. The relative proportions of the various particle sizes can vary greatly from one till sheet to another, or locally within a single till sheet. Individual till sheets, however, often exhibit a remarkable uniformity (e.g. White 1972; Scott, 1976). It is convenient to divide the grading into a fine (sand size and smaller) and a coarse fraction. It has been demonstrated (McGown and Derbyshire, 1977) that, for fine fractions above about 45% of the total dry weight, the fine fraction acts as a matrix holding the coarse fraction which tends to behave as discrete particles. A series of large scale triaxial tests on recompacted mixtures of gravel and sandy clay reported by Holtz and Ellis (1961) have shown that gravel contents of less than about 40% have little effect on shear strength. Thus till with a coarse fraction of less than about 40% is referred to as "matrix dominant" and it is the fine fraction that is of particular interest to the geotechnical engineer. Lodgement till with a higher coarse fraction is rare and in lowland areas is restricted to thin basal zones where the till directly overlies rockhead (deformation till: Fig. 1.4, Sect. 2.8.2). Such coarse lodgement tills behave essentially as dense granular materials are common on shield terrains (Fig. 1.3) and will not be considered further here.

The grading of matrix dominant tills is most usefully presented on a triangular (sand/silt/clay) diagram. Being well-graded, data points fall within broad bands as shown on Fig. 8.1a which is based on over 1000 determinations from published and unpublished work in Britain and North America. To demonstrate the uniformity usually exhibited by individual till sheets, typical discrete envelopes containing data points from different till sheets have been outlined.

The main factors which are believed to influence grading of lodgement till are:

a) Characteristics of the bedrock source (Scott, 1976)
b) Distance "downstream" from the bedrock source (Dreimanis and
     Vagners, 1971).
c) Incorporation of reworking of older sediments (Gillberg, 1977; Sect. 2.4)
d) Post depositional weathering (Sect. 8.2.4).

As a till sequence builds up, bedrock sources tend to become ´sealed´ and there is thus a tendency for material higher in a sequence to be finer grained. Individual till sheets, however, tend to fall within discrete envelopes on the triangular diagram which implies that this effect has more influence on the relative proportions of coarse and fine fractions than on the grading of the fine fraction itself.

In a study of the grading of Canadian lodgement tills, Scott (1976) demonstrated that the composition, strength and grain size of the bedrock source materials are dominant factors. He was able to distinguish seven till provinces in Canada on the basis of grading using triangular diagrams and to relate these provinces to bedrock source characteristics (Fig. 8.1b). For example the lodgement tills of the Canadian Shield, which are derived from gneiss and granites, generally contain less than 5% clay size material, and have a high coarse fraction. In contrast the lodgement tills of the Canadian Prairies which are derived from poorly consolidated sediments of Upper Cretaceous and Tertiary age with high clay contents, contain a high proportion of silt and clay size material and are almost stoneless with generally less than 5% coarse fraction.

8.2.2 Atterberg Limits.

When the plasticity indices (PI = Liquid Limit (LL) minus Plastic Limit (PL)) of tills are plotted against their liquid limits on the conventional plasticity chart, points fall close to a line parallel to but above the Casagrande ´A´ line. This line has been termed the ´T´ line (Boulton and Paul, 1976). Fig. 8.2 shows such a plasticity chart with typical data for lodgement tills from three sources in northeast Britain together with the ´T´ and ´A´ lines. Thus we can say for a till that:-

$$PI \simeq 0.73 \ (LL-11) \qquad (8.1)$$

Equation 8.1 implies a constant linear relationship between the plastic and the liquid limits. It should be noted that the ´T´ line is not unique to tills but is shared with several other soils of different geological origins.

Skempton (1953) defined "activity" as the plasticity index divided by the percent dry weight of clay sized particles. Unweathered lodgement tills are almost invariably "inactive", their activities being generally less than 0.75. This can

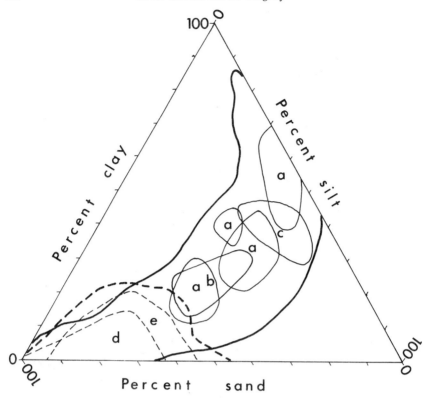

8.1a. Particle-size distribution envelopes based on percentage sand-silt-clay ratios of lodgement tills, (solid line), coarse-grained supraglacial diamicts (light dashed lines) and melt-out tills (heavy dashed line). Melt-out tills being derived from basal debris (Figs. 1.7, 3.2) occur in the same envelope as lodgement tills. Individual lodgement till sheets often fall into discrete envelopes (a, Ohio, U.S.A.; b, Northumberland, U.K.; c, Ontario, Can. From various sources). Regional variation in lodgement till grading relates to rock type and incorporation of preexisting sediments by the glacier (Fig. 8.1b). The better grading of coarse-grained supraglacial diamicts is offset by the larger size range of clasts present (Fig. 4.6); d, Iceland; e, Scotland.

be attributed to the composition of their fine fraction which is largely rock flour comprising finely ground quartz and carbonate and inactive clay minerals. Activity is essentially a function of the mineralogy of the clay fraction. Thus for a till sheet within which mineralogy and hence activity, does not vary greatly plasticity can be expected to be a function of grading. As an example Fig. 8.3a presents data from many samples of Devensian till from north-eastern lowland Britain where the tills are derived largely from Mesozoic and Palaeozoic sedimentary rocks which include abundant mudstones. It will be seen that plasticity index is not purely a function of the clay size content but also increases slightly with silt size content such that the activity of these sediments, which is generally within the range 0.60 to 0.80 is slightly higher

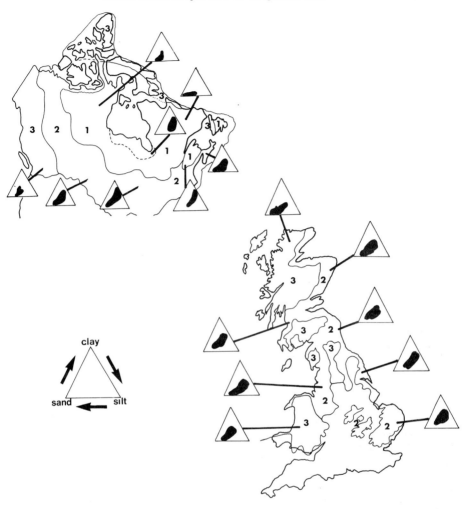

<u>Fig. 8.1b</u> Grading of lodgement tills in N. America and Britain.   1. Shield
terrain;   2. Sedimentary lowlands of subglacial terrain; 3. Glaciated valleys
(See Figs. 9.1, 9.2). Note the coarse sandy tills in glaciated valley and shield
terrain in North America except where fines have been dredged out of Hudson Bay.
The most matrix rich tills are found in the Great Lakes Basins and along the
St. Lawrence lowlands where lacustrine and marine sediments have been
incorporated (Sect. 2.4). Intermediate well-graded to matrix rich tills are
found on the remaining sedimentary lowlands. (After Scott, 1976). In Britain,
matrix rich tills are found in eastern England on sedimentary rock strata that
includes shales and mudstones; clast rich tills characterize glaciated valley
terrain in Wales, Scotland, the Pennines and Lake District.   (Based on various
published sources and unpublished site investigation data).

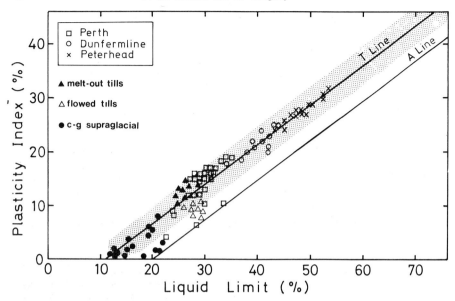

<u>Fig. 8.2</u> Plasticity chart for lodgement tills. Data points generally fall within the shaded area. Typical test results (Scotland) from unpublished site investigation reports. For comparison, typical test results for melt-out and flowed tills and coarse-grained supraglacial diamicts are also shown. Sources; Eyles, Paul, unpublished data.

where clay size content is low. By contrast Fig. 8.3b presents published data from some lodgement tills from Ontario, Canada. Whilst a similar trend of increasing plasticity index with clay and silt contents can be deduced activities are considerably lower, typically 0.25 to 0.45, so that for a given grading plasticity index is correspondingly lower. The marked differences between Figs. 8.3a and 8.3b can be attributed to differences in clay mineralogy, the Ontario tills being largely derived from limestones and dolomites and the crystalline basement rocks of the Canadian Shield (see Ch. 6 and Quigley <u>et al</u>, 1971).

<u>8.2.3 Natural Moisture Content And Liquidity Index</u>.

The natural moisture content (NMC) of a fine grained soil is a reflection of its state of consolidation. Unweathered lodgement tills are usually heavily overconsolidated, that is they have been subjected in the past to higher effective vertical stresses than exist at present. As a result the matrix moisture content of unweathered lodgement till is low, generally slightly below the plastic limit. The relationship between natural moisture content and the Atterberg Limits gives a useful insight into the consistency of fine grained soils. This relationship is most conveniently expressed by the liquidity index (LI = (NMC-PL)/PI). The liquidity index of unweathered lodgement tills is typically within the range -0.1 to -0.35. It should be noted that for soils of low plasticity the determination of liquidity index is particularly sensitive to errors in the determination of Atterberg index properties.

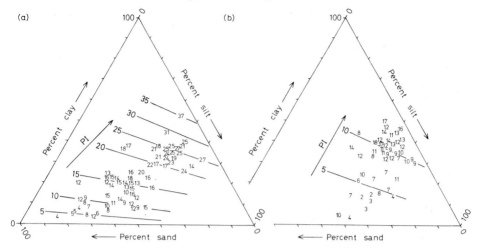

<u>Fig. 8.3</u> The relationship between grading and plasticity of lodgement tills.
Small figures are plasticity index and contours are general trend. (a) Eastern
lowland Britain - Sources: - Unpublished site investigation reports from the
Peterhead, Aberdeen, Perth, Fife, West Lothian, Northumberland and Durham areas.
(b) Some Canadian Tills - Sources: - Peters and McKeown (1976), Milligan (1976),
Cooper, (1979) and unpublished data.

Figures 8.5 and 8.12 show typical profiles of liquidity index versus depth in
lodgement tills from northeast Britain.

When a till contains an appreciable proportion of clasts (greater than sand size)
it is important, when calculating liquidity index, that the moisture content
determined is that of the matrix. Clasts within a matrix dominant till will have
individual moisture contents in general not equal to the matrix moisture content,
but usually lower. Hence if the moisture content is determined for the whole
sample, including clasts, it will generally be lower than the matrix moisture
content resulting in an underestimation of liquidity index. This problem can be
simply overcome by either;

1) Taking care that clasts are excluded from samples to be used for
    moisture content determinations or,

2) Determining the moisture content of the entire sample (M) and correcting
    for the coarse fraction to determine the matrix moisture content (Mm)
    as follows

$$Mm = \frac{M - Ac}{1 - c}$$
(8.2)

where

c = Proportion of the sample by dry weight greater than sand size.

A = Absorption Coefficient i.e. the average moisture content of the intact
      clast material.

´c´ can be determined by conventional sieving. ´A´ can be estimated by means of

Table 8.1. An engineering geological weathering scheme for lodgement tills.

WEATHERING STATE	ZONE	DESCRIPTION
Highly weathered	IV	Oxidized till and surficial material Strong oxidation colours High rotten boulder content Prismatic gleyed jointing Pedological profile usually leached brown earth Leached of most primary carbonate Increased clay content Increased activity
Moderately weathered	III	Oxidized till Usually dark brown or dark red brown Low rotten boulder content Little leaching of primary carbonate Increased clay content Increased activity Base commonly defined by fluvioglacial sediments
Slightly weathered	II	Selective oxidation along fissure surfaces where present otherwise as Zone 1
Unweathered	I	Unweathered till No oxidation Usually dark grey No post depositionally rotted boulders No leaching or primary carbonate

standard aggregate absorption tests (as for example) described in BS812 (1975). Unless ´c´ is large (more than about 20%) ´A´ can be estimated from visual examination of clasts without introducing appreciable error.

8.2.4 Post Depositional Modification.

The engineering and hydrogeological significance of weathering in tills has been reported both in North America (Quigley, 1975; Grisak et al., 1976; Quigley and Ogunbadejo, 1976) and Britain (Bonell, 1972a; Spears and Reeves, 1975; Eyles and Sladen, 1981).

Till at the surface weathers chemically through oxidation, hydration, leaching of soluble materials - mainly carbonates, and by mechanical disintegration of particles, changes of till structure by fracture and in some cases, by downward movement of very fine material (White, 1972; Leighton and MacClintock, 1962; Madgett and Catt, 1978).

A distinct sequence within the weathering process has been recognized by many workers (e.g. Christiansen, 1971). Oxidation takes place most rapidly and to the greatest depth, carbonates are leached to a lesser depth and further chemical alteration occurs to a still lesser depth. In situ alteration of feldspars to clay minerals occurs within the most highly weathered zones (Allen, 1959).

When considering the effects of weathering on the geotechnical properties of engineering soils and rocks it is convenient to zone such materials according to degree of alteration from the parent material. A fourfold weathering scheme has been developed for late Devensian lodgement tills in eastern lowland Britain (Eyles and Sladen, 1981) and is reproduced in Table 8.1. Subsequent studies by

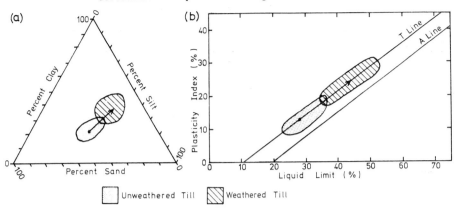

(a)

(b)

☐ Unweathered Till   ▨ Weathered Till

Fig. 8.4. Schematic illustration of the typical effects of weathering on Index properties (a) Grading (b) Atterberg Indices.

the authors have shown that this scheme can be applied in general terms, to tills in other lowland areas of Britain and North America. For single till sheets it is often possible to attribute ranges of variation of geotechnical properties to each weathering zone.

The depth of weathering and the character of individual zones vary greatly from area to area and depend on a complex inter-relationship of factors including mineralogy, climate, groundwater levels, drainage conditions (the inclusion of granular meltwater deposits within till sheets acting as "drainage blankets" has been found to have a marked effect on weathering depth; Fig. 2.11) and vegetation cover. A detailed discussion of weathering processes within lodgement till is outside the scope of this chapter. In the following section a few of the more important effects of weathering are briefly mentioned.

In areas of good drainage fine particles formed within Zone IV tend to migrate downwards into Zone III resulting in a general increased fineness within these zones. In areas of poor drainage, however, accumulation of fine particles within Zone IV can result in a highly plastic horizon, often termed ´Gumbotil´ in North America (Kay, 1916; Leighton and MacClintock, 1930). It is particularly important to note that the degree of clay mineral formation with weathering appears to vary greatly with till type and from area to area. For example Quigley and Ogunbadejo (1976) working in the Sarnia area of Ontario have reported a dramatic increase in activity from 0.4 in the unweathered zone to 1.0 in weathered zones of waterlaid tills (Sect. 2.9). By contrast an increase in average activity from 0.64 to 0.68 from unweathered to weathered zones is reported in lodgement tills, in Northumberland, England (Eyles and Sladen, 1981).

The proportion of carbonates in unweathered till appears to greatly affect the severity of changes brought about by weathering. Where the primary carbonate level is high changes in geotechnical properties resulting from weathering are greatest.

The net effects of weathering on the geotechnical index properties of lodgement tills can be summarized as follows:

J. A. Sladen and W. Wrigley

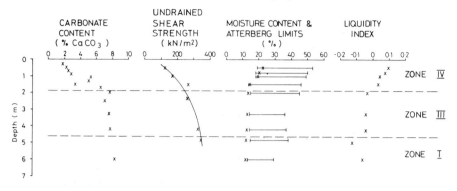

Fig. 8.5. Typical variation of some index properties with depth in weathered lodgement till in Northumberland, England (after Eyles and Sladen, 1981).

1   Increased clay and silt contents due to mechanical disintegration resulting in increased plastic and liquid limits and increased plasticity index.

2   Increased clay content and increased activity due to the formation of secondary clay minerals again resulting in increased plasticity.

3   Increased moisture contents. Moisture content generally increases at a greater rate than the plastic limit such that more weathered samples generally have higher liquidity indices.

These effects are illustrated in qualitative terms in Fig. 8.4. Figure 8.5 shows the typical variation in index properties with depth from a weathered till section in Northumberland, England.

8.3 Shear Strength

8.3.1 Drained Strength.

The strength of saturated soil in terms of effective stress is usually described by means of Mohr-Coulomb failure criteria, i.e. a drained friction angle (ϕ) and cohesion intercept (c'). The shear strength (S_f) is given by:

$$S_f = c' + \sigma'_n \tan \phi' \qquad\qquad (8.3)$$

where σ'_n is the effective normal stress on the failure plane. In terms of total stress equation 8.3 can be rewritten

$$S_f = c' + (\sigma_n - \mu) \tan \phi' \qquad\qquad (8.4)$$

where σ_n is the total normal stress on the failure plane and μ is the pore water pressure.

When soils are sheared to failure they usually exhibit a 'peak' drained strength, defined by ϕ' and c', followed at higher strains by a reduction in strength. At

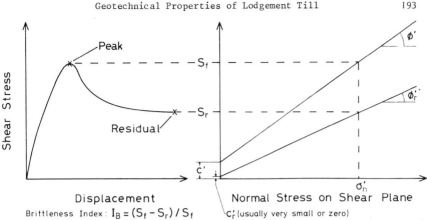

Brittleness Index: $I_B = (S_f - S_r) / S_f$

c_r' (usually very small or zero)

Fig. 8.6.   The Peak and residual strength of soils in terms of effective stress.

large strains the strength reaches a reasonably constant minimum value defined by the residual angle of friction ($\emptyset'r$); this concept is illustrated in Fig. 8.6 residual cohesion intercepts are usually so close to zero that they may be ignored (Skempton, 1964).

For fine grained soils there appears to be a general tendency for the peak friction angle ($\emptyset'$) to decrease with increasing plasticity and several correlations between $\emptyset'$ and PI have been proposed in the literature. Fig. 8.7 presents such a correlation (after Terzaghi and Peck, 1967) together with typical values of $\emptyset'$ for lodgement tills. Peak drained cohesion intercepts (c') for lodgement tills generally fall within the range 0-25kN/m^2. Thorburn and Reid (1973) from a series of tests on lodgement tills from the Glasgow area, Britain, have observed a tendency for c' to decrease with increasing $\emptyset'$. Their data are reproduced in Fig. 8.8, together with other typical values of c' and $\emptyset'$ for British lodgement tills. The more general applicability of this correlation, however, has not been tested. The peak strength of a soil will be influenced by any structure the soil may exhibit and its density prior to shearing, neither of which are reflected in the plasticity index, so that correlations such as indicated in Fig. 8.7 should be regarded as essentially qualitative.

Residual strength of an homogeneous soil on the other hand, is independent of structure and density prior to shearing (Bishop et al, 1971) and dependent primarily on grading and mineralogy (Kenney, 1967). As both these latter properties are reflected in plasticity index a correlation between PI and ($\emptyset_r'$) can be expected for soils of similar geological origin (Voight, 1973). Figure 8.9 presents published values of $\emptyset_r'$ for lodgement tills plotted against PI and tends to confirm the existence of this correlation. A noticeable feature of Fig. 8.9 is a marked drop in $\emptyset_r'$ at a plasticity index of about 20-25%. As can be seen from Fig. 8.7 the peak strength exhibits no such drop. There is thus a marked increase in drained brittleness, defined by the drop in strength post peak divided by the peak strength (Bishop, 1967) at a plasticity index of 20-25%. The engineering significance of this increasing brittleness, particularly in relation to slopes in lodgement till, has been discussed by Vaughan and Walbancke (1975). Lupini et al, (1981) have demonstrated that this marked drop in residual strength is the result of different modes of shear. Low plasticity soils behave

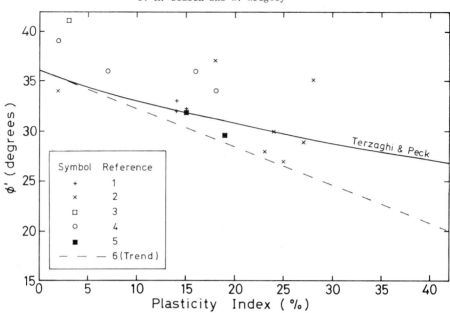

Fig. 8.7. The peak angle of internal friction (´) versus plasticity index for lodgement tills. Sources: (1) Skempton and Brown (1961). (2) Tarbet (1973). (3) Fookes et al. (1975). (4) Peters and McKeown (1976). (5) Eyles and Sladen (1981) and (6) Vaughan and Walbancke (1975).

essentially as granular materials in which residual strength is high, brittleness is low and is due predominantly to dilation in the failure zone within which no preferred particle orientation occurs. By contrast in soils with higher clay contents, or more strictly platy clay particles, and hence higher plasticity indices, a low strength shear surface of strongly orientated platy particles develops resulting in a marked drained brittleness. A transitional mode occurs in which residual friction angle is extremely sensitive to small changes in grading. The implications of this relationship are further discussed with reference to slope stability in section 13.5.

8.3.2 Undrained Strength.

The liquid and plastic limits of clay soils are perhaps most usefully interpreted as the moisture contents at two very crudely standardized remoulded undrained strengths. Thus the plasticity index is an indication of the rate of variation of remoulded undrained strength with moisture content. Undrained shear strength (C_u) of remoulded fine grained soils has been shown to be an inverse function of liquidity index (Skempton and Northey 1952). Wroth and Wood (1978) have demonstrated that this function can be approximated to a unique straight line for all fine-grained soils when the liquidity index is plotted against the logarithm of shear strength. This latter relationship is based on the observation that (i) shear strength at the liquid limit (LI = 1.0) varies from about 2.4kN/m^2, at a liquid limit of 30%, to about 1.3kN/m^2 at a liquid limit of 200% with a mean value of about 1.7kN/m^2 (Youssef et al., 1965) and (ii) the shear strength at the

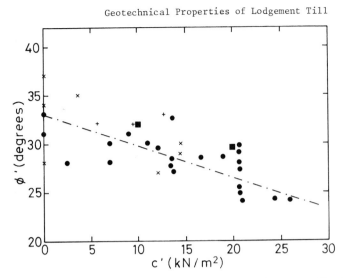

<u>Fig. 8.8</u>. c´ versus ´ for lodgement tills. Sources: Thorburn and Reid (1973), other symbols as Fig. 8.7.

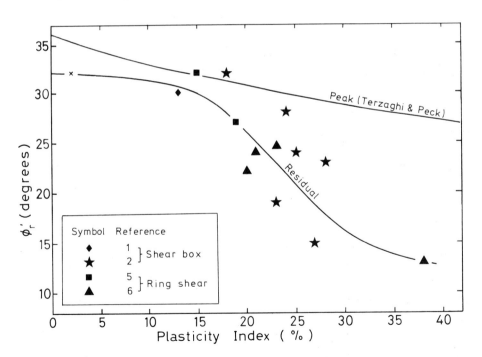

<u>Fig. 8.9</u>. The residual strength of lodgement tills. Sources: as for Fig. 8.7.

plastic limit is approximately 100 times that at the liquid limit (i.e. average about 170kN/m^2). The relationship proposed by Wroth and Wood can therefore be expressed as

$$LI \simeq 1.115 - 0.5 \log_{10} Cu$$

or

$$Cu \simeq 170 \exp (- 4.6 LI)$$

$$(8.5)$$

when Cu is measured in kN/m^2

As lodgement tills are almost invariably either insensitive (sensitivity = ratio of undisturbed to remoulded undrained shear strength) or of low sensitivity a reasonable correlation between their undisturbed undrained strength and liquidity index could be expected. Figure 8.10 shows typical results of undrained shear strength tests from a single site plotted against liquidity index together with the upper and lower bound lines from Skempton and Northey's work, extrapolated beyond 150kN/m^2, and Wroth and Wood's line. Some scatter is inevitable and can largely be attributed to errors in the determination of liquidity index which is sensitive to errors in the determination of liquid and plastic limits. It can be seen, however, that a reasonable correlation between liquidity index and undrained shear strength exists and that Skempton and Northey's lower bound line forms a sensible lower bound to undrained strength at a given liquidity index.

For liquidity indices greater than zero (i.e. for moisture contents above the plastic limit) points often plot above the lines predicted from the above work indicating sensitivities greater than unity.

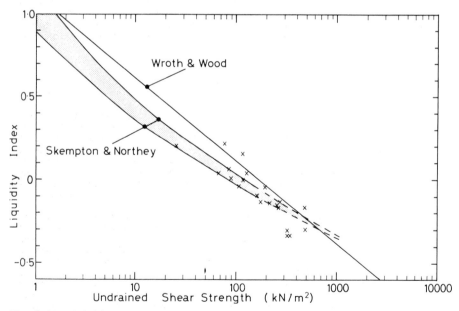

Fig. 8.10. Relationship between liquidity index and undrained shear strength for insensitive soils with some typical test results for lodgement till from Perth, Scotland.

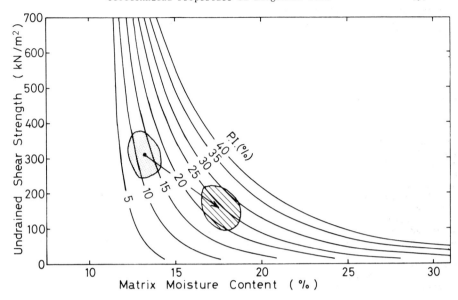

Fig. 8.11. Idealised relationship between moisture content and undrained shear strength for saturated insensitive soils of varying plasticity index whose Atterberg Limits plot on the 'T' line. Note, if points plot to the left or right of the 'T' line then strengths at a given moisture content would be lower or higher respectively. Shading illustrates typical effects of weathering on shear strength (as Fig. 8.4).

The authors have found that many test results which plot below the lower bound line, when liquidity index is based on the moisture content of the whole sample plot close to it when allowance is made for the lower porosity of the clasts in order to determine matrix moisture content (see 8.2.3 above). For low liquidity indices, however, corrected points frequently plot below the lower bound line. It will be seen from equation 8.4 that for high shear strengths at low total stresses very high negative pore pressures must be developed within the sample, these are caused by the tendency of a dense sample to dilate on the shear plane. Even during 'quick' undrained tests some migration of pore water towards the shear plane must occur resulting in a locally higher liquidity index and hence lower strength, the strength measured therefore may not be truly undrained.

In Fig. 8.11 undrained shear strength is plotted against moisture content for various values of plasticity index. These lines are based on the lower bound line from Skempton and Northey's work and on the interrelationship between index properties implied by the T line (equation 8.1) and can therefore be used to predict a sensible minimum value of undrained shear strength of a lodgement till given only basic soil index properties. Fig. 8.11 is also useful to demonstrate two important aspects of the engineering behaviour of lodgement tills. Firstly it can be seen that low plasticity tills are extremely sensitive to small changes in moisture content, for example an increase in moisture content of just 2% from 13% to 15% in a till with a plasticity index of 10% is associated with a drop in strength from 275kN/m^2 to 69kN/m^2. While this is an extremely important aspect of the behaviour of low plasticity lodgement tills it also limits the usefulness of such relationships between undrained strength and index properties, as shown in Fig. 8.11, for all but qualitative purposes as the accuracy to which moisture content and plasticity index can be measured is limited. Secondly Fig. 8.11 can be used to illustrate the effects of weathering on till behaviour where weathering results in an increase in plasticity and moisture content. Typical envelopes containing data points from weathered and unweathered zones of a single till section are shown. It will be appreciated therefore that if at a given site undrained shear strength is plotted against moisture content for all zones of weathering an apparent relationship is likely to be observed in which undrained shear strength appears to be less sensitive to changes in moisture content than it is in reality. This effect can have important consequences, as for example when lodgement till is used as a fill material if an upper level of moisture content is used as the criterion for acceptability of placed fill (Sect. 13.4).

Figure 8.12 presents a typical section showing the variation of index properties with depth in lodgement till from a site in Perth, Scotland together with measured values of undrained shear strength. Also shown are undrained strength versus depth patterns predicted from the liquidity index versus depth trend by Wroth and Wood's relationship (equation 8.5) and the upper and lower bound relationships from Skempton and Northey's work. It will be seen that there is reasonable agreement between the measured and predicted values.

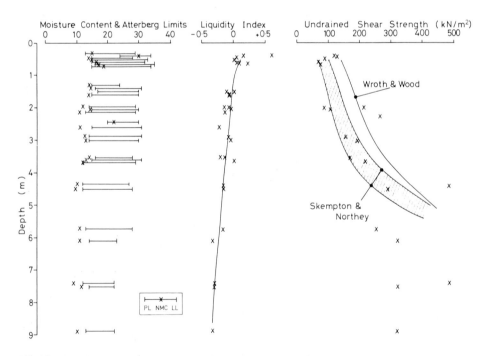

Fig. 8.12. Typical variation in lodgement till of Atterberg indices, liquidity index and undrained shear strength with depth together with strength versus depth relationships predicted from liquidity indices. Perth, Scotland.

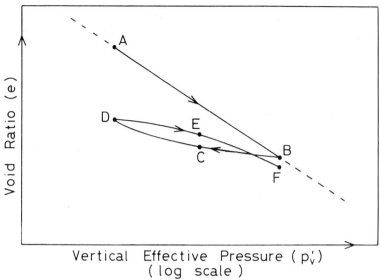

Fig. 8.13. Schematic illustration of volume changes during one dimensional compression and unloading ("e-log p" diagram). See text for details.

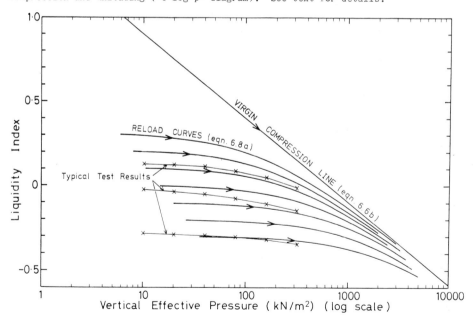

Fig. 8.14. Idealised relationship between liquidity index and vertical effective pressure during one dimensional compression for insensitive lodgement till, together with typical test results from Perth, Scotland.

8.3.3 The Influence Of Discontinuities.

The above discussion of drained and undrained strength parameters is based
largely on studies of the behaviour of relatively small laboratory specimens
which can be regarded as representing the "intact" strength of lodgement tills.
Correlations with index properties assume that the soil can be approximated to an
homogeneous mass. Lodgement tills have been observed to exhibit well developed
sets of joints, fissures and bedding discontinuities (e.g. Kazi and Knill, 1973;
McGown and Radwan, 1975; McKinlay et al, 1975; Al-Shaikh-Ali, 1978). It has
been suggested that discontinuities can result in important discrepancies between
the intact behaviour and en-masse performance in the field (e.g. McGown et al.,
1974). Much work remains to be done in studying the possible implications of
discontinuities on en-masse behaviour of lodgement tills (Sects. 11.9, 12.2).
Great care should be taken when comparing field and laboratory observations to
ensure that field drainage conditions are known, field conditions in lodgement
till for example can rarely be considered to be truly undrained. Before

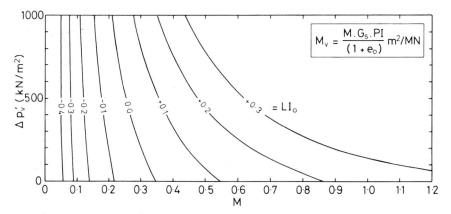

$$M_v = \frac{M \cdot G_s \cdot PI}{(1 + e_o)} \ m^2/MN$$

Fig. 8.15. Idealised relationship between the coefficient of compressibility
(Mv) and index properties for insensitive soils.

discrepancies between field and laboratory behaviour are attributed to the
influence of fissures careful consideration should be given to all other factors.

Hight et al, (1979) have published a useful review of the likely effect of
fissures on the shear strength of sandy clays and their conclusions can be
summarised as follows:

(i)   Unless discontinuities represent local variations in composition
        or density (i.e. water content) their presence is unlikely to
        have a significant effect on en-masse behaviour by lowering
        shear strength parameters.

(ii)   Locally higher moisture content along fissures is unlikely to
        have a pronounced effect on the drained strength of sandy clays
        of low plasticity but may affect undrained behaviour on the large
        scale (cf. equation 8.5).

(iii) The presence of more plastic coatings on fissures within a clay
       of otherwise low plasticity may have a marked effect on the
       drained and undrained en-masse behaviour.

It is also important to note that fissures that are more permeable than the soil
mass will   act   as preferential drainage paths so that their presence will allow
any excess pore water pressures, created for example by any given loading or
unloading condition, to dissipate more quickly than would be the case in an
homogeneous soil mass, markedly reducing the length of time which any analysis
based on undrained strength parameters could be considered valid.

McGown et al, (1977), working in the Glasgow area, Scotland, have reported that
within a lodgement till, with a matrix plasticity index of 15%, bedding plane
discontinuities exist with coatings with measured plasticity indices of up to
30%.   Such coatings would likely have a considerably lower residual angle of
friction ($\emptyset_r{}'$) than the matrix so that, if adversely orientated, they could
influence long term slope stability. Detailed reliable documentation of observed
in situ effects attributable to the influence of such features in lodgement till
are,   however,   uncommon.   Lodgement till slopes are discussed further in
Sect. 13.5.

8.4 Compressibility

One dimensional compression of soil is modelled in the laboratory in the
conventional oedometer apparatus in which load is applied vertically and lateral
strain is prevented. In the field these conditions are approximated where the
loaded area is large in relation to the sediment thickness such as during
deposition of a marine or lacustrine clay, beneath a large ice sheet or beneath a
wide building supported by a raft foundation.

The change in volume which occurs during one dimensional compression is
illustrated schematically in Fig. 8.13 in which void ratio (e) is plotted against
the logarithm of effective vertical pressure (pv'). Such a plot is often referred
to as an "e-logp" diagram. During virgin (normal) compression the state of a
soil sample can be represented by a point on the "virgin compression line" which
is approximately linear on the e-logp diagram. As the effective pressure is
increased from A to B the void ratio decreases. If load is then removed, as
would occur in nature for example by erosion of overlying sediments or melting of
ice, the state of the soil is represented by points on the unload or "swelling"
line B-C-D.   The initial part of the swelling line, B to C on Fig. 8.13, is
usually approximately linear but at low values of p the swelling line is often
curved as shown by C to D. Soil in a state represented by points on the virgin
compression line are referred to as "normally consolidated" whereas points on a
swelling line are referred to as "overconsolidated". The degree of
overconsolidation is generally expressed by the "overconsolidation ratio" (OCR)
which is defined as the maximum past vertical effective pressure ($pv'm$) divided
by the current value of vertical effective pressure.

If the soil is reloaded from point D the state of the soil is represented by
points along the reload curve such as E and F. The reload line is generally
curved on the e-logp diagram, is initially above the swelling line, crosses the
swelling line and tends towards the virgin consolidation line at high effective
vertical pressures.

The equation of the virgin compression line is given by:

$$e = e_1 + C_c \log_{10} pv'$$

(8.6)

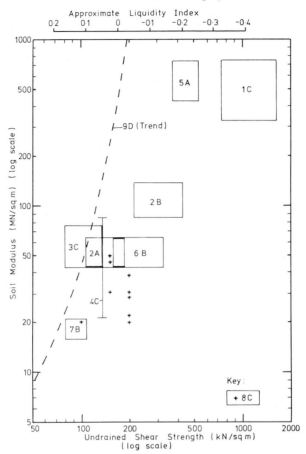

Fig. 8.16. Relationship between soil modulus determined by various *in situ* means and undrained shear strength for lodgement tills. Key: A - Back analysis of settlement records; B - Pressure- meter tests; C - Plate bearing tests; D - Back analysis of pile load tests. Sources: - (1) Radhakrishna and Klym (1974) (2) Crawford and Burn (1962), (3) Soderman *et al*, (1968), (4) Trow (1965), (5) Dejong and Harris (1971), (6) and (7) Milligan (1976), (8) Marsland (1977), (9) Weltman and Healy (1978). From Milligan, 1976 with additions. (See also Table 12.3).

where C_c is termed the compression index and e_1 is the void ratio on the virgin compression line at unit pressure (i.e. when $\log_{10} pv' =$ zero). C_c and e_1 are unique for a given soil.

Because it is approximately straight the swelling line can be expressed as

$$e = e_2 + C_s \log_{10} \frac{pvm' + \Delta pv'}{pvm'} \qquad (8.7)$$

where C_s is termed the swell index, e_2 is the void ratio prior to unloading, at $pv'm$, and pv' is the reduction in effective vertical pressure. C_s is reasonably constant for a given soil.

Reload curves are often assumed to be approximately parallel to swelling lines but from a study of many one dimensional consolidation tests carried out on lodgement till samples the authors have concluded that reload curves are better modelled by the equation

$$e = e_1 - C_c \log_{10} (pv' + c)$$ (8.8)

where C is a constant in the same units as pv' for each reload curve depending on the values of e and pv' prior to reloading.

Wroth (1979) has shown that, during virgin one dimensional compression, for a given level of vertical effective stress the liquidity index of all insensitive soils is approximately equal. Consequently the virgin compression line (equation 8.6) is approximately unique for all insensitive soils if it is rewritten in terms of liquidity index rather than void ratio; from equation 8.6:

$$C_c = \frac{\Delta e}{\Delta \log_{10} pv'}$$ (8.9)

The void ratio of a saturated soil is related to moisture content (m) by:

$$e = m G_s \text{ , hence } \Delta e = \Delta m G_s$$

where Gs is the specific gravity of the soil particles, Liquidity Index is defined as:

$$LI = \frac{m - PL}{PI} \text{ , so that } \Delta LI = \frac{\Delta m}{PI}$$

where PI, PL and m are expressed as fractions Equation 8.9 can therefore be rewritten

$$C_c = G_s PI \frac{\Delta LI}{\Delta \log_{10} pv'}$$ (8.9a)

so that equation 8.6 becomes:

$$LI = LI_1 - \frac{C_c}{G_s PI} \log_{10} pv'$$ (8.6a)

where LI_1, is the liquidity index at unit pressure. From Wroth's work it can be demonstrated that for insensitive soils $Cc/GsPI \simeq 0.5$ and for pv' in the units of kN/m^2, $LI_1 \simeq 1.4$ so that equation 8.6a becomes

$$LI \simeq 1.4 - 0.5 \log_{10} pv'$$ (8.6b)

Equation (8.8) defining the infinite family of possible reload curves can similarly be rewritten:

$$LI \simeq 1.4 - 0.5 \log_{10} (pv' + C)$$ (8.8a)

where the value of the constant c can be simply determined by inserting the values of LI and pv' prior to reloading (the "in situ" values in the case of field conditions). Equation 8.8a describes the behaviour of overconsolidated samples in one dimensional compression.

Equation 8.6b, the virgin consolidation line, and some of the family of reload curves, equation 8.8a, are plotted as LI versus $\log_{10}$ pv' in Fig. 8.14 together with the results of a typical series of one dimensional compression tests carried out on undisturbed samples of lodgement till.

When analysing settlements due to consolidation resulting from loading of over consolidated soils it is often convenient to make use of a soil property known as the "coefficient of compressibility" (Mv) defined as the decrease in unit volume per unit increase in pressure.

In terms of void ratio and vertical effective pressure Mv is given by the expression;

$$Mv = \frac{1}{(1 + e_o)} \cdot \frac{\Delta e}{\Delta pv'} \qquad (8.10)$$

where e_o is the void ratio prior to loading. Now

$$\Delta e = G_s \, PI \, \Delta LI$$

so that

$$Mv = - \frac{G_s \, PI}{(1 + e_o)} \cdot \frac{\Delta LI}{\Delta pv'} \qquad (8.11)$$

From equation 8.11 and 8.8a it is therefore possible to estimate Mv values from simple soil index properties as presented in Fig. 8.15 in which LI and e_o are the liquidity index and void ratio respectively of a soil layer prior to loading and pv' is the increase in effective vertical pressure applied to the layer.

It will be noticed in Fig. 8.15 that for low values of LI (typical of lodgement tills) values of Mv do not vary greatly with increasing $\Delta pv'$ which implies that the stress strain response is close to being linear. It has thus become common practice (e.g. Stroud and Butler 1975) when estimating settlements of foundations on lodgement till to use a quasi- elastic method with a "drained" value of Young´s Modulus (Ev'). Ev' should not be confused with the so-called "undrained elastic modulus" derived from the stress-strain curves of quick undrained triaxial tests.

Ev' is related to Mv for isotropic soils by the expression

$$Ev' = \frac{1}{Mv} \frac{(1 - 2\upsilon')(1+\upsilon')}{(1-\upsilon')} \qquad (8.12)$$

where υ' is Poissons ratio in terms of effective stress.

Atkinson (1975) has shown that heavily overconsolidated clay is better modelled as an anisotropic elastic material. The relationship between Mv and the elastic constants for the anisotropic case is given by:

$$Ev' = \frac{1}{Mv} \left[1 - \frac{2\,(v'_3)^2}{n\,(1 - v'_1)} \right] \tag{8.13}$$

where

$E'v$ is Young's modulus in the vertical axis, v'_1 is the Poisson's ratio effect of vertical strain on horizontal strain, v'_3 is the Poisson's ratio effect of horizontal strain on complimentary strain and n is the ratio of the vertical to the horizontal Young's Modulus (Ev'/Eh')

Equation 8.13 reduces to equation 8.12 when $v_1{'} = v_3{'}$ and n = 1.

For the values of Poisson's ratio typically assumed for lodgement till, 0.15 to 0.25, equation 8.12 yields

$$Ev' = \frac{0.95}{Mv} \quad to \quad \frac{0.83}{Mv}$$

Atkinson (1975) showed for London Clay that n = 0.5, $v'_1 = 0$ and $v'_3 = 0.19$ so that from equation 8.13

$$Ev' = \frac{0.86}{Mv} \quad and \quad Eh' = \frac{1.71}{Mv}$$

Thus it can be concluded that the effect of anisotropy on the relationship between Ev' and Mv is unlikely to be large and that the relationship is not sensitive to variations in Poisson's ratio within its likely range of values. $Ev'.Mv$ cannot be greater than unity.

For reasons discussed below, and in Sect. 12.2, it has been found for lodgement till that computations based on laboratory determinations of Mv and Ev' generally lead to overestimates of settlements. Large scale in situ load tests and back analyses of settlement records of completed structures are therefore often used to provide estimates of the compressibility of lodgement till.

In Fig. 8.16 published values of Young's Modulus for lodgement tills derived by various in situ means are plotted against undrained shear strength.

Without knowing stress levels and liquidity indices no direct comparison between Figs. 8.16 and 8.15 can be made but inspection will show that Fig. 8.15 consistently predicts higher compressibilities for stress levels typical in civil engineering. For example consider a till with the following properties:

PI = 25% = 0.25

LI = -0.15

e_o = 0.44

Gs = 2.7

v' = 0.20

subjected to an increase in vertical effective stress (Pv') of 200kN/m^2. We would expect from Fig. 8.15 a value of Mv of approximately 0.075 m^2/MN and from equation 8.12 a value of Ev' of approximately 12MN/m^2. From equation 8.5 we would expect an undrained shear strength of approximately 340kN/m^2 which allows a comparison with Fig. 8.16 to be made where it will be seen that a modulus of 12MN/m^2 is considerably lower than indicated by _in situ_ tests at this shear strength.

It is useful to enumerate the probable causes of this apparent anomaly.

(i) None of the _in situ_ tests can strictly be considered to represent one-dimensional compression. The modulus value is usually derived by assuming that isotropic conditions exist and back analysing using appropriate elastic theory. In reality the load settlement performance will be a complex function of the geometry of the loaded area and soil layer, and of the vertical and horizontal profiles of moduli. In the case of pressuremeter tests it is largely the horizontal modulus which is measured. For overconsolidated soils the horizontal modulus is often appreciably higher than the vertical modulus, approximately twice as high for example for London Clay (Atkinson, 1975).

(ii) Fig. 8.15 is based on an empirical relationship (equation 8.8a) derived from a study of oedometer test results which are known to be sensitive to disturbance during sampling and specimen preparation resulting in overestimation of compressibility.

(iii) Plate load tests may not have been carried out sufficiently slowly to ensure drained conditions pertain. Pressure-meter tests are usually carried out quickly. In both cases the modulus measured is likely to be below the undrained but above the truly drained value.

(iv) Unless load cells are incorporated in foundations, errors can be introduced when back analyzing settlement records. Design loads are typically considerably higher than actual structural loads and their use in back analyses can lead to overestimates of modulus. In addition redistribution of load within a structure subjected to differential settlements complicates estimation of foundation stresses.

(v) Undrained strengths in Fig. 8.16 may have been underestimated for the reasons outlined in section 8.3.2. above.

The authors believe the combined effects of the above factors adequately account for the apparent anomaly but they are difficult to evaluate individually. It is certainly true that (i), (ii) and (v) above are likely to have more marked effects for higher shear strength at which anomalies appear greatest.

For foundation design purposes (Ch. 12), at the present state of knowledge, values derived from _in situ_ studies, such as those presented in Fig. 8.16, probably represent the most realistic estimates of compressibility. Care should be taken, however, when interpolating from such tests that loading conditions are comparable. Further, laboratory test results should not be discounted as erroneous purely on the basis of an apparent discrepancy with _in situ_ observations.

8.5 In situ Stresses

At their maximum development continental ice sheets were several kilometres in thickness and most lodgement tills have been consolidated under large vertical effective stresses. Effective stresses at the base of an ice sheet cannot be simply related to the thickness of superincumbent ice but are also a function of piezometric pressures which can be high (Figs. 2.1, 2.12; Boulton and Paul, 1976). For this reason the attempt by Harrison, (1958) to estimate former ice

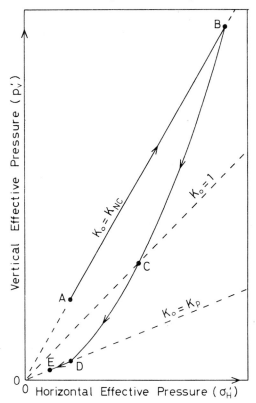

Fig. 8.17. Schematic illustration of the relationship between vertical and horizontal effective stresses during one-dimensional compression and unloading.

thicknesses by reference to consolidation characteristics of overridden silts were unrealistic. Nevertheless an estimate of the maximum ice thickness allows an upper limit to be set on the past vertical effective stress. For example it has been estimated that the maximum thickness of the Devensian (last glaciation) ice sheet in Britain was about 1700 to 2000 m (Boulton et al, 1977) so that the upper limit to the maximum effective vertical stress is therefore of the order of 17,000 to 20,000 kN/m^2. In North America maximum ice thicknesses of greater than 4000 m have been calculated for the Laurentide Ice Sheet (Sugden, 1977; Fig. 9.9). It can be demonstrated that shear stresses at a glacier sole remote from the ice margin remain relatively low and within the narrow range 50-150 kN/m^2 (Sect. 4.2). Thus it can be concluded that at least when lodgement tills were subjected to maximum vertical effective stresses the major principal stress direction was close to being vertical.

For the till sheet to fail in shear and hence be remoulded shear stresses must exceed the value given by equation 8.3. As the potential shear plane is subhorizontal, σ'_n is approximately equal to the effective vertical pressure. For typical strength parameters for lodgement till of $c' = 20$ kN/m^2 and $\phi' = 30°$

and an average shear stress of 100 kN/m^2 it will be seen that shear failure can only occur when σ'_n is less than about 140 kN/m^2.

If lodgement till sequences were sheared at low effective stresses after deposition then their shear strengths during shearing must have been less than about 150 kN/m^2 (the maximum likely shear stress). From equation 8.5 this implies a liquidity index of more than about 0.027. As the liquidity index of unweathered lodgement tills is almost invariably lower than this value and usually considerably lower it can be inferred that if they have indeed been sheared subglacially then they have subsequently been reconsolidated by high effective vertical stresses.

The above arguments are only valid upglacier from ice margins. Near to the periphery of ice sheets under conditions of high subglacial water pressures and decreasing effective stresses previously deposited lodgement till is frequently sheared (Boulton et al, 1974). This shearing affects only a thin ($\cong$1.5m) surface layer making a deformable bed where drumlins and flutes may be moulded from pre-existing lodgement till by ice flow (Figs. 2.1, 2.12, Sect. 2.3.1). In many cases the thin post-depositionally remoulded layer exposed at the margins of modern glaciers semi-liquid, has a high void ratio (.8), is matrix enriched as a result of shearing and is rapidly destroyed by solifluction, periglacial freeze drying and deflation (Boulton, et al, 1974).

As the lateral extent of lodgement till sheets is usually large in relation to their thickness and till plains are typically sub-horizontal it is not unusual to assume that the stress history of lodgement tills can be modelled by one-dimensional loading and unloading. This is almost certainly a considerable over-simplification for the following reasons:

(i) As mentioned above some shearing is likely to have occurred during ice retreat or readvance.

(ii) It is possible that due to changes in ice thickness and/or subglacial water pressure some cycling of load has occurred.

(iii) Near the surface high horizontal stresses may have been reduced by sub aerial weathering.

(iv) Following ice retreat periglacial conditions are likely to have occurred. Thus most terrestrial lodgement till sequences have probably been subjected to repeated freeze thaw cycles. Although some work has been carried out to investigate the effect of freezing and thawing on stress history and in situ stresses (Nixon and Morgenstern, 1973; McRoberts and Morgenstern, 1974) at present no accurate estimate of the net effect of an unknown number of freeze thaw cycles can be made (Sect. 5.4.1).

While recognizing these limitations it is nevertheless of interest to estimate likely in situ stresses of lodgement tills and to do so at present it is necessary to assume the simplified stress history outlined above.

Jaky (1944), Bishop (1958), and Brooker and Ireland (1965) have all shown that the ratio of horizontal effective stress (σ'_H) to vertical effective stress (pv') (= Ko) is given with reasonable accuracy for a wide variety of normally consolidated soils (Knc) by Jaky's semi-empirical expression.

$$Ko = Knc = 1 - Sin \; \emptyset' \qquad\qquad (8.14)$$

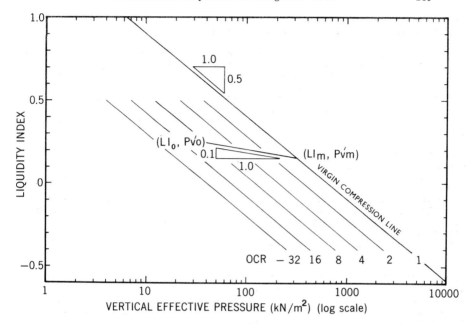

Chart for estimating overconsolidation ratio (OCR) from current values of liquidity index and vertical effective pressure (LI , pv').

Typical results of one dimensional consolidation tests in which the lateral stresses have been measured are illustrated in Fig. 8.17. During normal consolidation, denoted by the portion of the curve A-B, the vertical and horizontal effective stresses are in constant proportion as indicated by equation 8.14. On unloading from a normally consolidated state such as B, the value of Ko increases reaching a value of unity at point C, with an overconsolidation ratio (OCR) typically between 4 and 5. This initial portion of the curve (B-C, where Ko < 1.0) is approximately linear. Further unloading of the soil from point C on the stress path results in a curved stress path C-D over which σ'_H is greater than pv'. When point D is reached passive yield occurs and on further unloading the curve follows the passive yield envelope D-E.

The curve B-C-D is defined by the expression

$$Ko = Knc \ (\ OCR \ )^m \tag{8.15}$$

where m is a constant. Meyerhof (1976) has suggested $m \simeq 0.5$ and Knc is determined from equation 8.14. Along D-E, Ko equals the coefficient of passive resistance (Kp) which is the maximum possible value for Ko and is given by

$$Kp = \frac{\sigma'_H \ max}{pv'} = \frac{1}{pv'} \left[pv' \ \tan^2 \ (\ 45° + \frac{\phi'}{2}) + 2 \ c' \ \tan \ (\ 45° + \frac{\phi}{2}) \right] \tag{8.16}$$

Thus if ϕ' and c' are known and the maximum preconsolidation pressure is known or can be estimated, a value of Ko for any current value of pv' on an unloading path can be estimated.

Wroth (1979) has shown that an estimate of maximum preconsolidation pressure to which an insensitive soil sample has been subjected can be made given its current liquidity index and the current vertical effective overburden stress. The idealized relationship between liquidity index and vertical effective stress for one dimensional <u>normal</u> consolidation of insensitive soils was given in equation 8.6b and is shown graphically in Fig. 8.18. An idealized equation for the unload (swelling) line was given in equation 8.7 which can be rewritten

$$LI = LI_m + \frac{C_s}{GsPI} \, \log_{10} \frac{pvm' + \Delta p}{pvm'} \tag{8.7a}$$

where LI_m is the liquidity index at P'_{vm} on the virgin compression line and p is the reduction in P'_v.

The slope of this line $\left(\frac{C_s}{G_sPI}\right)$ depends on the plasticity index and Wroth suggests a variation from 0.085 at PI = 15% increasing only to 0.17 for PI = 100%. For typical lodgement tills the gradient of the unload line would therefore be close to 0.10.

If LI_o and pv' are the <u>in situ</u> values, pvm' can therefore be estimated by solving the equations:

$$LI_m = 1.4 - 0.5 \log_{10} pvm' \tag{8.17}$$

$$LI_o = LI_m + 0.1 \, \log_{10} \frac{pvm' + \Delta p}{pvm'} \tag{8.18}$$

$$pv' = pv' - pv'_o \tag{8.19}$$

Alternatively the point (LI_o, pv' can be plotted on Fig. 8.18, a line of slope -0.1 drawn through it and the point of intersection of this unload line with the normal consolidation line provides an estimate of the preconsolidation pressure. Fig. 8.19 presents an example of the use of this construction and shows values of OCR determined from the liquidity index versus depth data presented in Fig. 8.12 together with corresponding values of Ko determined from equation 8.15 and Kp from equation 8.16. The average effective strength parameters determined from drained triaxial tests on specimens from this site were $C'=10$ kN/m^2, $\phi=30°$. Ko can be seen to approach Kp at high values of OCR although the good agreement shown in Fig. 8.19 may be slightly fortuitous.

Had overconsolidation been due to a known uniform excess pressure applied at ground level and subsequently removed, then the OCR versus depth profile could be predicted, and would be similar to that drawn on Fig. 8.19 for an excess pressure of 2000 kN/m^2. That most of the data points fall close to the 2000 kN/m^2 line suggests that this is likely to have been the maximum excess effective pressure applied to the till. The four points below 6.0m depth plot significantly to the right of this line possibly indicating that the deeper parts of the section have been subjected to higher excess pressures. That the near surface points fall to the left of this line is probably the result of weathering or dilation due to passive failure.

This estimated excess pressure (2000 kN/m^2) is considerably less than the upper bound predicted by ice cap models (17,000-20,000 kN/m^2), possibly due to either the continuous existence of high subglacial water pressures during deposition or that areas adjacent to ice cap centres are usually characterized by glacier scour (Figs. 2.1, 9.2, 9.9). Matrix-rich lodgement tills are generally limited to

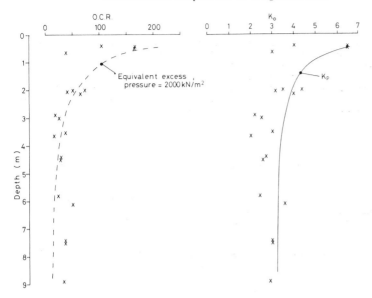

<u>Fig. 8.19</u>. Typical profiles in lodgement till of OCR and Ko versus depth estimated from the data presented in Fig. 8.12; Perth, Scotland.

outer zones of glaciation where ice thicknesses were considerably reduced (Eq. 2.1; Figs. 2.1, 2.3).

It should be noted that this method of estimating preconsolidation pressure, even if the assumptions concerning stress history are not unreasonable, is at best approximate. It is particularly sensitive to errors in estimating the slope of the swelling line (taken here for simplicity as -0.1), further the swelling line is not in reality straight, as noted in Section 8.4 above, and is markedly non linear when unloading causes the soil to approach passive failure. In addition the virgin compression line, while approximately straight within stress levels typical in geotechnical engineering, must at high stress levels be asymptotic to the horizontal line representing zero moisture content i.e. the line given by LI = -PL/PI. Both these effects lead to considerable underestimates of preconsolidation pressure when the preconsolidation pressure was high.

The authors have found, however, that for lodgement tills this method gives generally more consistent results than methods based on oedometer tests (Casagrande 1936, Schmertmann 1953) as these depend for accuracy on relatively little sample disturbance and on the e-logp plot exhibiting a reasonably marked break in slope, neither of which are typical of lodgement till. It is also pertinent to note that, if the purpose of estimating preconsolidation pressure is to enable an estimate of <u>in situ</u> horizontal stress to be made, as Ko approaches Kp its estimation becomes less sensitive to errors in estimating O.C.R.

8.6 Concluding Remarks

In this chapter the authors have shown how simply determined index properties can be used to relate the geology of lodgement tills as established in Chapter 2, to

their principal engineering properties. A framework has been provided which can
be used for the critical examination of laboratory test data. Correlations with
simple index properties have been developed as they provide a useful means of
demonstrating the interrelationships of many geotechnical properties. Such
correlations should be regarded as at best approximate and sometimes merely
qualitative so that while these correlations may be used for initial assessments
they can never replace careful laboratory testing. They have proved particularly
useful, however, in areas where it is difficult to obtain good quality samples,
such as offshore work.

We are aware that not all engineering properties have been discussed and the most
notable omissions are permeability and time-dependent consolidation behaviour,
and stress-strain behaviour in triaxial compression and simple shear. As
lodgement tills are often relatively permeable the time-dependency of
consolidation settlements is often of minor importance. Stress-strain behaviour
in triaxial compression and simple shear, however, have been omitted as an
adequate account of these important aspects of soil behaviour could not be given
within the framework of this short chapter. Readers will find useful discussions
in several soil mechanics text books for example Lambe and Whitman (1969) and
Atkinson and Bransby (1978).

Finally this chapter has been concerned specifically with one particular but
widespread till type. In the authors´ experience in lowland glaciated areas
lodgement till is usually the predominant till type represented in the field
though extreme care is needed in areas of postglacial lakes to discriminate
lodgement tills from other massive diamicts (Sect. 2.9). Many of the arguments
presented in this chapter, however, can be applied with modifications to any well
graded insensitive soil.

CHAPTER 9

The Distribution of Glacial Landsystems in Britain and North America

N. Eyles, W. R. Dearman and T. D. Douglas

INTRODUCTION

The dominant Quaternary event in the mid-latitudes has been that of episodic ice sheet growth with a maximum southward extent in Britain during the oldest (Anglian) glaciation (300-250,000 y.b.p.). The limit is marked approximately by a line connecting the River Avon in the west to the River Thames in the east at about 51 N. latitude (Fig. 9.1).

In North America (Fig. 9.2) the southward limit of glaciation runs from Washington in the west, cutting across the Dakotas to Kentucky at 38 N. latitude, north to Pennsylvania and then east (Fig. 5.3). In both North America and Britain, within the area of the last glaciation (the Wisconsin and Devensian respectively) freshly constructed, relatively undissected glaciated terrain contrasts with an outer zone of fluvially and periglacially dissected glaciated terrain extending to the limit of glaciation. Within the younger glaciated terrains it is possible to portray the distribution of glacial landsystems (Figs. 1.3, 1.4, 1.5, 1.6).

9.1 Timing Of Quaternary Glaciations and Glacial Deposition

During the last decade, a substantial revision of knowledge about the timing of Quaternary glaciations has occurred. The Quaternary period varyingly represents the last $1.7 - 3.0 \times 10^6$ years of geological time: opinions differ as to the definition of the start and the need for such (Bowen 1978). The Quaternary is characterized by climatic variations which in the mid-latitudes have meant the alternation of cold stages (glacials) and warm stages (interglacials). The only environment which is likely to yield a continuous record of sediments throughout this period is the deep ocean floor because repeated glaciation on land and on the continental shelves has eroded older sediments.

Sediment cores, recovered from the deep ocean floors, preserve a record of changing climates which can be unravelled by reference to changing oxygen isotope concentrations ($^{18}O/^{16}O$) in microfossils within the sediment. The ratio of ^{18}O to ^{16}O changes, with every 0.0023 C., and thus is temperature dependent. Free swimming microfauna, such as foraminifera, that live at shallow water depths fix oxygen in their calcareous skeletons. When these organisms die, the skeletons are precipitated onto the ocean floor, as part of the total pelagic (suspended) sediment flux and a record of the isotopic composition of the ocean surface water is preserved. The ratio between ^{18}O and ^{16}O can be determined by mass-spectrometry from samples at different levels in a sediment core to obtain a picture of changing surface water temperature, and, as a very crude extrapolation, changing climatic conditions. Fig. 9.3 shows the oxygen isotope record from the 14 m long core V8-238 taken from the Pacific Ocean and which has been used as a standard for the latter part of the Quaternary (Shackleton and

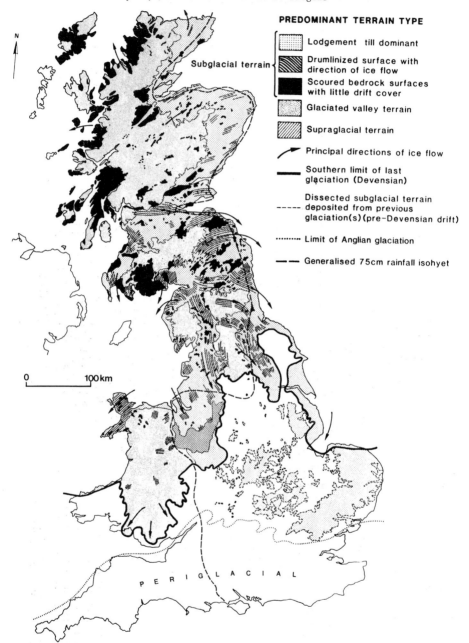

PREDOMINANT TERRAIN TYPE

Subglacial terrain

Lodgement till dominant

Drumlinized surface with direction of ice flow

Scoured bedrock surfaces with little drift cover

Glaciated valley terrain

Supraglacial terrain

Principal directions of ice flow

Southern limit of last glaciation (Devensian)

Dissected subglacial terrain deposited from previous glaciation(s)(pre-Devensian drift)

Limit of Anglian glaciation

Generalised 75cm rainfall isohyet

0   100 km

N

P E R I G L A C I A L

Fig. 9.1. Glacial landsystems in Britain.   After Eyles and Dearman (1981).

Fig. 9.2. Glacial landsystems in North America (see also Fig. 2.21).

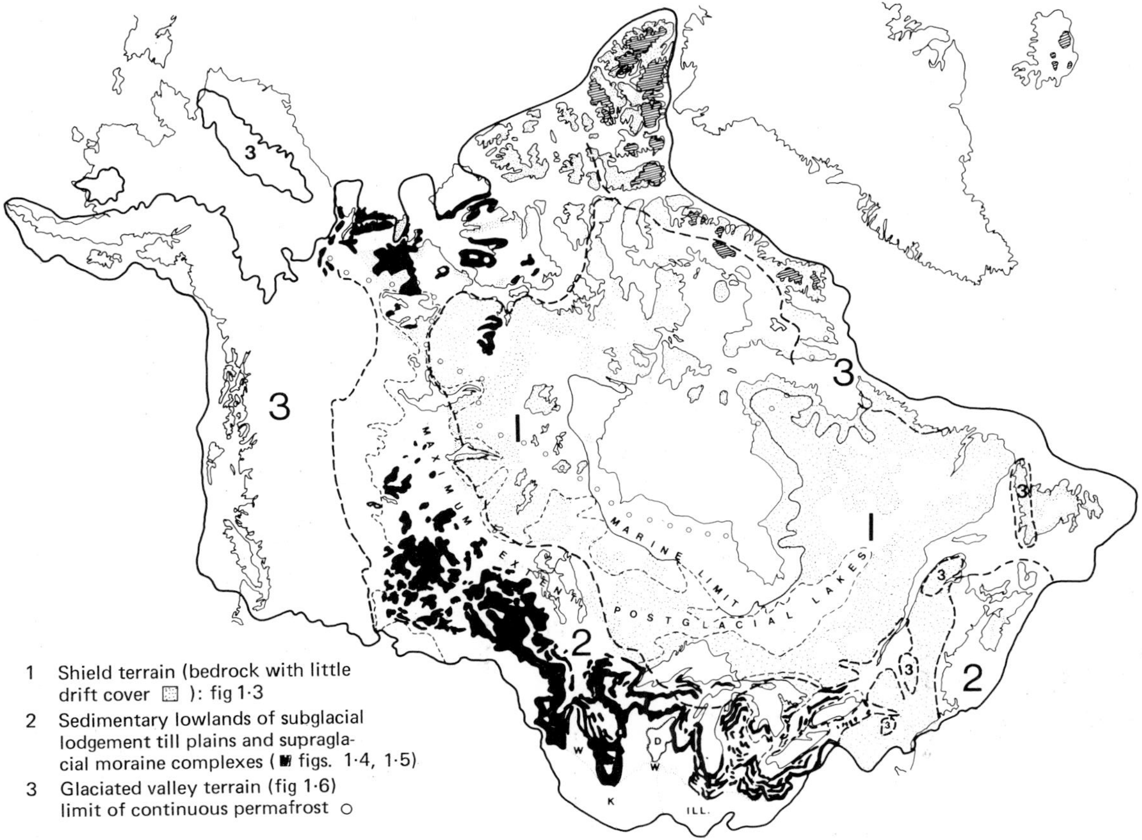

Glacier ice cover at the present day is only shown for Arctic Canada. D = driftless area. W = limit of Wisconsin glaciation. Ill = Illinoian. K = Kansan. — = limit of Nebraskan glaciation.

1 Shield terrain (bedrock with little drift cover ▣): fig 1·3

2 Sedimentary lowlands of subglacial lodgement till plains and supraglacial moraine complexes (▨ figs. 1·4, 1·5)

3 Glaciated valley terrain (fig 1·6) limit of continuous permafrost ○

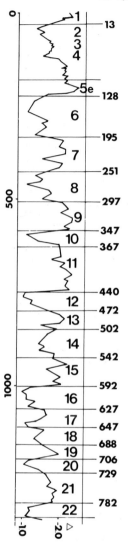

Fig. 9.3. Deep sea core V-28-238 from the Solomon Plateau, Pacigic Ocean (After Shackleton and Opdyke,1973). Core dep4h (cm) appears on the left, age in thousands of years on the right. The middle numbers refer to stages or periods. Oxygen isotope variations (△) are expressed as parts per thousand; a deviation of -2.0 approximates to interglacial conditions though direct paleotemperature analysis is not possible. Stage boundaries are mid-points between minima and maxima and are dated by assuming constant sedimentation rates from a dated paleomagnetic reversal at 12 m depth. Stages 1 - 5 correspond to the last glaciation (Devensian; U.K., Wisconsin, N.A.), 5e is the last interglacial (Ipswichian, U.K., Sangamon, N.A.). Glaciations are even numbered. Note that 8 major glaciations alone may be recorded after 700,000 y.b.p. There is some suggestion that the length of interglacials as represented here is exaggerated, (Bowen, 1978). The oxygen isotope record provides a good fit with calculated variations in solar radiation generated by long term periodicities in the earth's orbit - such astronomical variations were first suggested as generating recent ice ages by Milankovitch in the 1920's (Imbrie and Imbrie, 1979). For a recent review see Budd (1979).

Opdyke 1973). Cores showing a similar pattern but of varying duration have been recovered from both the Atlantic, and Indian Oceans, and from the Greenland and Antarctic ice-caps; ice also provides a good record of changing oxygen isotope ratios (Bowen, 1978).

The climatic ´norm´ in the mid-latitudes for the majority of Quaternary time has been the cold stage or glacial (Fig. 9.3). Characteristically, each cold stage lasts about 80,000 - 90,000 years and each interglacial only 10,000 - 15,000 years. Superimposed on this simple rhythm are cold periods (stadials) and warm periods (interstadials) of short duration. Correlation of the terrestrial record in Britain and North America with the ocean oxygen isotope stratigraphy is difficult (Bowen, 1978) and the number of cold stages evident from the ocean record is far more numerous than previously considered. This contrasts greatly with the classic four glaciations originally recognised in the forelands of the European Alps and formerly used as a world-wide norm (Gunz, Mindel, Riss, Wurm; Sect. 4.7). Kukla (1977) demonstrated that the fourfold sequence in the Alps may be a result of repeated tectonic uplift cycles not widespread climatic change per se.

Similarly, it is now realised that the classic Nebraskan tills, traditionally thought to be the oldest glacial units in continental North America, are younger than 1 million y.b.p. whereas the earliest continental ice sheet reached its maximum around 2.5 m.y.b.p. doing so again, but with a more restricted extent from 1.9 to 1.5 m.y.b.p. (Boellstorff, 1978). Thus post-Nebraskan deposits, earlier thought to record the entire Quaternary Ice Ages in continental North America, in fact only record the last 40% of the time. Much stratigraphic confusion also surrounds many type sites (Hallberg and Boellstorff, 1978; Hallberg, 1980). Perhaps the most complete glacial record of the Late Cenozoic Quaternary ice ages is found in the Yakataga Formation of the Gulf of Alaska which almost spans the entire last 15 million years (Plafker, 1981). Again however the record (largely offshore glaciomarine) is likely to record the tectonic evolution and uplift pattern of the area rather than world-wide climatic change (e.g. Kukla, 1977).

The oxygen isotope stratigraphy and the wider application of radiometric dating thus has thrown into sharp relief the problems of correlating terrestrial glacial events and it is quite clear that the glacigenic depositional record on land is only a partial one. However with the apparent 'development upon the continents of the many glaciations suggested in the ocean results' demonstrated by Ford et al (1981 p. 8) a time framework for both the modification of relief by glacier erosion and glacial deposition in the mid-latitudes seems a real possibility in the not too distant future.

9.2 The Quaternary of Britain

For many applied purposes and for general small-scale geological mapping the Quaternary of Britain can be divided into two basic groups:

    (i)   the non-glacial Quaternary, and
    (ii) the glacial Quaternary.

9.2.1 Non-glacial Quaternary

The non-glacial group comprises a distinctive shallow water marine, estuarine and freshwater sequence of Lower and Middle Quaternary age (older than 300,000 y.b.p.). These occupy a basin developed on the surface of the underlying Cretaceous Chalk. The edge of the basin fill is exposed onshore in East Anglia but deepens to the east below the North Sea where the total thickness of Quaternary sediments is over 1 km (McCave et al, 1977). The succession comprises the famous 'crag' sequences of shelly calcareous sands found in Norfolk and Suffolk. The non-glacial group includes soils deposited in cold periglacial environments but there is no direct evidence of glaciation. As a whole the group comprises shelly mud to sandy sequences, and is almost entirely overlain by glacial sediments.

9.2.2 Glacial Quaternary

Of fundamental importance are sediments that were deposited during a number of glacial and interglacial cycles. Deposits of three distinct glaciations can be recognized which in order of approximate ages are the Anglian (300-250,000 y.b.p.), Wolstonian (200-150,000 y.b.p.) and the Devensian (120,000-10,000 y.b.p.; Fig. 9.3). The southward extent of the Anglian glaciation is the greatest and that of the Devensian the least. As a result, a large tract of Anglian and Wolstonian sediments are preserved as interfluve cappings in dissected terrain south of the Devensian limit (Fig. 9.1). The limit of the Wolstonian glaciation is poorly defined and in what follows glacial sediments of Wolstonian and Anglian age are referred to collectively as Pre-Devensian.

The gross distribution of landsystems resulting from successive glaciations in Britain reflects topography and precipitation. The 75 cm rainfall isohyet divides a wetter highland portion of Britain from drier eastern lowlands (Fig. 9.1). The highland area is dominated by crystalline and metamorphic rocks, heavily disected by glacially overdeepened valleys (Fig. 9.4). The degree of dissection in the Scottish Highlands, North Wales and the English Lake District reflects repeated valley glaciation (Boulton et al, 1977). Drier eastern lowlands are underlain by sedimentary rocks where the erosive effects of large ice sheets have been more limited and where glacial deposition has been dominant resulting in thick glacial sequences (e.g. Holderness, East Anglia). In this zone, deeply weathered granites survive in the Cheviot area and in eastern Scotland (Phemister and Simpson, 1959) and are found in association with drainage basins that have been little affected by glacial erosion (Fig. 9.4; Linton, 1959).

While regional patterns of erosion and deposition are of interest, local conditions are more important and here the generalizations implicit in Fig. 9.4 break down. Thus drift-filled glacially overdeepened valleys are in fact ubiquitous in lowland areas of low regional erosion with depths of 160 m below sea level being reported for some valley floors (Anson and Sharp, 1960; Al-Saadi and Brooks, 1973; Anderson, 1974). In many estuarine areas the bulk of infilling sediments are glaciofluvial or marine origin (Fig. 4.9).

The deposition of thick glacial sequences on extensive marine surfaces planated by high sea-levels during the previous interglacial (Figs. 15.1, 15.13) results in high rates of coastal erosion and large coastal sediment fluxes. This is emphasized by continued isostatic uplift in Scotland and tectonic sinking in the southern North Sea Basin.

9.3 The Distribution of Glacial Landsystems in Britain

Complexly layered soils and ground conditions encountered in glaciated and periglaciated terrains are too variable to present on a map at the scale of Fig. 9.1. Certainly the general distribution of superficial engineering soils can be depicted, as on any geological map, but this gives no information on likely subsurface ground conditions and soil types. Similar comments can be directed to the glacial geology maps produced by Clayton (1963), Unesco (1967), West (1968), Derbyshire (1975) and the Institute of Geological Sciences (1977).

For each of the landsystems identified in Figs. 1.3, 1.4, 1.5 and 1.6, the areas for which these terrain models provide a representative survey of ground condition has been located on Fig. 9.1.

Relatively fresh glaciated terrain and extensive glacial sediment cover north of the Devensian ice limit contrasts with older disected and subdued glaciated terrain to the south. The importance of this distinction is reflected in the

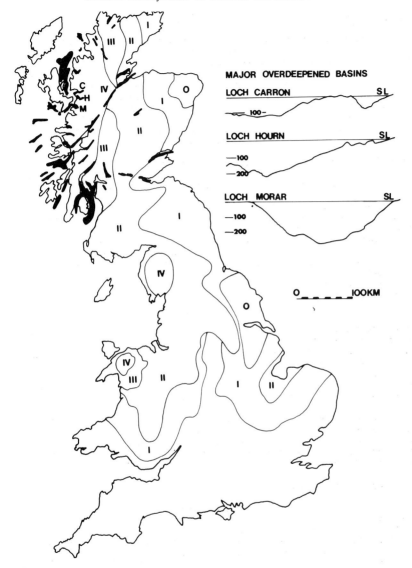

Fig. 9.4. Generalized glacial erosion zones in Britain and location of major
overdeepened basins (SL = sea level); 0, (unmodified lowlands) to IV
(extensively modified glaciated valleys). After Clayton (1974).

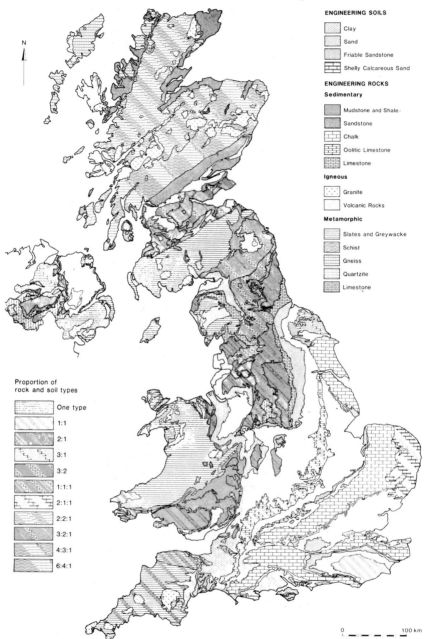

ENGINEERING SOILS

- Clay
- Sand
- Friable Sandstone
- Shelly Calcareous Sand

ENGINEERING ROCKS

Sedimentary

- Mudstone and Shale
- Sandstone
- Chalk
- Oolitic Limestone
- Limestone

Igneous

- Granite
- Volcanic Rocks

Metamorphic

- Slates and Greywacke
- Schist
- Gneiss
- Quartzite
- Limestone

Proportion of
rock and soil types

- One type
- 1:1
- 2:1
- 3:1
- 3:2
- 1:1:1
- 2:1:1
- 2:2:1
- 3:2:1
- 4:3:1
- 6:4:1

0   100 km

Fig. 9.5. Engineering geological map of the solid, or bedrock, of Britain (After Dearman and Eyles, 1982).

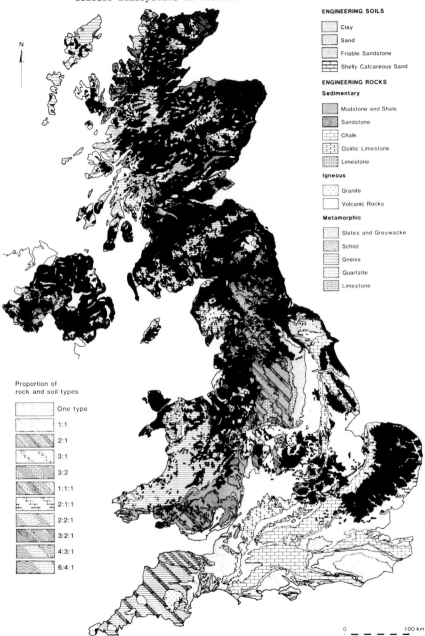

ENGINEERING SOILS

Clay

Sand

Friable Sandstone

Shelly Calcareous Sand

ENGINEERING ROCKS

Sedimentary

Mudstone and Shale

Sandstone

Chalk

Oolitic Limestone

Limestone

Igneous

Granite

Volcanic Rocks

Metamorphic

Slates and Greywacke

Schist

Gneiss

Quartzite

Limestone

N

Proportion of
rock and soil types

One type

1:1

2:1

3:1

3:2

1:1:1

2:1:1

2:2:1

3:2:1

4:3:1

6:4:1

0   100 km

Fig. 9.6. Till cover (black) with rock types depicted either as engineering rocks or soils.

long standing concept of Newer and Older Drift (Charlesworth, 1957). To the
south of the Devensian limit, in the absence of original surface topography, the
landsystem approach, with its emphasis on landform recognition as a guide to
subsurface sediment type and geometry, cannot be directly employed. Nonetheless
sufficient detail has been presented with regard to the likely geometry and type
of subsurface stratigraphy in each landsystem to enable interpretation of
borehole logs and or open sections.

As related in Sect. 2.9.3, it is not yet possible to identify areas of
glaciomarine deposition exposed along the British coastal margin with anything
like the accuracy for North America (Fig. 2.21). This awaits detailed work.

9.3.1 Subglacial Landsystem

Within the area of the subglacial landsystem, the dominant glacial sediment is
often lodgement till, the 'boulder clay' of countless publications. At many
sites two or more distinct till units can be mapped in section. Near the former
margins of the Devensian glaciers, where the depth of glacial erosion has been
more limited compared to the upland and highland parts of Britain (Fig. 9.4),
organic material and non-glacial sediments lying between tills indicate the
presence of either pre-Devensian glacial deposits or readvancing ice lobes
(Mitchell, et al, 1973). In drumlinized terrain ubiquitous 'tripartite'
stratigraphies composed of lower till(s), middle sands, gravel and laminated
clays and upper till(s) have been used extensively as mapping units and taken to
indicate multiple glaciation. Many multiple successions can probably be
explained by the former presence of subglacial melt streams coupled with shifting
ice divides and changing ice flow directions (Fig. 2.11). Drumlinized subglacial
terrain is found most widely in lowlands below 300 m in elevation. The Lake
District and the northern Pennines in England are surrounded by streamlined
topography comparable to the classic areas of the Central Lowlands of Scotland
and the Tweed valley (Fig. 9.1). The dissection of the till cover south of the
limit of the last glaciation should be noted (Fig. 9.1). The extensive dissected
till sheets in the East Midlands and East Anglia ('chalky boulder clays')
suggests that these may be the remnants of lodgement till plains, (Perrin, et al,
1979), and they have thus been included in the subglacial terrain group. The
scoured metamorphic and igneous bedrock surfaces in Scotland described as 'knock
and lochan' by Linton (1963) are comparable to shield terrains in North America
(Sugden, 1977; Gordon, 1981).

9.3.2 Supraglacial Landsystem

Large areas occupied by supraglacial terrain are limited to arcuate morainic
belts marking the position of Devensian ice lobe margins in the
Cheshire-Shropshire lowlands (Fig. 3.4). Drumlinized terrain emerges from under
the hummocky cover indicating deposition of flowed tills and outwash over
subglacial surfaces (Figs. 1.5, 3.4). This distinctive terrain type is also
encountered, on too small a scale to be depicted on Fig. 9.1, in the Vale of York
and other areas (Boulton, 1972; 1977). Economically such areas are major
aggregate sources. The 'trapping' of meltwater streams within belts of ice-cored
ridges along former ice margins (Fig. 3.5) accounts for the large volume of
glaciofluvial sediments associated with this distinctive terrain type.

9.3.3 Glaciated Valleys

Hummocky morainic topography along valley floors and complex lateral
accumulations on valleysides is a ubiquitous terrain type in upland and highland
Britain (Fig. 9.1). The latest phase of mountain glaciation, with a major
ice-cap in the western Highlands and smaller corrie and valley glaciers in the

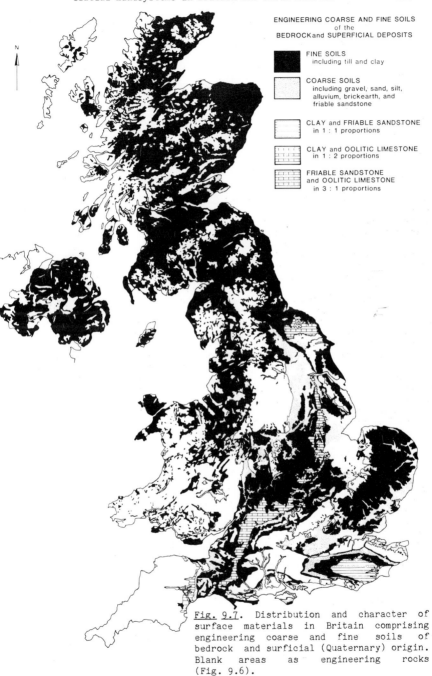

ENGINEERING COARSE AND FINE SOILS
of the
BEDROCK and SUPERFICIAL DEPOSITS

FINE SOILS
including till and clay

COARSE SOILS
including gravel, sand, silt,
alluvium, brickearth, and
friable sandstone

CLAY and FRIABLE SANDSTONE
in 1 : 1 proportions

CLAY and OOLITIC LIMESTONE
in 1 : 2 proportions

FRIABLE SANDSTONE
and OOLITIC LIMESTONE
in 3 : 1 proportions

Fig. 9.7. Distribution and character of
surface materials in Britain comprising
engineering coarse and fine soils of
bedrock and surficial (Quaternary) origin.
Blank areas as engineering rocks
(Fig. 9.6).

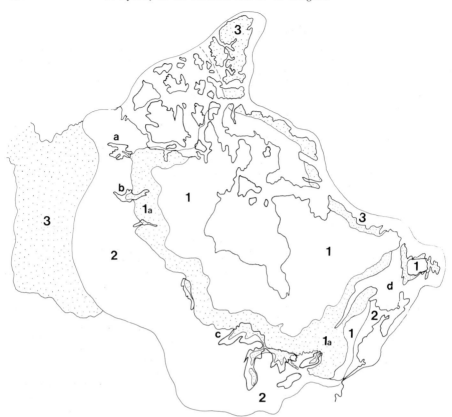

<u>Fig. 9.8</u>. Generalized glacial erosion zones in North America. Modified from Sugden (1977). 1. Shield terrain, areal abrasion dominant with moderate surficial soil cover; 1a, areal scour and selective linear erosion with restricted and shallow surficial soil cover. 2, sedimentary lowlands, with thick surficial cover from multiple glaciations frequently obscuring glacitectonized rockhead. 3, intensively eroded and overdeepened glaciated valleys; fiords along coast. a. Great Bear Lake, b. Great Slave Lake, c. Lake Superior, d. Gulf of St. Lawrence.

Lake District and North Wales, developed after the decay of the main Devensian ice sheet. This is referred to as the Loch Lomond Stadial, lasting from c 11,000 to 10,000 y.b.p. (Sissons, 1976, 1979, Fig. 5.2). The extensive tracts of ´hummocky moraine´ found in Scotland within the limit of the Loch Lomond Stadial Ice Cap may be the result of supraglacial deposition by valley glaciers (Figs. 1.6, 4.5; Eyles 1983). Specific engineering problems related to road, dam and reservoir construction in these valleys are discussed in Chapters 13 and 14.

9.4 The Importance Of Bedrock Lithology

Within the sedimentary strata of Britain many of the rock units must be considered as engineering soils as they are weathered and can be disaggregated in water. Therefore a review of the distribution of surficial soils must also emphasize the areal extent of bedrock ´soils´ as their treatment in excavation, testing and design is the same. Unfortunately the normal geological map does not show the distribution of engineering bedrock ´soils´ and rocks; the two are not distinguished and are grouped together as ´solid´. Lithology and lithological variations are not always given and the name given to a deposit may give no indication, and indeed may be positively misleading, as to the lithology of a given map unit. A much quoted example in this context is the London Clay (Burnett and Fookes, (1974). Without considerable background knowledge or reference to a comprehensive memoir it is usually impossible to determine with certainty the lithological types of rock and soil present on the conventional geological map.

Fig. 9.5 is an engineering geology map of Britain that is based on bedrock lithology. Sixteen lithologies have been recognised; four engineering soils, five sedimentary rocks, two igneous rocks and five metamorphic rocks. In addition, the key to the problem of depicting lithological changes along a single geological outcrop lies in the adoption of a stripe system by which eleven different proportions of one, two or three types can be clearly differentiated.

Igneous rocks are limited to granite and volcanic rocks. Basic intrusives such as sills and dykes, that could only be represented on a map by a line, are omitted. Pyroclastics and fine-grained igneous rocks and volcanic glasses are included with volcanics. No contact metamorphic rocks associated with large igneous intrusions are shown nor are tectonic breccias and mylonites. Other groups of metamorphic rocks are well represented. The choice of sedimentary rocks has been selective; mudstone and shale, sandstone, and three types of limestone are all that are represented. Greywacke, elsewhere regarded as a sedimentary rock, is portrayed here as a metamorphic rock because of its frequent association with slate; one of the mapping units is the lithological complex ´slates and greywacke´.

Fig. 9.5 show the distribution of solid rocks and soils in Britain as if the Quaternary superficial soils were stripped off. Fig. 9.6 shows the extent of till cover in Britain. The final stage is to produce a map showing the distribution of engineering soils in Britain combining both ´solid´ and superficial soils (Fig. 9.7). Note that the superficial soils have been divided into coarse and fine soils. Figure 9.7 emphasizes the large areal extent of engineering soils in Britain of both surficial and ´solid´ origin.

9.5 The Distribution of Glacial Landsystems in North America

The general distribution of landsystems is depicted in Fig. 9.2. Coupled with the block diagrams illustrating typical ground conditions likely to be encountered (Chapter 1), it gives a schematic representation of the distribution of glacial sediments. The limit of glaciation on Fig. 9.2 is a maximum reconstruction; the recent identification of ice-free areas would suggest a reduced ice cover in eastern Canada and the Arctic Islands (Denton and Hughes, (1981). The basis for identifying ice-free areas is the recognition of weathering zones; old highly weathered surfaces at elevation, surrounded by younger glaciated terrains. These weathering zones should not be confused with the weathering zones indentified by engineering geologists for soil description (Sects. 5.2, 8.2.4).

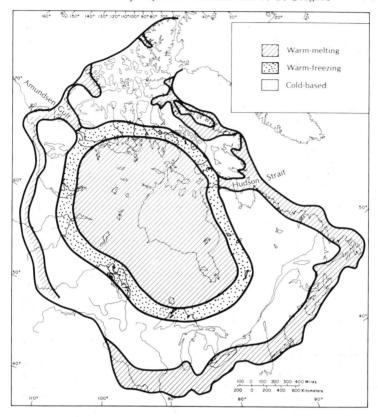

Fig. 9.9. Basal thermal regime of the Laurentide Ice Sheet (From Sugden, 1977).
Maximum surface altitude was about 3500 m with a thickness of 4200 m allowing for
isostatic subsidence. Temperatures are calculated using estimated values of
parameters discussed in Sect. 2.1. Annual accumulation (in H_2O equivalents;
Sect. 4.2) ranges from 80 cm near the margin to 5 cm near the centre. Velocities
increase from 10 to 200 m aong flow lines with highest ice velocities where the
ice sheet discharged through valleys (Fig. 9.2). The zonation in temperature is
hypothetical but provides an illustration of the importance of thermal regime in
determining the gross distribution of erosion zones and landsystems (Fig. 9.2).
Areal abrasion characterized the wet based conditions on the shield (Zone 1a,
Fig. 9.8) with debris being entrained into the ice base by refreezing of basal
waters on the upglacier margins of the freezing zone (Fig. 2.1). The enhanced
erosion of the outer shield (Zone 1a Fig. 9.8) possibly results from the
streaming of ice flow around freezing or frozen zones (Sect. 2.1, Fig. 2.2).
There is no exact correlation of reconstructed basal thermal regime and
landsystems in Zone 2 of Fig. 9.2 largely because the reconstruction appertains
to the ice sheet at its maximum. During retreat and thinning the extreme ice
margin was likely frozen with complex thermal mosaics upglacier depending on bed
elevation (Fig. 2.2).

Figs. 9.2 and 9.8 show a concentric arrangement of landsystems and erosion zones within the area covered by the former Laurentide Ice sheet (Fig. 1.1). The shield type subglacial landsystem (1,1a) characterises the inner zone reflecting both resistant crystalline rocks of the Canadian Shield and areal scour (Fig. 2.1) rather than selective glacial overdeepening (Sugden, 1977).

9.5.1 Shield-Type Subglacial Landsystem (Sect. 2.8.1)

The boundary of the shield with sedimentary lowlands (1,1a/2: Fig. 9.8) underlain by Palaeozoic and Mesozoic carbonate and clastic strata, demarcates the outer zone of thicker soil cover where lodgement tills form extensive plains. The wide exposure of bedrock on the Canadian Shield and the depression of Hudson Bay were originally attributed to glacial overdeepening at the supposed ice-sheet centre. It is now known that the ice sheet was multicentred (Andrews, et al, 1983) and that the depression of Hudson Bay is structurally controlled and enhanced by unrecovered isostatic recovery. Recent estimates of erosion over the Canadian shield are as low as 10 m for each major glaciation (Kasycki and Shilts, 1979) though debate continues (Gravenor, 1975; Andrews, 1982 and refs. therein) fostered by possible subsurface 'disposal' of nuclear waste in shield granites. The general absence of deeply eroded and drift-filled glacial valleys and the similarity of shield topography beneath Lower Palaeozoic sedimentary rocks indicates the great antiquity of the subdued shield surfaces, the general ineffectiveness of glacial erosion and the predominance of areal scouring (Sugden, 1977; Fig. 9.9). The most typical scoured shield terrain is found at the outer margin of the zone (1a, Fig. 9.8); enhanced linear erosion has occurred across a circumferential belt where the shield disappears under the Lower Palaeozoic cover. The Great Lakes Basins, Lakes Winnipeg, Great Bear and Great Slave Lakes and the outlet of the St. Lawrence River (1a, Fig. 9.8) all lie in overdeepened basins. The bedrock floor of Lake Superior for example (Fig. 2.21) at its deepest point is over 200m below sea level. The north facing escarpments of the south dipping sedimentary sequences that lie on the shield margins have been locally modified by glacial erosion in the form of deep narrow 'through' valleys occupied by finger lakes (Clayton, 1965; Straw 1968). Many escarpment faces have been glacitectonised and are intimately associated with moraine complexes of the supraglacial landsystem (Figs. 2.18, 2.19). Toward Hudson Bay, repeated marine incursions and the formation of large postglacial lakes, results in a thicker soil cover (Chapter 6) compared to the rest of the shield (Fig. 9.2) and is obscured over large areas by peaty organic terrains (Radforth and Brawner, 1972). Drift thicknesses in general, range between 2 - 10 m on the shield and typical sequences are portrayed in Fig. 1.3B. The economic importance of this glacial sediment cover are discussed in Sect. 2.8.1.

9.5.2 Sedimentary Lowlands: Lodgement Till Plains And Supraglacial Moraine Complexes

Zone 2 on Fig. 9.2 is identified by restricted bedrock outcrop obscured by thick (100 m) stratigraphic sequences resulting from multiple glaciations. The juxtaposition of streamlined lodgement till plains, developed upglacier of wide belts of hummocky supraglacial moraine complexes ('end moraines'), indicates a systematic distribution of landsystems with respect to ice margins and flow directions. It was in the Western Prairies of the United States that the landsystem, or morphostratigraphic, approach to the regional mapping of glacial landscapes and underlying sediments was first clearly demonstrated (Clayton and Moran, 1974).

The pattern of reconstructed basal thermal conditions (Fig. 9.9) is difficult to apply to the distribution of landsystems in this zone (Fig. 9.2) as detailed work shows the importance of complex local thermal mosaics. In the western plains the association of supraglacial moraine complexes and thrusted glacitectonised rock masses has resulted in a model of an outer ice margin, frozen to a deforming substrate, giving way upglacier to a sliding bed below wich is deposited lodgement till (Sect. 2.8.2; Moran et al, 1980). In several instances supraglacial deposition can be related to thrusting against escarpments and upland margins developed in the underlying Mesozoic sedimentaries. Debate continues as to the status of hummocky, apparently supraglacial terrain, in the northern sector of zone 2 north of Great Bear Lake (Figs. 9.2, 9.8; Sect. 3.3).

In the eastern half of zone 2 the major control on the location of supraglacial complexes results from the lobation imposed on ice sheet margins by the Great Lakes Basins. The parallelism of moraine complexes and lake basins can be clearly seen (Fig. 9.2). The low subdued relief on many of the moraine complexes is the result of several factors including the reworking of fine-grained lacustrine sediments, incapable of supporting steep slopes, modification by post-glacial lakes and glacial deposition in the subaqueous environment below floating ice margins. In general the implications of supraglacial and subaqueous deposition are only just being examined in this area; stratigraphic complexity within such terrain that is currently regarded as evidence of multiple glacial events, may be more realistically exained in terms of such deposition. The considerable area affected by proglacial and postglacial lakes and marine overlap is shown in Fig. 2.21 and discussed in Sect. 2.9 and Ch. 6. Note the outer circumferential distribution of glaciated valleys in Fig. 9.2.

CHAPTER 10

Engineering Geological Mapping in Glaciated Terrains

A. D. Strachan and W. R. Dearman

INTRODUCTION

Mapping for engineering geological purposes has the aim of representing the ground conditions, their uniformity or variability within specified limits, in sufficient detail to enable the engineer to achieve a safe and economical design for the proposed works.

Apart from the single large project in rural and remote areas, e.g. reservoirs, dams, airports, or pipelines, most construction is associated with urban areas and communication between them by roads, railways and pipelines. It can be demonstrated that many of the main centres of population in Britain for example are associated with both coalfields and glaciated terrain (Fig. 10.1). Problems in mapping and representation of glacigenic sediments for engineering purposes are thus largely associated with mapping in an urban environment where exposure is minimal but a vast amount of subsurface data is usually available. By contrast mapping for engineering purposes outside urban areas commonly involves air photograph interpretation of largely virgin terrain with limited ground truth.

The main contrasts in mapping in the two types of area, rural and urban are reflected in the major division of engineering geological maps, in terms of scale purpose and content (Fookes, 1969; Anon. 1972; Dearman and Fookes, 1974), into Engineering geological maps produced for planning on a regional scale, usually 1:1000 or smaller and Engineering geological plans, produced on a larger scale for specific engineering purposes. In Britain there is little need for small-scale engineering geological maps. Since 1858 the primary geological survey of the country has been carried out at a basic scale of 1:10560 (1:10000 since metrication), and maps may show the solid formations, solid-with-drift, or the drift deposits alone. Derived from the 1:10000 geological maps that cover at least 85 per cent of the country is a whole range of maps at various smaller scales from 1:2500000 to 1:50000, not all of which show the distribution of glacial deposits in drift editions. These are all geological maps which can be interpreted for engineering use. Eckel (1951) showed how published geological maps could be interpreted in terms of construction materials, foundation and excavation conditions, surface and groundwater, and soil conditions. This is important given that in other countries Quaternary and solid mapping is usually carried out separately and in some cases the area of the country mapped is low requiring initial small-scale terrain evaluation for many engineering projects.

Recently, recommendations have been made for the supplementation, by addition of engineering data, of general geological maps of the types produced by government agencies (Anon., 1972). Particular emphasis should be paid to the need for detailed descriptions of both bedrock and superficial deposits as far as possible in conformity with a unified code. Codes for engineering descriptions

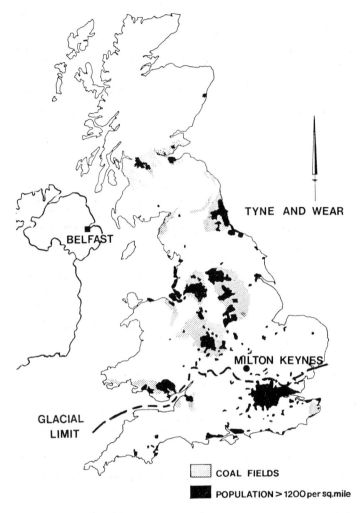

TYNE AND WEAR

BELFAST

GLACIAL
LIMIT

MILTON KEYNES

☐ COAL FIELDS

■ POPULATION > 1200 per sq.mile

Fig. 10.1. Coalfields and population centres in Britain (After Price, 1971). Localities are those mentioned in the text.

(Fig. 10.6) are discussed below and in Sect. 11.9, codes for detailed sedimentological descriptions are discussed in Sect. 1.4 and Ch. 7.

10.1 Engineering Geological Maps and Plans

The major division between engineering geological <u>small-scale</u> <u>maps</u>, generalized in content and made for regional and planning purposes and <u>large-scale plans</u>, detailed in content and made for specific civil engineering purposes (Anon 1972) is followed in this chapter. Figure 10.2 summarizes aspects of the

	Typical Scales	Information Shown	Prepared by	Method	Engineering Use	Examples
A. Geological Maps						
	1:2000000	General distribution of glacigenic sediments and exposed solid formations categorised in lithological terms	University	Assessment of published small-scale geological maps	Very general planning at small scale	Engineering geological map of Britain (Figs. 9.1, 9.5, 9.6, 9.7)
	1:625000	Sand and gravel resources	I.G.S.	Selection from 1:50000 I.G.S.	Very general resource and land use planning at small scale.	I.G.S. Sand and Gravel Resources. 2 sheets.
	1:625000	Distribution of Quaternary deposits	I.G.S.	Abstraction from 1:50000 I.G.S. maps and other published maps	Very general planning at small scale	I.G.S. Quaternary maps. 2 sheets.
	1:50000	Distribution of glacigenic sediments in a limited number of categories	I.G.S.	Reduction from published and manuscript 1:10500 and 1:10000 geological maps	Regional planning and general assessment of engineering sites. Needs skilled interpretation for engineering use.	I.G.S. 1:50000 Sheet 21: Sunderland
B. Engineering Geological Maps						
	1:50000	Distribution of glacigenic sediments in four thickness categories	University	Assessment of all available data	General assessment of engineering conditions over the map area	Tyne and Wear Map of superficial deposits. (Fig. 10.7)
	1:21120	Distribution of Drift deposits including Boulder Clay and Glacial Sand-and-Gravel. Rockhead contours for central city area; selected engineering parameters.	I.G.S.	Reduction from 1:10560 geological maps and assessment of borehole records	General assessment of engineering conditions over the map area.	I.G.S. Belfast Engineering Sheet.
	1:25000	Distribution of Boulder Clay, Glacial Lake Deposits, Sand-and-gravel	I.G.S. Geol. Unit	Assessment of basic geological mapping site investigation boreholes	New town planning	Engineering Geological map of Milton Keynes (Fig. 10.3)
C. Engineering Geological Plans						
(i)	1:2500 to 1:10000	Distribution in three dimensions of fill and glacigenic sediments above rockhead; rockhead contours; thickness categories or actual spot thicknesses of the mapped units; rockhead lithology.	University	Assessment of all available site investigation and other data; limited field assessment	Detailed Planning and reconnaissance; desk study phase of site investigation; index to site investigation data bank	Tyne and Wear Survey. (Fig. 10.11)
(ii) Reconnaissance	1:500 to 1:10000	Surface topography, geomophology, location of site investigation boreholes and pits; mapping in terms of descriptive soil classification.	Engineering consultants, site investigation contractors.	From air photography and walk-over survey	Detailed planning and reconnaissance	
(iii) Site investigation	1:20 to 1:5000	As above	As above	As above plus instrument assisted techniques	Site investigation	Taff Vale Nantgarw (Fig. 10.5a)
(iv) Construction	1:100 to 1:1250	As above	As above	From photogrammetry, walk-over surveys plus instrument assisted techniques	Investigation and recording during construction	Taff Vale Nantgarw (Fig. 10.5b)

Fig. 10.2. Summary data on available geological maps and engineering geological maps and plans of glacigenic soils in Britain.

classification of maps and plans and is an amplification of the classification given in Dearman and Fookes (1974), specifically related to glacigenic sediments.

In Section A of Fig. 10.2 geological maps, ranging in scale from 1:2000 000 to 1:50000, are listed that show the general distribution of glacigenic sediments in Britain. Engineering uses vary from very general planning to regional planning and general assessment of engineering sites (e.g. Figs. 9.5, 9.6, 9.7). To a very great degree the usefulness of such maps depends on the skill and background knowledge of the user.

10.1.1 Engineering Geological Maps.

There is only one published example of an engineering geological map of a till covered area in Britain. This is the 'Geology of Belfast and District: Special

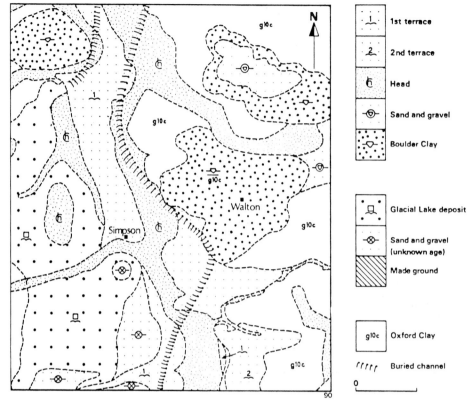

Legend:
- 1st terrace
- 2nd terrace
- Head
- Sand and gravel
- Boulder Clay
- Glacial Lake deposit
- Sand and gravel (unknown age)
- Made ground
- $g^{10}c$ Oxford Clay
- Buried channel

Fig. 10.3. Part of the engineering geological map (1:25,000) produced by the Eng. Geol. Unit of the I.G.S. After Cratchley and Denness (1972).

Engineering Geology Sheet, Solid and Drift' (1971) published by the Institute of Geological Sciences. The choice of 1:21120 scale was determined by the availability of a suitable topographic base-map, but as the reduction was prepared from 1:10560 maps engineering sites can be located precisely (Bazley, 1971).

Overprinted on the topographical map is a conventional coloured solid-and-drift map. As the region is largely drift covered, apart from the few relatively small outcrops of bedrock formations, the inferred boundaries of the solid are given beneath the drift. Glacial sand and gravel and boulder clay (lodgement till) are differentiated by distinctive colours and symbols, but no other information is given on the printed map. However, on the reverse side of the sheet is a summary account of the geology and a separate map showing contours on rockhead for the central part of the city. Rockhead contours combined with the isopachytes of the estuarine clay (sleech) and topographical contours enable the thickness of the glacial deposits to be estimated. Some 2000 borehole records were used in the preparation of the map. The sites of 60 selected boreholes with endorsed records of the sequence of superficial deposits are located on the map and provide a useful indication of the variability of the glacial deposits. A separate table

	Superficial-Clayey Till
4. Other features:	
a) Range of Geotechnical	
Parameters: LL%	40–60
PL%	20–30
PI%	20–30
LI	(−0.1)–(0.2)
SG	2.45–2.6
γ(Mg/m³)	1.8–2.0
Intact c_u (kN/m²)	50–150
pH	
Sulphate (% init. dry wt.)	Probably 5
b) Collective implications of 4a	Good foundation and fill
c) Variability and openness	Variable and closed
of fissuring	
d) Implications of 4c	—
e) Bedding orientation	—
11. Foundation stability	
a) Light loads (<150 kN/m²) on	
thick deposit: i) Loading rate	Rapid
ii) Loading	
intensity	High
iii) Foundation	
type	Any except point or strip
b) Heavy loads (>150 kN/m²) on	
thick deposit: i) Loading rate	Slow
ii) Loading	
intensity	Moderate
iii) Foundation	
type	Raft
c) Susceptibility to moisture	
content change during and	
after construction	Moderate
d) Metastability	None
e) Swelling pressure	Low
f) Pre-existence of slip surfaces	Possibly abundant
g) Strength variation	High laterally and vertically
h) Compressibility variation	High laterally and vertically
i) Special conditions	Differential settlement near surface slip zones

Fig. 10.4. Extract from table of
engineering geology characteristics to
accompany 1:25,000 Engineering geology
map of Milton Keynes.

of 'Soil and Rock Characteristics' summarises the chief geotechnical properties
and other information on the mapping units and it is this that gives the map its
engineering flavour, differentiating this sheet from the standard solid-and-drift
1:50000 sheets produced by the Institute of Geological Sciences.

It must be accepted that the Belfast sheet is not a true engineering geological
map but is an example of a basic geological map supplemented with additional
engineering geological information as recommended in Anon. (1972).

A second example is a map with a series of overlays, e.g. those produced by the
Engineering Geology Unit of the Institute of Geological Sciences of the new town
of Milton Keynes, situated 80 km northwest of London adjacent to the M1 Motorway
(Fig. 10.1). Unfortunately the maps have not been published, but some idea of
their style and content can be gained from the account published by Cratchley and
Denness (1972), and a single illustration (Fig. 10.3).

Initially a new geological survey of the whole area was made in terms of
stratigraphical units at a scale of 1:25000. A geotechnical survey was then

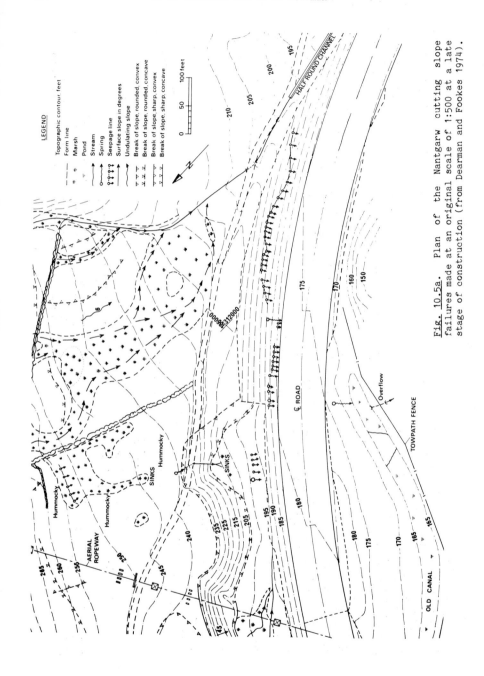

Fig. 10.5a. Plan of the Nantgarw cutting slope failures made at an original scale of 1:500 at a late stage of construction (from Dearman and Fookes 1974).

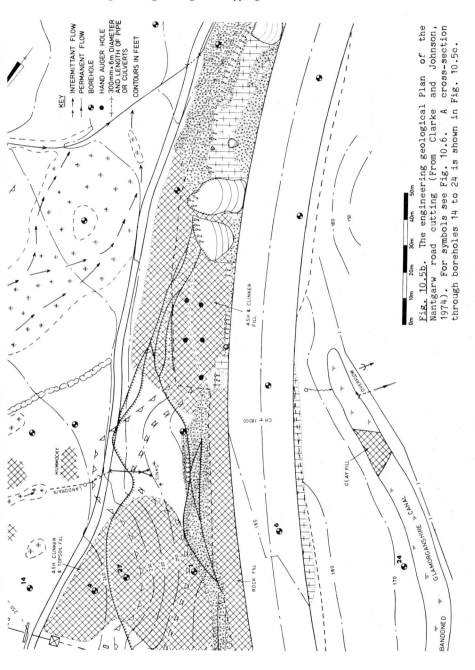

Fig. 10.5b. The engineering geological Plan of the Nantgarw road cutting (From Clarke and Johnson, 1974). For symbols see Fig. 10.6. A cross-section through boreholes 14 to 24 is shown in Fig. 10.5c.

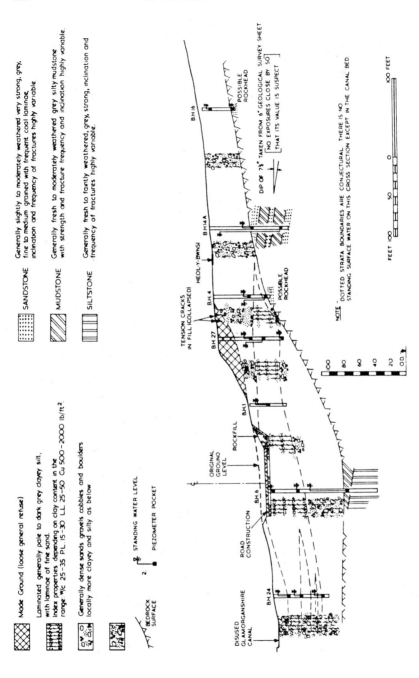

Fig. 10.5c. Cross-section of the Nantgarw road cutting failure (from Fookes et al, 1972).

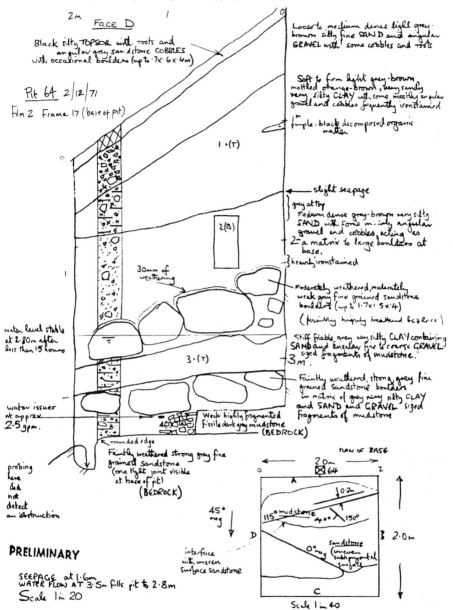

Fig. 10.5d.   Trial pit log, Taff Vale trunk road (After Fookes et al, 1972).

carried out and from visual examination of the borehole samples and assessment of relevant index properties the stratigraphical units were regrouped into a number of sensibly uniform engineering geology units regardless of their age. Accompanying the new 1:25000 engineering geological map was a chart (Fig. 10.4) of quantitative engineering properties and interpreted behavioural characteristics.

Sand and gravel, boulder clay, glacial lake deposits, and head are the glacigenic deposits represented on the engineering geological map (Fig. 10.3). Boulder clay (lodgement till) is the main superficial deposit at Milton Keynes, and a limited local investigation in a trench section suggested that the geotechnical properties were sufficiently variable to justify the use of in situ testing of these deposits. Representation on the resultant map is by standardized symbols for the different stratigraphical units, united regardless of geological age under the same colour into lithological units having the same engineering characteristics on the derived engineering geological map.

10.1.2 Engineering Geological Plans.

Most engineering geological mapping on the plan scale is carried out by commercial organizations. A single case history will suffice to illustrate the mapping methods adopted at each of the Reconnaisance, Site Investigation and Construction stages of the development of a civil engineering project.

Mapping methods are simple and are geared to a very rapid assessment of terrain or an individual aspect of construction. Plans are made by a small number of highly trained staff using simple survey techniques that have been adequately described elsewhere (Anon. 1972). The mapping method consists essentially of recording observations and points of measured data (e.g. the log of a hand-augered hole) on an existing or specially prepared topographic plan, or on squared paper using a survey grid, or on a photograph. Location is by chainage-and-offset measurements, simple instrument surveying or inspection. By far the most difficult aspect of mapping is the insertion on the plan of geological boundaries which are located by observations of field changes in geomorphological form (Cooke and Doornkamp, 1974), in rock or soil characteristics, or by interpretation based on judgement. For investigations at the plan scale ground conditions are normally investigated by means of boreholes and/or pits and trenches where either the conditions are critical and need to be viewed in three-dimensions, or boring is difficult (Section 11.4). Representation of the geology is then usually made by means of sections.

A single example will suffice. The Taff Vale trunk road between Cardiff and Merthyr Tydfil in South Wales is a dual carriageway designed and built as an improvement to the main north to south road network through the coalfield (Fig. 10.1). Severe physical constraints were imposed by the narrow valley with its glacially oversteepened slopes, glacial, fluvioglacial, periglacial and talus superficial deposits as well as an extensive legacy of industrial activities (Brunsden et al., 1975).

North of Nantgarw Colliery the road lies in a deep cutting for some 750 m through a kame terrace (Fig. 1.6) on the east side of the valley. As soon as excavation was complete, mud runs started over about half the length and large tension cracks developed in and beyond the cutting slopes. A site investigation was put in hand and geotechnical mapping was undertaken of a large area behind and adjacent to the slip at a scale of 1:500 (Fig. 10.5a). These plans are reproduced in Dearman and Fookes (1974) and larger versions with indications of soil types in the cut slopes in Clarke and Johnson (1975). A small part of the large version is shown in Fig. 10.5b as an example of the use of standard symbols

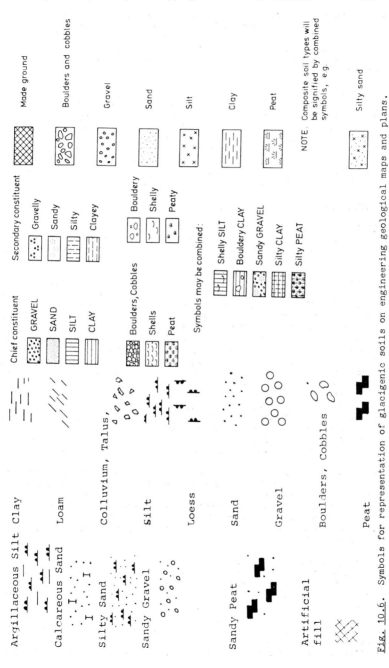

Fig. 10.6. Symbols for representation of glacigenic soils on engineering geological maps and plans.

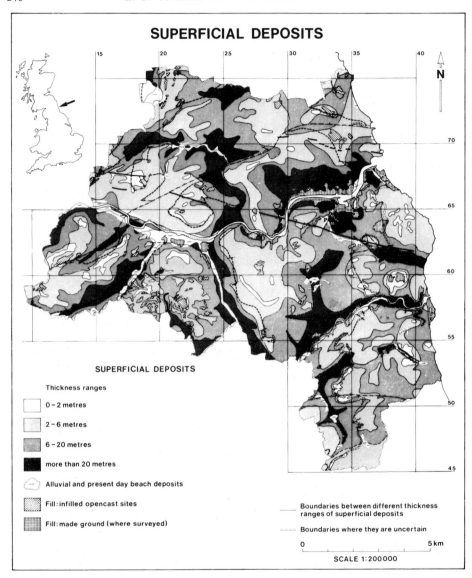

Fig. 10.7. The 1:50,000 engineering geological map of Tyne and Wear (Fig. 10.1) reproduced here at 1:200,000 showing the distribution of superficial deposits in four thickness range classes (from Dearman et al, 1979).

in the representation of different lithologies (Anon. 1972).

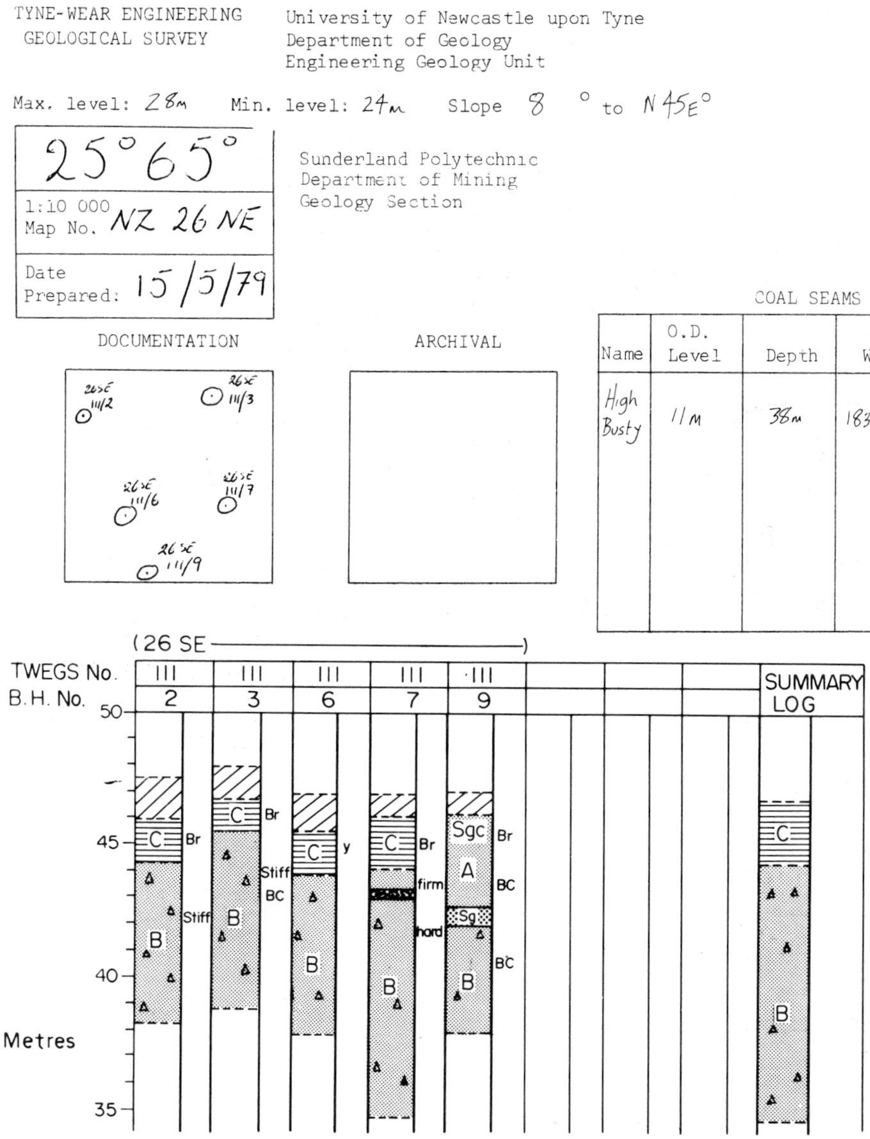

Fig. 10.8. Data summary sheet for a single 100 m square, Tyne and Wear Engineering Geological Survey.

Plotting of lithological boundaries was based on the use of chainage-and-offset techniques from the road ground control. Offsets were measured up the cutting face at intervals of approximately 3 m and all changes of lithology recorded along these to an accuracy of 0.1m and reduced to plan distances before plotting.

This investigation can also be used to illustrate the use of geological cross-sections drawn through a series of boreholes. Again standard symbols are used in columns slightly offset from the borehole positions except below rockhead (Fig. 10.5c).

An essential part of site investigation is pitting, and it is standard practice to record the lithologies and structures present in the base of the pit and on at least one of the faces. A typical example of a field-sheet record of trial pit in glacial sediments is shown in Fig. 10.5d.

At the plan scale the whole range of illustration techniques has now been presented, and central to this is the use of standard symbols and combinations of symbols to represent the varied soil types formed in glacigenic sediments.

10.2 Symbols For Glacigenic Soils

Symbols given in Fig. 10.6 are those recommended for use in Britain as well as those that are internationally accepted. The symbols are simple and distinctive and they combine easily into symbols for composite types of soils and clastic rock.

While the proposed symbols are primarily intended for use in logs and sections, usually on a large scale, they can also be used for plans and maps. For the latter it may be advantageous to lighten the ornament by spacing it more widely or by using thinner lines.

Symbols are given for the four divisions of soils based on particle size. Each symbol has two variants, one for use when the material is the chief soil constituent, the other for use when it is the secondary constituent.

International symbols, have been adopted by the Internal Association of Engineering Geologists. Mapping Commission (Anon. 1982).

10.3 Preparation Of Engineering Geological Maps And Plans From Site Investigation Data

This section describes how a map should be compiled in an urban environment and shows some of the problems likely to be encountered.

In the urban environment there is usually a wealth of site investigation data that is uncollected, stored in many places and is consequently inaccessible. There is a strong case to be made for some form of central registry for such data (Legget, 1973). Such a project has been undertaken in the Tyne and Wear Metropolitan County in North-east England and illustrates the use that can be made both of site investigation data and a variety of published sources of information. A data bank of about 30,000 borehole logs has been amassed, with the main object of the research being to produce derivative engineering geological maps and plans; the direct utility to the planner and engineer unversed in conventional geological interpretation can be stressed. Because of the variety of glacigenic sediments in the area, description of the mapping and compilation techniques adopted provides an effective conclusion to this chapter.

10.3.1 An Engineering Geological Map Of Tyne And Wear Area.

In the Newcastle area (Fig. 10.1) coal measures have been worked both by deep mining and by opencast methods and the worked sites have been infilled and restored. Outcrops of superficial deposits include fill, till, and sand-and-gravel. Sand-and-gravel also occurs locally as part of the till

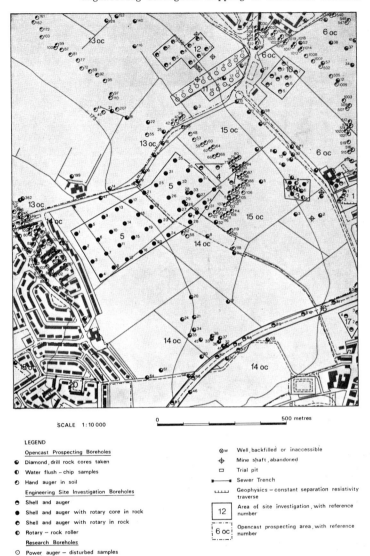

SCALE 1:10 000

0   500 metres

LEGEND

Opencast Prospecting Boreholes

- Diamond, drill rock cores taken
- Water flush – chip samples
- Hand auger in soil

Engineering Site Investigation Boreholes

- Shell and auger
- Shell and auger with rotary core in rock
- Shell and auger with rotary in rock
- Rotary – rock roller

Research Boreholes

O Power auger – disturbed samples

⊗ w   Well, backfilled or inaccessible
⊕   Mine shaft, abandoned
□   Trial pit
▪—▪   Sewer Trench
⊔⊔⊔   Geophysics – constant separation resistivity
      traverse
[12] Area of site investigation, with reference
        number
[6 oc] Opencast prospecting area, with reference
         number

Fig. 10.9. Documentation map recording the location and nature of archival information for a small area of North Tyneside (Fig. 10.7). From Dearman et al (1977).

sequence.

One of the main problems is how to represent variable thickness, multilayer superficial deposits resting on bedrock. A solution to the two-or three layer

deposit on bedrock was available in the elegant stripe method first developed in
Czechoslovakia in 1947 (Pasek 1968) and illustrated by Matula (1969, 1971). On
the otherhand a well-designed engineering geological map (Varnes, 1974) can also
meet these multilayer requirements (Dearman and Matula, 1977). A broad picture
of the glacial deposits, reproduced here at 1:200000 (Fig. 10.7) has served as
the basis for more detailed mapping at scales of 1:2500 and 1:10000 in the
Newcastle area. Construction of the map involved three steps:

i)   data collection and location
ii)   data evaluation and preparation of summary logs
iii) compilation of the map.

Data Collection And Location. Borehole records from site investigation reports
remain the prime source of information on ground conditions. Reports are
located, then borrowed or copied, and the individual borehole logs microfilmed.
The microfiche collection comprises the main data bank once the boreholes have
been located on borehole documentation 1:2500 plans.

Data Evaluation And Preparation Of Summary Logs. Mapping units for the
superficial deposits were selected on the basis of ease of recognition in the
field with distinct engineering properties. The following units were chosen:

Symbol	Soil Type	Colour or Pattern
A	Weathered till - Till A	Yellow
B	Lodgement till - Till B	Brown
C	Clay	Blue
L	Laminated clay	Blue
S	Sand-and-gravel	Green
M	Multilayer	Purple
P	Peat	Blue
-	Fill	Standard symbol or red
R	Rock	Black diagonal stripe

Compliation Of The Map. A mapping technique was devised to show the distribution
of up to four layers (including solid at rockhead) from the surface down,
excluding fill, with an accompanying letter notation to identify the materials.
Surface deposits were represented by a distinctive colour or pattern over each
square, with the underlying layers represented by a bar-strip across the middle
of the square (Fig. 10.10). This mode of representation follows mapping
convention in showing the lithological unit that crops out, and the bar-stripes
are read from top of the map downwards as a stratigraphic sequence. Rockhead
contours at 10 m intervals are drawn on the plan.

An assumption was made, based upon the borehole evidence, that the most likely
surface material was weathered till. Different surface deposits are likely to be
of limited extent, and were thus enclosed by a boundary. The exact shape of the
boundary depended upon how many boreholes indicated that particular material.
For materials found in only one borehole, an arbitrary circular boundary was

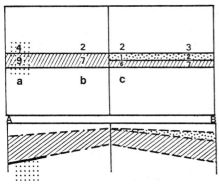

Fig. 10.10 Modified stripe method for characterization of individual 100 m squares on the 1:2500 engineering geological plan of Tyne and Wear. <u>Above</u>: two plan squares with single and double stripes. At (a) a 4 m thick layer is underlain by a 9 m layer resting on sandstone bedrock. At (b) the thicknesses vary and bedrock has not been reached. At (c) a 2 m thick surface layer is underlain by a 1 m layer, in turn underlain by a 6 m thickness of a third layer. <u>Below</u>: interpretative cross-section along A-B.

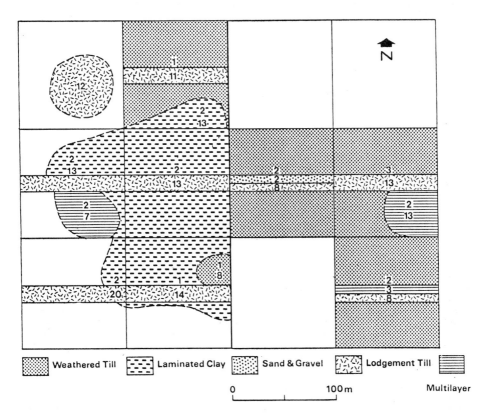

Weathered Till   Laminated Clay   Sand & Gravel   Lodgement Till   Multilayer

0   100 m

Fig. 10.11. Extract from the 1:2500 Engineering geological plan of Tyne and Wear. In the absence of data, blank areas are filled in as weathered till.

drawn round the borehole position.

An example of a group of squares on the completed 1:2500 plan is given in Fig. 10.11. Compilation of 1:10000 plans was either by direct reduction of the 1:2500 sheets or by the same method applied to a larger basic square size. The latter is appropriate where boreholes are sparse.

Apart from the fact that weathered till underlies most of the present land surface (Eyles and Sladen, 1981), the units have no definite stratigraphical implications. The 'multilayer' unit was adopted to deal with sequences in which more than three units were detected.

It was decided that the basic map unit would be defined by National Grid lines on the plan and would cover an area of 100 square metres. All the borehole logs for a single square are gathered together, plotted on a data summary sheet in relation to ground level (Fig. 10.8) and the different strata noted by a colour code on the log itself. The aim of this comparative exercise is obtained at a single summary log that could be considered to be representative of the superficial deposits within that particular 100 m square. The method of plotting and colour coding brings out any obvious correlations between boreholes. Judgement has to be made of the quality of data from individual boreholes, and where there is doubt the information is discarded. Documentation maps have also been produced for larger areas with reference to individual construction or mining projects (Fig. 10.9).

It was frequently the case that individual 100 m squares were too large as the basic mapping unit, because of the wide range of variation in ground conditions and the easily defined but limited extent of individual units in the drift sequence. In such cases judgement had to be exercised in the next stage of the investigation, the compilation of the map.

10.3.2 Use Of The Computer

The great mass of data from boreholes, each of which can be located by grid references, and the sequence of lithological soil types coded in terms of standard soil types with depths to boundaries between types specified with reference to Ordnance Datum, is an obvious candidate for computer analysis, manipulation, and display. Some progress to this end has been made (Reekie, Coffey and Marsden, 1979). An excellent general review of data compilation in urban geology projects can be found in Utgard et al, (1978). National computer registers of geotechnical and site investigation data are likely developments in the future, and computer graphics offer exciting developments (Walton, 1978; Radbruch-Hall, et al, 1979).

CHAPTER 11

Site Investigation Procedures and Engineering Testing of Glacial Sediments

S. Somerville

INTRODUCTION

For any new civil engineering project, some form of site investigation is
required even if this only involves the site of a single-storey domestic
dwelling. As a project increases in size and complexity, the need for an
investigation becomes increasingly more obvious not only to provide the Client
with information, to ensure the feasibility of his scheme and to allow it to be
designed and costed but also to permit the Contractor tendering for the work to
design any temporary works required for construction and arrive at a competitive
tender price. It also reduces to a minimum the risk of a contractor making
subsequent claims on the basis of being provided with inadequate ground
information in the tender documents.

Site investigations are carried out for many reasons including the determination
of ground conditions for the design of new works, checking the safety aspects of
existing structures such as dams and reservoirs or investigating causes of their
collapse, and providing information to support legal claims.

In addition, the results of a site investigation can be used to determine whether
a change in the location of the works is warranted - either to another part of
the same site or to a completely different location, and may also provide
information on possible changes that may take place in the ground (e.g. a change
in the ground-water regime), or in the environmental condition that the proposed
works may bring about.

The wide variation in composition and consistency of sediment (engineering soil)
sequences in glaciated terrains has had a considerable influence on the evolution
of site investigation techniques. Site investigations are generally undertaken
by commercial firms relatively few being carried out by Government departments,
Universities and other research organizations. Thus the cost effectiveness of
the equipment purchased and its employment in the field has had a high priority.
The tendency has been for Companies to purchase or develop equipment which can be
used in a variety of ground conditions to allow a reasonable cross-section of
their geotechnical properties to be assessed regardless of what might be
encountered at any one site. As Terzaghi (1943) wrote - "There is no glory in
the foundations" - and this attitude has always been reflected to some extent on
a resistance by developers to spend money on something that is subsequently
hidden in the ground. Thus limited money has been available by the commercial
companies to finance research of new methods of investigation and most of this
work has been carried out by the Universities and Government Research
organizations.

Of the many hundred or so companies that undertake site investigations, the vast
majority operate only one or two rigs and less than a dozen are big enough to be

internationally or even nationally known. In general, it is these larger firms
who have the range of field and laboratory equipment to undertake the more
sophisticated investigations where a high degree of technical expertise is
required. Naturally, such companies tend to be more expensive but as with most
things in life 'you get what you pay for' or perhaps to be more precise 'you
don't get what you don't pay for'.

In general, the variable clast content of glacial soils is a cause of a high
degree of boring and sampling disturbance and this makes it very difficult to
obtain reliable geotechnical data for engineering interpretation. These problems
limit the suitability of any one system of site investigation technique.
Frequently more than one glacial soil type can be experienced at a site and the
method of investigation will generally be a compromise dependent on the size and
technical complexity of the proposed project together with the available finance
for its undertaking. For any development, it is important to obtain as much
information as possible about the site prior to undertaking the field work. Much
useful information can be obtained from published topographic and geological maps
and memoirs and many sites can be examined on stereo pairs of aerial photographs.
A call or letter to the local Authority's Engineering Office can often yield
useful locally known knowledge of the site and any previous uses to which it may
have been put. Any such information obtained from these an other sources,
together with a walk-over of the area, can provide an intelligent assessment of
the likely ground conditions and the possible geotechnical problems that may
arise. At some future date, it may be that access will be available to national
registers of site investigation reports (Sect. 10.3.2). A trial collection of
such reports on microfiche has already been made for a 12,000 km^2 area in Southern
England and is reported in CIRIA Report No. 70, 1977. The scale of the
subsequent field work will depend on the size and type of project and may vary
from a few shallow mechanically excavated trial pits where only light industrial
or housing development is planned, to something far more sophisticated where say
a major bridge or dam is proposed.

However, whatever the size of the project, it is important to maintain a flexible
approach so that the scope of the investigation and techniques used can be
modified in the light of the information obtained. Thus it is often more
economical to use one of the larger contractors who have the inhouse experience
and alternative equipment available.

In the past, marine site investigations have been mainly confined to coastal
developments such as ports, jetties, tanker terminals and water intake works, and
the techniques required have not differed greatly from those used on land.
However, the more recent investigations required for offshore gas and oil
structures - particularly in the hostile environment of the North Sea and
Beaufort Sea - have accelerated the development of new techniques in boring,
sampling and in situ testing.

11.1 Desk Study And Site Walk-Over

For the average sized site investigation contractor, a site investigation
contract is normally limited to field work e.g. boreholes, trial pits and field
tests together with the subsequent laboratory testing and reporting. However, an
initial desk-study and a walk-over of the area form distinct and important phases
of the overall investigation which together can significantly effect the scope of
subsequent field work.

For the larger projects the appointed consulting engineers normally undertake a
desk-study and walk-over as part of their design brief. However, for smaller
projects where a site investigation contractor is appointed by say a small

authority or building firm, these investigatory stages may be omitted or at least only performed perfunctory if undertaken by personnel inexperienced in this type of operation and the outcome may be that some vital piece of information may be missed which could seriously affect the scheme. The value of appointing a site investigation company having the necessary geotechnical engineering expertise to be able to advise their client on desirable changes in scope, is therefore obvious.

In Britain the Code of Practice for Site Investigations (BS15930:1981) includes Appendices setting out the general information required for a desk study and associated sources of information, and also gives notes on site reconnaissance which is basically a memory aid for use in site walk-over surveys. Briefly, the code lists the following main headings for a desk study:

General Land Survey including site location on published maps and charts, air survey, site boundaries, contours, drainage, above and below ground obstructions e.g. transmission lines and sewers, survey stations, bench marks and meteorological information.

Permitted Use and Restrictions including any planning and statutory restriction by Planning acts, government regulations, land use zoning, rights of way, tunnels, undermining and ancient monuments, etc.

Approaches and Access by road, rail, water and air.

Ground Conditions - obtained from geological and soil maps, etc.

Sources of Material for Construction including natural, imported and tipped materials.

Drainage and Sewerage

Water, Electricity, Gas and Telephone Services

Heating - availability of suppliers.

Sources on information referred to include topographic maps, Geological Survey, Soil Survey and Water Resource maps and memoirs, admiralty charts and hydrographic publications, meteorological information, air photographs and seismological information. Information can also usually be obtained from the engineers and planning offices of local or regional government. The geology departments of the local University or college probably have glacial geologists with detailed knowledge of the regional glaciogenic stratigraphy. The University library will also contain many sources of geological information.

During the subsequent walk-over of the site area, information should be added to the site plan which might affect foundation design and construction. This should include notes on topography, valley courses, springs, ponds, marshy areas, streams, rivers with rate of flow and direction, etc., trees and hedges (size and type), type of vegetation cover, soil/rock outcrops and exposures on slopes and cuttings, access routes for site investigation, construction vehicles and plant.

It will be appreciated that a great deal of advance information can be obtained about a site from these various sources which together with a walk-over of the area can provide a fairly clear idea of the suitability of the site for the proposed development and often a reasonable guide to the anticipated ground conditions and problems likely to arise during the construction phases of the works.

11.2 Boring Techniques

The object of sinking boreholes is to identify strata boundaries, perform in situ tests, monitor groundwater conditions and recover samples of the ground in a state as undisturbed as possible in order that the materials can be identified and that laboratory testing will provide results representative of their in situ condition. The recovery of ideal information has to be tempered by cost and overall value related to the proposed works.

The standard equipment used in Britain for investigating soils is the so-called Shell and Auger (drop-tool) or percussive boring rig with which borings can be sunk in all types of soil. Mostly, these rigs are of the semi-mobile type (Fig. 11.1b) in that they have wheels but require towing for movement between borehole positions and are usually referred to as Trailer Mounted Rigs (TMR). Heavier rigs for boring deeper than about 60m and diameters greater than about 0.5m are usually of the static type. The semi-mobile rigs normally carry a standard set of equipment comprising two sets of lining tubes and sets of tools allowing boring to about 15m with 20cm diameter equipment. The basic tools used with this type of boring are the clay cutter and the shell. The cutter consists of a heavy open-ended steel barrel with a cutting edge or a 3-vaned cutter which is suspended on a wire line and repeatedly dropped down the hole, the clay being cleaned out after the tool is withdrawn from the hole. The shell (or baler) is similar to the clay cutter but has a flap valve (clack) of steel or leather on the bottom. This tool is pumped into granular soils and when full is raised and up-ended on the ground to empty whilst being flushed with water to clean the valve. Other tools used include the clay auger. This was the original tool used for boring in clay soils and causes much less disturbance to the soil than the clay cutter but has to be run down the hole on rods and to be rotated by hand and so is far slower and more costly. Chisels are used to break up obstructions such as rock, cobbles and boulders and to penetrate very strong ground, the fragments then being baled out using a shell. As boring is advanced, the boreholes are stabilised in loose ground and groundwater is sealed off by lining tubes (often referred to as casing). These are advanced and withdrawn by percussion methods but sometimes require to be jacked out in difficult ground conditions.

Small percussive bored piling rigs, normally used for constructing bored piles of about 0.5m diameter, are sometimes utilized for site investigation work to penetrate cobble and boulder beds. Such rigs are more expensive than standard heavy duty percussive boring rigs but may, even so, be more economical as greater and faster penetration can be achieved. Large diameter bore piling rigs are also occasionally used to sink shafts and permit in situ down hole inspection and testing to be carried out.

When boring has to be continued below superficial deposits into rock, an alternative method of boring is usually necessary since penetration by percussion methods (chiselling) is slow, laborious and expensive. Even when it is only required to prove that bedrock has been reached, it is necessary to penetrate about 2m to 3m into rock to ensure that it is in fact bedrock and not a large bedrock raft (Sect. 14.5.1). Chiselling can take several days in very hard rock and an alternative is to continue with rotary core drilling using either separate purpose-made rigs (e.g. Fig. 11.1a) or specially designed rotary drilling attachments that are suspended from the legs of the shell and auger rig and can be swung aside when lifting and lowering the drill string.

The purpose of this type of drilling is to obtain relatively undisturbed core of the rock material in order to examine the geological succession beneath the site. The borehole is advanced by rotating a core barrel and allowing the grinding

a

Fig. 11.1a. Rotary drilling rig. The rig illustrated is a Hands-England Drillmast 36 fitted with a mast having a rotary head travel of 7m and is mounted on a detachable skid for "on the ground" operations. The sub-frame mounted drill can easily be mounted on a truck or trailer.

Fig. 11.1b Pilcon Wayfarer 1500 Shell and Auger (drop-tool) rig.

action of an annular bit set with diamonds or other hard material to cut into the rock. Unfortunately the parts of the rock mass tending to govern its mechanical behaviour are the weakest, and the most likely to be lost or disturbed by the drilling process. The correct use of first class equipment is therefore important to obtain good quality undisturbed core suitable for geotechnical purposes. In general the greater the core diameter the greater the chance of obtaining good core recovery, however this is offset by an equivalent increase in cost.

A wide range of rotary core drilling equipment is available from several manufacturers which can be mounted on skids, trucks, crawlers, tractors or as mentioned above housed in separate attachments, and can be used to prove bedrock

Fig. 11.2.   Atlas   Copco   lug   mounted
overburden drill.

at the base of the superficial deposits or for drilling individual boreholes.

When detailed information of the ground is not especially required, rapid
low-cost boring can be achieved by Rotary Probe Drilling. However, the method is
noisy, can give rise to dust problems and is least effective in sticky cohesive
soils. Basic methods in this type of drilling include Rotary Percussive Drilling
whereby penetration is achieved by a slowly rotating hard metal chisel or
cross-chisel bit, so that the full face of the hole is cut, the cuttings being
blown back to the surface. One machine widely used for this type of drilling is
the Atlas Copco Overburden Drill (Fig. 11.2) which can drill a cased hole through
bad ground.

Rock Bit Drilling is another technique which uses a tricone roller bit
(Fig. 11.3b) in which toothed wheels run around the full face of the borehole
removing rock cuttings. Flushing media may be water, air or mud and can be used
in all materials. A heavy rig is required to drill through rock and the method
is often used to penetrate obstructions in difficult ground. Dual tube reverse
circulation drills are used extensively in drift prospecting projects in
glaciated shield terrains (Fig. 11.3; Sect. 2.8.1).

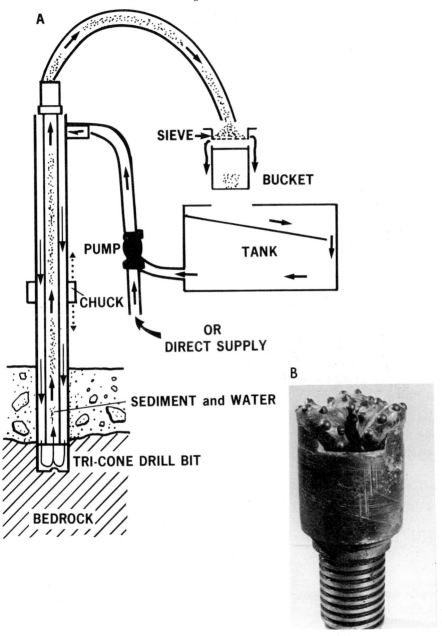

A

SIEVE →

BUCKET

PUMP

TANK

CHUCK

OR
DIRECT SUPPLY

SEDIMENT and WATER

TRI-CONE DRILL BIT

BEDROCK

B

<u>Fig. 11.3.</u> A) Dual-tube reverse-circulation drill as commonly used in drift exploration (Sect. 2.8.1).   B) Tri-cone drill bit

C

Fig. 11.3c Resonant drill system; Hawker Siddeley Superdrill 150 (Series 2).
The   system   is   also   available   as   a   helicopter   portable   ´Helisonic´   rig.
Photograph courtesy of D.R. Dance.

A further technique is termed Full Face Diamond Drilling where a bit has a full
face of diamonds and permits open-hole drilling with a medium weight rig using
water, air or mud flush. An advantage is that it can be integrated with core
drilling when using a medium weight rig and thus avoid the use of more than one
driller on a small site. It is often useful for forming regular holes for in
situ rock mechanics testing, e.g. in situ stress measurements.

In general air flush methods are always preferred because the high uphole
velocities ensure that cuttings are recovered from the material actually being
drilled at the time.   It also provides a more sensitive indication for
determining the water table and any cavities or fissures. It is also usually
easier to tow a compressor than a water bowser around a site.   Wash boring

methods, widely used in the United States, in which soil is loosened by a high pressure jet of water from a pipe lowered down the borehole, also allows rapid and low cost penetration of suitable superficial deposits. Disturbance of the soil below the bottom of the borehole is generally less than with other forms of boring but the 'washings' recovered are usually too disturbed for anything other than a rough identification of the soils penetrated. In the U.S., the borehole size is normally of 2.5in (63.5mm) diameter to allow the entry of Standard Penetration Test equipment from which spoon samples are recoverable (Sect. 11.4).

Wash probings that adopt a similar principle for penetration but without the use of casing can be employed in suitable conditions to determine the thickness of soft or loose superficial deposits overlying a more competent stratum such as lodgement till or rock.

Simple lightweight rotary auger drills (Minuteman, DIG-R-MOBILE, etc.) and percussive drills (e.g. Wacker) are frequently used; these have the advantage of being cheap to purchase and easy to operate in remote areas but they are stopped by coarse-grained sediments. These methods are useful for rapid penetration of superficial deposits to determine and prove bedrock but are limited where the overburden consists of stiff cohesive soils.

There is considerable interest at present in sonic and resonant drill systems that produce clean cores from sediments of a wide variety. Sonic drills (e.g. Vibrocorer) employ ultra-high frequency vibrations to drive an open core barrel through sediments. The vibration is produced by a motor at the top of the barrel driven by compressed air, hydraulic power or electricity. The principle is the same as that employed in vibro-compaction of cement and sediments and for pile driving i.e. local fluidisation and displacement of soil particles. Sonic drills presently available are generally of small diameter (4 cm), lightweight and easily portable but are only suitable for fine-grained sediments being stopped by cobbles, dry sands and lodgement tills. These systems operate best in wet sediments to about 25 m depth (Anon, 1981; Thompson and Wadge, 1981).

Resonant drill systems (e.g. Hawker Siddeley Superdrill; Fig. 11.3c) operate by combining the actions of high frequency vibration with conventional rotary motion. Vibrations are generated by a mechanical oscillator with the frequency tunable to that of the natural frequency of the drill pipe. This action is reinforced by a pulldown force at the winch head and produces a system unmatched for the penetration and coring of sediments and soft rocks. Penetration rates of 30 cm/sec for sands and gravels and 7 cm/sec for lodgement tills are common (Dance, 1981). The system requires no drilling fluids and yields high quality core allowing detailed analysis of sediments and soft rocks for geotechnical and drift exploration study (e.g. Sect. 2.8.1).

Whereas site investigations carried out on land and in shallow-water offshore sites conventionally employ shell and auger rigs to investigate soils and to some extent to prove the upper horizon of the bedrock, for deeper water offshore investigations these techniques are strictly limited and the majority of offshore investigations are now undertaken using wireline rotary core drilling and sampling methods. With the wireline system of rotary core drilling, a single outer drill string combines the functions of casing and drill rod and contains an outer core barrel and drilling bit, the whole assembly being run down to the base of the borehole on a wireline. An inner core barrel assembly which can be of the rotating or non-rotating type passes down through and latches into the outer barrel. The system can be used for boring in sediments or rock (See Bjerrum, 1973 and Focht and Kraft, 1977 for review).

11.3 Trial Pits and Trenches

Pits and trenches are normally the cheapest method of exploration to shallow
depth and can be excavated by hand or mechanically dug by, for example, a tractor
mounted hydraulic backhoe. They allow an examination and assessment of the
ground and although this method of exploration is normally limited to a depth of
about 4m, it can be used to advantage when used in conjunction with boreholes to
ascertain ground conditions at depth. It is possible to obtain a picture of
complex stratification and the presence of lenses of weaker material, and this
can be important where the site has been overlain by fill materials, when voids
or loosely deposited layers of deleterious materials can be easily identified and
reliably measured, or when tracing slip planes in unstable ground (Fig. 10.5).
With an experienced observer, the dip of the strata, degree of weathering and
frequency of joints and fissures can be determined, as can the soil/rock
strength, ground-water seepages and inflows, and stability of the sides as it
will effect the excavation of service trenches, basements and foundation bases.
The depth of the agricultural topsoil can be recorded (often missed in
boreholes), as can the immediate subsoil which is important in relation to the
support of road pavements and floating floor slabs for houses and warehouses,
etc. Where required, in situ tests such as soil density and California Bearing
Ratio (Fig. 13.2) can be carried out and vane and hand penetrometer tests made to
determine shear strength. Trial pits are particularly useful for obtaining
samples such as block samples encompassing slip planes, or gouge material scraped
from such horizons. However, it is emphasized that wherever entry to pits is
required, side support must be provided where the depth of excavation exceeds
about 1.2m to ensure the safety of personnel.

11.4 Sampling Equipment

In parallel with the evolution of drilling rigs, sampling equipment has evolved
which can be used in a variety of ground conditions. In the UK it is common
non-marine practice to obtain undisturbed samples of cohesive glacial material
from boreholes with the U4 open-drive sampler (or U100 in metric units;
Fig. 11.4). In the past, it has been normal practice to take samples at about
1-1/2m intervals, however BS15930:1981 recommends a closer interval of 1m and in
view of the likely rapid variation in type and consistency of glacial soils, the
closer spacing should be adopted. The sampler consists essentially of a 100mm
diameter tube open at one end and fitted at the closed end with means for
attachment to the drill rods. Sample and rods are lowered to the base of the
borehole and driven into the soil using a jar-hammer fitted just above the
sampler or preferably pushed into the soil by jacking the rods against a reaction
at ground level. A non-return valve allows the escape of air or water as the
sample enters the tube and assists in retaining it when the tube is withdrawn
from the ground (Fig. 11.4). After withdrawal, the adapter containing the
non-return valve together with the cutting shoe is removed from the sample tube
and the ends of the sample are trimmed to a depth of about 25mm, removing any
obviously disturbed material, and then sealed with micro-crystalline wax and
capped. The sample is then labelled clearly showing the site location, borehole
number and its depth below ground level and is transported with as little
vibration as possible to the laboratory for examination and testing. Providing
proper care is taken with these procedures, the sample will remain at its in situ
moisture content and will not deteriorate for several weeks.

The fundamental requirement of any sampling tool is that on being forced into the
soil it should cause as little remoulding or disturbance as possible. The degree
of disturbance is controlled basically by the design of the cutting shoe which
controls the area ratio of the sample (i.e. the volume of soil displaced by the
sampler in proportion to the volume of the sample recovered) and to some extent

the inside wall friction, and the design of the non-return valve.

Ideally, a sampler would have an area ratio of zero but in practice it must be thick and strong enough to meet the requirements of average soil conditions. In the UK, the standard U(100) sampler has an area ratio of about 30 per cent and is strong enough to be used in clays having a range of consistency from very soft to hard, and in weak rocks such as chalk and marine clays. It can cope with quite stony soils providing damage to the cutting shoe is acceptable. In very soft clays, either two or three barrels may be screwed together to increase the chances of recovery and special core-catcher shoes can be fitted to retain loose material. It is normal practice to couple two tubes together when sampling stiff glacial soils to ensure that the material retained in the lower tube is beyond the range of disturbance affected by boring.

To remove samples from the sampling tubes, thin walled steel or plastic liners re sometimes used within the main barrel. These are relatively cheap and can be cut away to expose samples that may be sensitive to the normal methods of extraction. However, as the linings are themselves rather flexible, care must be exercised to avoid sample disturbance when handling. Also, the area ratio of the sampler is increased - to around 40 per cent - which tends to increase sampling disturbance which may offset any advantage gained from their use.

In North America and increasingly in Britain, use is made of Shelby tubes. These are constructed of high tensile steel with diameters ranging from 50mm to 250mm and lengths from 300mm to 1400mm. The tubes have lower area ratios than U(100) samplers (from about 8 to about 20 per cent) and because of their higher strength and lower resistance lend themselves to jacking or thrusting-in rather than being driven in by percussion methods, thereby providing a higher class sample.

The lighter sizes are generally limited to clays containing little or no stone content and up to around stiff consistency. For stronger clays, the heavier gauge tubes are used, the 70mm diameter tubes being a common size. This has an area ratio of about 19 per cent and is designed to pass down a diamond drill hole of 76mm diameter. Shelby tubes are used in Britain almost exclusively in marine investigations where they are driven by percussion methods using wire-line operated down-the-hole hammers. Thrust driving methods have been recently introduced as for example on the Fugro Wipsampler wherein the sample tube is attached to a drill string which can be latched into the borehole casing to allow the sampler to be thrust ahead into the soil.

The American Pitcher Sampler allows samples of stiff sandy and stony tills to be obtained by rotary drilling methods. This sampler has an outer rotating barrel fitted with a cutting bit to allow it to drill into the soil, and an inner thin-walled stationary barrel similar to a Shelby tube. Water is used to assist penetration and the sample, about 1m long by 74mm diameter remains in the inner barrel until required for testing.

The recovery of samples of soft clays and silts typically encountered within normally consolidated glacial lake deposits has always been difficult and many samplers incorporating fixed or sliding pistons have been introduced to avoid the sample being sucked out of the sample tube as it is withdrawn from the ground. Such samplers vary in size from about 150mm to 260mm diameter and up to 1m long and a typical fixed piston type is illustrated in Fig. 11.4. Similar difficulties exist in the recovery of clean sand below the water table and the modified Bishop sampler introduces air at the base of the sample to overcome these problems (Fig. 11.4). Both this type and the piston samplers are generally limited to sampling soil with little or no stone content. For very hard clays, clays with considerable stone content and cemented granular deposits, it is not normally possible to recover more than very short lengths of sample using conventional samplers and the material that is recovered is often in a shattered condition and useless for laboratory testing. Even so, it is always better to get some sample rather than none at all as coupled with results of in situ testing it is likely that a reasonably good engineering assessment of the ground conditions can be made. Sampling in these harder soils can be carried out using rotary coring methods with various specialist core barrels such as the triple tube barrel with which good recovery can be obtained (often 100 per cent) of a quality suitable for testing.

A continuous disturbed sample of soft clay, silt or fine sand can be obtained using equipment developed by the Delft Soil Mechanics Laboratory in Holland for the weak alluvial and marine soils generally met in that country. A continuous length of up to about 18m of material can be sampled by pushing either a 29mm or a 66mm sampler into the ground using respectively either the 2Mg or the 20Mg Dutch deep sounding machine. The sample is automatically fed into a nylon stockinette sleeve made impervious by a vulcanising process. On withdrawal from the ground the sample can be split and the complete ground sequence examined and logged and as a preliminary investigation can thus serve a useful purpose in determining the best location for conventional undisturbed sampling from adjacent boreholes.

Sampling in marine investigations can be undertaken using the wireline boring techniques with a suitable inner core barrel assembly. Alternatively, drive samples (e.g. thin walled Shelby tubes) can be lowered down the outer barrel of the wireline equipment and either driven or jacked into the soil. Various methods are available for obtaining samples from the sea floor such as the Shipek sediment sampler which basically comprises two concentric half cylinders, the inner of which rotates as the sampler reaches the sea bed and engulfs the scooped material, and various gravity samplers such as the Moore free-fall corer, Kullenberg sampler and Cambridge sampler, which comprise weighted sample tubes with hardened metal core cutters at their bases to facilitate penetration, that are dropped onto the sea bed.

Other samplers are available for obtaining a limited length of core from below the sea bed. These include the Stingray developed by McClelland Engineers, and the Vibrocorer and the Superdrill (Sect. 11.2).

11.5 Sample Quality

The action of taking a sample of soil from the ground causes some disturbance of the material however carefully the sampling procedure is carried out. Disturbance is caused by a number of factors including the release of in situ stress which allows the soil to expand and swell, the development of shear strain as a result of stress anisotropy and the action of pushing the sample tube into the soil which tends to compress material in its path. In addition to the physical problems associated with recovery of the soil from its environment, the quality of a sample is also affected by the soil fabric which greatly influences the engineering properties of soils (Rowe, 1972). In this respect, the majority of glacial soils are fissured and this feature has been shown to be the dominant factor rather than the presence of stones in the selection of a representative sample size. McKinley et al. (1975) suggest that U(100) samples are a reasonable engineering compromise for the size of samples necessary in standard site investigations and provides consistent data for determining shear strengths of tills having light to moderate loadings (Sect. 12.2).

Although field sampling techniques in glacial deposits are generally determined by the presence of strong stony soils, in some situations it may be possible to

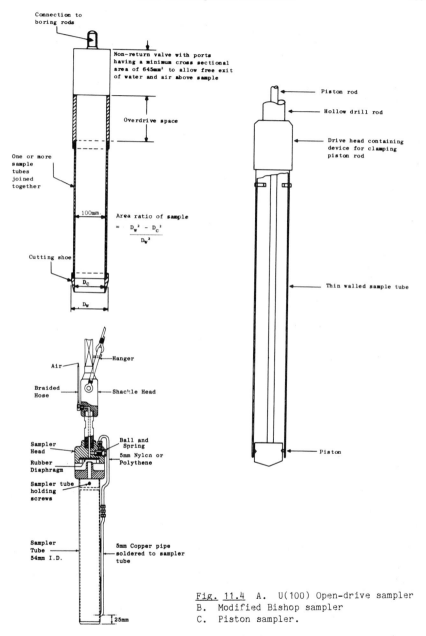

Connection to
boring rods

Non-return valve with ports
having a minimum cross sectional
area of 645mm² to allow free exit
of water and air above sample

Overdrive space

One or more
sample
tubes
joined
together

100mm

Area ratio of sample
= $\dfrac{D_w{}^2 - D_c{}^2}{D_w{}^2}$

Cutting shoe

D_C

D_W

Piston rod

Hollow drill rod

Drive head containing
device for clamping
piston rod

Thin walled sample tube

Piston

Air

Hanger

Braided
Hose

Shackle Head

Sampler
Head

Ball and
Spring

5mm Nylon or
Polythene

Rubber
Diaphragm

Sampler tube
holding
screws

Sampler
Tube
54mm I.D.

5mm Copper pipe
soldered to sampler
tube

25mm

<u>Fig. 11.4</u> A. U(100) Open-drive sampler
B.   Modified Bishop sampler
C. Piston sampler.

select the sampling procedure in accordance with the quality required to determine particular geotechnical data. British Standard 5930 (1981) provides a useful classification of sample quality (Table 11.1).

Table 11:1 A classification of sample quality (after Idel et al, 1969)

Quality*   Properties that can be reliably determined

Class 1   Classification, moisture content, density,
             strength deformation and consolidation
             characteristics

Class 2   Classification, moisture content, density

Class 3   Classification and moisture content

Class 4   Classification

Class 5   None; sequence of strata only

*Class 1 is considered as an undisturbed sample. In certain
 circumstances, whatever sampling method is used, samples
 may only be of Class 2 quality if sample size is too small
 to be representative of the soil fabric governed by
 discontinuities.
 Classes 3, 4 and 5 generally regarded as disturbed samples.

In non-brittle glacial cohesive soils containing little or no stone content, samples obtained with the standard U(100) sampler should be of Class 1 quality provided good practice is followed by the drilling crew. However, as discussed above, in hard brittle cohesive soils or fissured or stony clays, it is likely that samples of only Class 2 will be recovered regardless of the care taken in the sampling procedure. A somewhat higher degree of quality is obtained using Shelby tubes although similar limitations apply where the soil is closely fissured or stony. Samples of high quality can be obtained using the Rowe 250mm diameter sampler which is particularly useful for sampling glaciolacustrine deposits. The sampler comprises a galvanized steel piston tube that is jacked into the ground against kentledge reaction. Special arrangements for handling the samples are taken to avoid air being drawn into the soil by transporting them in sealed water jackets and extracting them in the laboratory under water. The cost of such sampling is very expensive but the value of the information gained can be proportionally higher than with conventional sampling techniques.

Thrust driven thin-walled piston samplers can recover Class 1 samples in cohesive soils of up to stiff consistency whilst Class 1 to 2 samples are obtainable with the larger diameter Delft sampler. However, with the small diameter Delft sampler samples generally no better than Class 2 to 3 are usually obtainable, whilst with soil samples recovered by rotary drilling methods Class 2 are obtainable in cohesive soils but only about Class 4 in non-cohesive material. Material recovered in the split tube of the SPT sampler would not be expected to have a sample quality greater than Class 3.

11.6 Field Testing Techniques

11.6.1 Standard Penetration Test.

Along with the use of the standard open-drive U(100) sampler to obtain relatively undisturbed samples of cohesive glacial soils, the Standard Penetration Test is

CROSS SECTION OF SAMPLER

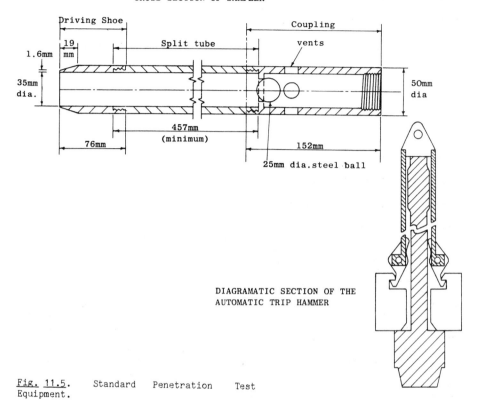

DIAGRAMATIC SECTION OF THE
AUTOMATIC TRIP HAMMER

Fig. 11.5.   Standard   Penetration   Test
Equipment.

normally used to provide an assessment of _in situ_ strength of granular materials.
An increasing use of the test is also being made to provide an assessment of
strength and compressibility of cohesive soils.

This test was initially introduced by the Raymond Concrete Pile Company in the
U.S.A. as a prepiling test to give an indication of the _in situ_ density of sand
and the basic procedure published by Terzaghi and Peck (1948) was adopted as a
British Standard test (BS 1377 : 1975 Test 19).

The equipment which has the merit of being robust and simple to operate comprises
a 50mm diameter split spoon sampler of standard dimensions (Fig. 11.5). This is
driven into the soil at the bottom of the borehole using an automatic trip hammer
which allows a 65Kgf weight to drop through a height of 760mm. The number of
blows is counted to give a penetration of 300mm after the tool has been driven
through an initial seating depth of 150mm assumed to have been disturbed by the
boring tools, this number being referred to as the N value. Reduction
coefficients to allow for boring disturbances in granular soils are shown in
Fig. 12.18. During boring and carrying out the test, it is important to maintain
a water level in the borehole higher than the ground-water level to eliminate or
reduce to a minimum the unbalanced water head and the consequent tendency for the

material at the base of the hole to ´boil´ and loosen.

On withdrawal of the sampler, the split tube can be opened to expose the sample which although disturbed, allows description of the material actually penetrated by the tool; it can also be tested to determine its grading and can often provide information on properties such as permeability from an inspection of the soil fabric. In gravelly soils and weak rocks, the test is similarly carried out except that the driving shoe is replaced by a 50mm diameter 60 degree apex solid cone, the test then being referred to as a ´standard penetration test (cone)´.

Terzaghi and Peck (1948) correlated allowable bearing pressures for different width footings founded in sand with N values. It was apparent that the SPT carried out at very shallow depths gave misleading values of relative density. Sutherland (1963) drew attention to the work of Gibbs and Holtz (1957) who had concluded that penetration resistance increases with increase in either relative density or overburden pressure. Hence, since the principal object of the standard penetration test in non-cohesive soils is to evaluate relative density, the effect of overburden pressure at the depth of the test must be taken into account. Subsequently, Thorburn (1963) presented the graphical relationship (Fig. 11.6) between the value of N, overburden pressure and the relative density for non-cohesive soils based on field tests and building performance within the district of Glasgow, Scotland. It is now standard practice to modify the field value of N in accordance with Thorburn´s curves and to use the corrected value in Terzaghi and Peck´s relationship to assess the design value of allowable bearing pressure (Sect. 12.1.6).

Because of the difficulty of obtaining undisturbed samples of hard and stony cohesive soils, considerable attention has been given to establishing a relationship between $\underline{in\ situ}$ shear strength and the SPT. Stroud (1974) when he compared standard penetration test data obtained from a large number of boreholes sunk in various soil types including London Clay, Boulder Clay, Keuper Marl, Oxford; Kimmeridge, Lias and Frankfurt clays, Flinz and Devonian Marl, with standard undrained triaxial compression tests made on 100mm specimens. The general conclusions were that a simple correlation existed in many insensitive clays and weak rocks between N values and $\underline{in\ situ}$ shear strength of the form $C = f_2 N$: f_1 varies from about $4kN/m^2$ in materials of high plasticity to about $6kN/m^2$ in material of medium to low plasticity. Values between 2.5 and 3.5 kN/m^2 are identified for matrix rich lodgement tills by Weltman and Healy (1978). The results of Stroud (1974) also indicated a correlation existed for the coefficient of volume compressibility of the form $m_v = \frac{1}{f^2 N^2}$ with f^2 increasing from about $400kN/m^2$ for highly plastic material to over $600kN/m^2$ for materials with plasticity indices of less than 20 per cent. Evidence was given to suggest that

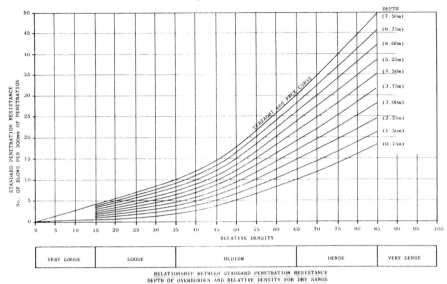

Fig. 11.6. Thorburn Relationship.

overconsolidated materials. Evidence was also given indicating a relationship
between E_v/N and q_n/q_f where q_f is the ultimate bearing capacity estimated for
local shear failure beneath the foundation and q is the net bearing pressure
(Chapter 12).

It is interesting to note that the relationship between N values and unconfined
compression strength published by Terzaghi and Peck give shear strengths almost
identical to those obtained by Stroud using his value of f_1 = $6kN/m^2$.   It is
worthwhile bearing this in mind when dealing with soils and weak rocks in which
it is neither possible to obtain representative samples nor good core recovery
when using rotary drilling methods. Thus in such cases, the adoption of _in situ_
standard penetration tests with U(100) samples should be carefully considered.

11.6.2 Cone Penetration Test.

The cone penetration test (CPT) is variously known in the past as the Static
(cone) Penetration Test, Quasi Static Penetration Test and Dutch (Deep) Sounding
Test.

The test (Fig. 11.7), is used in the design of pile foundations placed in
fine-grained granular deposits and consists of a steel cone of standard
dimensions fitted to a string of rods which is pushed into the ground whilst
measuring the end resistance to penetration of the cone and in addition, the
resistance due to local friction between the soil and the penetrometer. In many
areas soil conditions are not amenable to this _in situ_ testing technique as it is
not usually possible to penetrate far into strong soils or where cobbles and
boulders exist.   However, it has been used for some time where suitable
conditions have existed, initially with hand operated (2Mg) machines and later
with 10Mg and 20Mg engine driven units which have allowed increasingly greater
penetration of the stronger soils. Because of its limited use the equipment has

not gained much favour and is held by only a handful of Site Investigation Companies.

In general, the mechanical penetrometer as described by Vermeiden (1948) has been used in the test procedure following standard Dutch practice. Also, its use has been confined to land sites where the necessary reaction could be obtained using screw anchors or kentledge, or in marine situations where the sounding rig could be set up on a firm platform such as on staging or on a bottom-out barge.

More recently, with the introduction of the electric cone (Zuidberg and Windle, 1979; and the development of wire-line drilling equipment, an increasing use of the cone has been made in marine investigations and particularly so for investigating the special problem of piling in the North Sea for oil and gas platforms where considerable thicknesses of glacial soils exist (Fig. 11.7A, 2.23).

Basically, the electric cone penetrometer operates through the drill string latching into an outer barrel assembly and comprises an hydraulic cylinder and piston which pushes the cone into the soil ahead of the borehole using the weight of the drill string as reaction for the piston thrust. On completion of each test, the hole is drilled deeper and the test is repeated. The penetrometer is connected to the surface by an umbilical cable containing the hydraulic hose and the electric conductors, the data being automatically recorded on a pen recorder and in digital form on a data logger.

The main difficulties with this type of test is in handling the heavy equipment used and in being able to compensate for the wave action of the sea. Normally, only a few tests would be possible in a working day but with specialist drill ships and sophisticated heave compensators, quite a high rate of testing can be achieved. With the early mechanical cones the end resistance and local friction values were measured separately by first advancing the cone together with a following sleeve and then pushing the cone ahead of the sleeve. In the electrical penetrometer, the cone resistance and local friction are recorded simultaneously by means of load cells installed in the penetrometer. It has been found possible to infer the soil type from the ratio of cone resistance to local friction which is of considerable help in interpreting the results where it is not possible to retrieve samples of the soil (Sanglerat 1972; Fig. 11.7A).

As discussed by Weltman and Healy (1978) difficulties arise in assessing the in situ cohesion of glacial soils from the pressure on the cone at failure. C , mainly due to the small size of the cone not reflecting the total effect of the soil fabric, e.g. fissures, and because the cone resistance is known to vary both with the rate of penetration and the soil type.

11.6.3 Plate Bearing Tests.

Plate bearing tests have the merit of giving direct and positive measurements of the strength and deformation properties of the ground providing the formation on which they are carried out has not been disturbed by the excavation of boring process. However, they are very expensive compared to most other forms of in situ testing and it must be remembered that they stress only a short depth of material immediately below the plate so that if the soil is heterogeneous as is usual with most glacial deposits, they may be of limited use. The value, however, of undertaking such tests on large diameter plates has been shown by Marsland (1977) who compared the results of laboratory tests with loading tests on 865mm diameter plates; laboratory determined undrained shear strengths may be up to twice the operational value obtained from plate tests whereas the values of modulus of elasticity obtained from the plate tests may be up to 6.5 times the

1. Conical point ($10cm^2$)
2. Load cell
3. Strain gauges
4. Friction sleeve
5. Adjustment ring
6. Waterproof bushing
7. Cable
8. Connection with rods

CROSS-SECTION OF FRICTION
CONE PENETROMETER

Fig.11.7   Cone Penetrometer Equipment.

values obtained from laboratory triaxial tests.

11.6.4 Pressuremeter Tests.

These are used to determine strength and deformation characteristics of soils and
rocks and basically consist of a probe which is inserted into a borehole to the

required depth and expanded laterally against the sides of the hole. The required information is obtained from measurement of the applied pressures and the resulting deformations. The types in use are basically similar to the original type developed by the late French engineer Louis Menard (Baguelin et al 1978). The instrument can only be used in unlined boreholes and so is not suitable where there is any tendency for the sides of the hole to slump, although

lf-boring type - the Camkometer - developed at Cambridge University has overcome this problem. As yet, however, its usefulness in stiff glacial till has still to be determined. The test is useful for determination of soil deformation especially at depth where it would be costly to undertake in situ plate bearing tests, although results obtained have not always been appropriate to the behaviour of piles (Weltman and Healy, 1978). McKinley and Anderson (1975) have summarised the use of the test and have commented on the influence of size and soil fabric in assessing the results; they stress the need for more information and cross reference to other types of testing and full scale loading tests.

11.6.5 Groundwater Measurement And Field Permeability Tests.

The measurement of standing groundwater level, or water-table, and the permeability of the ground is a fundamental requirement in the analysis of most engineering problems and its knowledge is vital to the contractor for design of any excavation and temporary works and in groundwater abstraction and waste-disposal projects (Sect. 15.2).

During boring in high permeability soils, i.e. gravels and clean sands, any ground-water met will rapidly rise to the true groundwater level (Fig. 11.8). However, as the permeability decreases, the time taken for equilibration of the water-table will be proportionally greater and for clays may be of the order of several months. In the more permeable materials, it is usually sufficiently accurate to measure the groundwater level by plumbing with an electrical dip-meter in the borehole. However, for the less permeable soils and where it is required to record possible fluctuations in level over a long period of time, small diameter tubing (e.g. 19mm ID PVC tubing) is placed in the borehole and surrounded by a graded gravel filter (gravel pack) before withdrawal of the borehole casing. The tube is perforated over its basal 2 metres and the length of filter media provided is termed the response zone or section. Where this zone extends to ground level, the tube is known as a standpipe, but where only a limited response zone is required, an impermeable seal (bentonite or cement grout) is placed above the filter layer and the system is termed a piezometer (Fig. 11.8).

In soils of very low permeability, there may be insufficient groundwater to fill the tube up to the piezometric level and therefore the pore water pressure in the ground is measured instead. This is performed by placing a special piezometer tip of the hydraulic, pneumatic or electrical type, at the base of the borehole which is connected to the surface by single or twin small-bore nylon tubes to allow the pressure to be monitored at ground level.

Because of the heterogeneity of surficial soils in situ permeability is difficult to assess from simple falling or rising head tests carried out in single boreholes. Consideration should be given to measuring bulk permeability by full scale pumping tests with measurement of drawdown in boreholes sunk on radiating lines from the pumping well. The pumping well should be provided with a well screen (slotted or perforated casing) surrounded by a gravel pack of suitable grading whereas the observation boreholes may be conveniently placed site investigation boreholes fitted with standpipes.

Three observation holes on a line radiating out from the well are normally

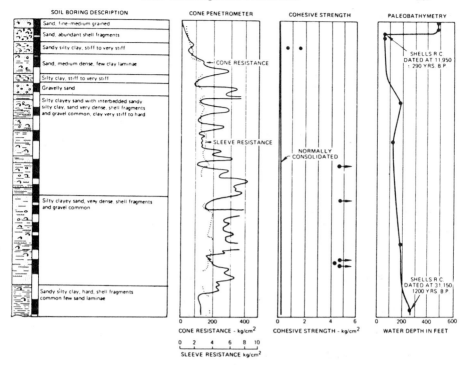

Fig. 11.7A Log from electric cone penetrometer shown alongside other data from site in the North Sea (from Milling, 1975).

sufficient;   more than one set of holes may be necessary if very variable ground conditions exist. The holes should preferably be set at logarithmic distances from the well for convenience in plotting the drawdown curve on semi-log paper. Pumping is generally continued until steady seepage conditions are reached when the permeability can be calculated for the condition shown in Fig. 11.8 from the equation

$$K = \frac{Q \log{r^2}/_{r1}}{\pi(h^2_2 - h^2_1)}$$

or from $K = \dfrac{Q \log \dfrac{r^2}{r^1}}{2\pi m(h_2 - h_1)}$ for artesian flow

condition where m is the aquifer thickness.

11.7 Geophysical Exploration

In general, geophysical methods do not lend themselves readily to exploration of heterogeneous surficial deposits and are therefore not often used during land site investigations. Hammer selsmic equipment is used in areas of thin drift

cover;   gravimetric and resistivity techniques are frequently used for tracing
buried drift-filled valleys.

However, such methods are frequently used for off-shore investigations to obtain
a generalized picture of the substrata and provide a reasonable idea of the
thicknesses of recent sediments and soils overlying bedrock.   The principal
method used is the Continuous Seismic Reflection Profiling method, a technique
which produces along the track of the survey vessel, a continuous printed record
of the seabed and sub-seabed discontinuities. Acoustic pulses of short time
duration, reflected from the seabed and underlying substrata are detected by a
synchronized recording device which displays the seismic trace as a sub-bottom
geological profile showing the two-way travel times to different reflective
horizons (e.g. Lineback, et al, 1971; Holmes, 1977).

The type of equipment used to provide the acoustic energy depends on the survey
objectives and may be airguns (air discharge); sparkers (electrical discharge);
gas guns (chemical combustion); boomers (electromagnetic induction); and
pingers (piezoelectrical or magnetostrictive).   In general, pingers and high
resolution boomers are suitable for resolving near surface layering; standard
boomers and sparkers for coarser and thicker over-burden and air guns for high
resolution shallow water profiling.

11.8 Laboratory Testing

Glacial soils present especial difficulties when it comes to obtaining good
representative undisturbed samples. It is often difficult enough to recover any
sort of sample let alone one that is truly representative.   However, sampling
disturbance is not a significant factor in the accuracy of many laboratory tests,
such as classification tests including moisture content, density, liquid and
plastic limits, sieve and mechanical analyses and specific gravity, and strength
tests including compaction, California Bearing Ratio (CBR; Fig. 13.2) and shear
strength of remoulded material.   For many small developments, involving only
light to moderately loaded foundations, the results of such tests together with a
visual examination of the ground can often provide sufficient information to
allow an experienced geotechnical engineer to design the foundations.

For prestigious and important structures, it is necessary to undertake
appropriate laboratory testing.   The high cost in obtaining large diameter
representative samples of the soil or undertaking special in situ testing such as
large diameter plate loading tests, may well be warranted. It must be remembered
that any test result - whether laboratory or in situ - is of limited value in
itself and must be considered in relation to the likely degree of sampling
disturbance and representative nature of the amount of soil tested. A knowledge
of the available geological conditions, information from boreholes, shafts, trial
pits and natural exposures coupled with the known engineering performance of any
structure in similar ground conditions is essential.

Standard routine laboratory tests include classification tests such as plasticity
index tests and particle size grading tests which allow the basic soil type to be
identified. Shear strength of very soft to soft cohesive soils are often
determined by vane tests carried out within the actual sample tube, however,
where specimens can be satisfactorily extruded from the sample tubes, then shear
strength is generally obtained by the triaxial compression test.

In this test either 40mm or preferably 100mm diameter specimens are enclosed
within a cell and subjected to an all-round pressure before being sheared by an
axially applied deviator stress. Basically, three types of test can be carried
out to determine shear strength parameters, the relevant test being dependent on

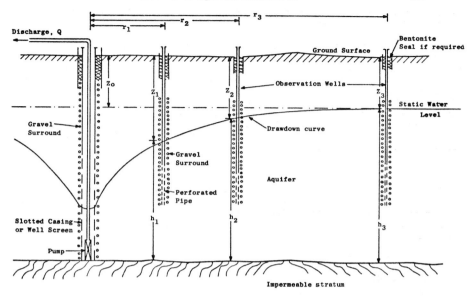

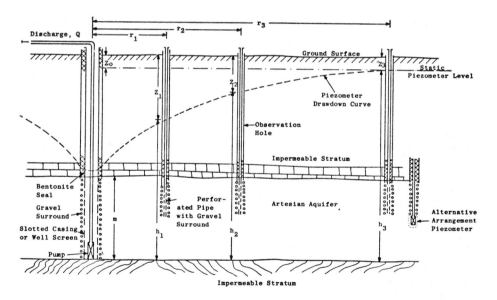

Fig. 11.8. Pumping test arrangement.
A: water table condition
B: artesian condition.
See also Fig. 15.4.

the drainage conditions of the soil under stress in the ground, i.e.

   (i)   Undrained test
  (ii)   Consolidated-undrained test with pore pressure measurement
 (iii)   Drained test

For undrained conditions such as those applying where a cohesive soil is stressed by a foundation (bearing capacity analysis), the immediate condition where soil is removed from within a supported or unsupported excavation, or the short term stability of a clay cutting, test (i) will provide the appropriate shear strength parameters in terms of total stress (c,). However, where drainage of the soil can take place and the loading conditions are such that the pore pressures gradually dissipate and the effective stress (the difference between total stress and the pore pressure) changes, conditions that apply, for example, to the long term stability of slopes and the foundations of slowly constructed structures such as earth dams, then tests (ii) or (iii) should be carried out to provide shear strength parameters in terms of effective stress (c´, 0´). These tests and their application to engineering problems are fully discussed by Bishop and Henkel (1957), Bolton (1979) and in Chapter 8.

Shear strength of the more sandy soils can be determined by the laboratory shear box test. This is an old established test which at one time was ousted by the triaxial test but subsequently regained favour in view of its ability to allow multiple reversals of stress and thus provide shear strength parameters for both peak and residual strain conditions. The earlier 60mm square test equipment was later supplemented by a 300mm square shear box allowing coarser material to be tested. Recently introduced ring shear apparatus is also used for the determination of residual parameters (Fig. 8.9).

Soils undergo a reduction in strength when subjected to cyclical stress variations. This phenomenon is relevant in certain foundation conditions such as when foundations are subjected to vibrations from reciprocating machinery or from earthquake shocks, and for sea bed structures e.g. offshore oil gravity structures, due to wave action (Martin et al, 1980; Rahman et al, 1980). In the tests, specimens are subjected to cyclic stresses over a range of frequencies most likely to include the natural frequency of the foundation/soil mass, before determination of its shear strength.

Apart from soil strength, other major engineering parameters of soils include their consolidation characteristics. These are measured in a consolidation machine or oedometer. In this test, 75mm or 100mm diameter specimens are consolidated under various increments of pressure, the specimen being allowed to expel or suck in water to equilibrate with the ambient conditions. The pressure - voids ratio relationship obtained allows the amount and rate of settlement of a structure to be estimated (Chapter 12).

11.9 Description of Soil and Rock

It is essential for soils and rocks to be described in a standard manner. A full soil description gives detailed information on the mass characteristics including strength, bedding, discontinuities and state of weathering, plus material characteristics including colour, particle shape, composition, grading, plasticity and soil name (Fig. 10.6). The Code of Practice for Site Investigation (B.S.5930: 1981), provides detailed information of the descriptive system currently in use in Britain and typical descriptions given on a borehole record would be:

     Stiff closely fissured grey sandy CLAY of low to medium

plasticity with fine to coarse gravel, cobbles and
occasional boulders (Lodgement Till)

Dense yellow fine to medium SAND with thin lenses of
soft grey silty clay (Recent Alluvium)

It should be noted that there is a difference between the engineering description
(as outlined above) and the classification of soils. Soil classification
provides a concise and systematic method of grouping different types of soil by
reference to their main characteristics and is primarily for soils to be used as
constructional materials. The standard classification system used in the UK is
the British Soil Classification System for Engineering Purposes (BSCS) which is
fully described in British Standard 5930:1981. The North American equivalent is
published by AASHTO (1979).

The current British system is not too dissimilar to that used in North America
because they are both based on the original Casagrande System of Soil
Classification. In the system, soil can be placed in one of a number of soil
groups on the basis of its grading and plasticity. Main group terms of G
(gravel) and S (sand) are given for the coarse components of the soil with
qualifying terms of W (well graded), P (poorly graded), Pa (uniform) and Pg (gap
graded). The fine components are similarly differentiated by, for example, F
(fine soil), M (silt), and C (clay) with qualifying terms indicating the degree
of plasticity.

In general, these classificatiions are similar in nature to sedimentological
classifications and coding schemes (e.g. Fig. 1.8, Table 7.1; Sect. 1.4) but the
latter are in many cases too elaborate for engineering application.

A Working Party of the Geological Society of London proposed various parameters
including colour, grain size, texture and structure, discontinuities, weathering,
rock name, strength and permeability using standardized scales and descriptions
(Anonymous, 1977). British Standard 5930 (1981) also sets out a preferred method
of rock description but is based on an earlier Working Party report. Currently,
the International Society for Rock Mechanics has set up a number of Working
Parties to report on the classification and characterization of rocks. A
universally accepted system of description of both rock and soil may follow.

11.9.1 Reporting.

There is little point in undertaking a site investigation and putting effort into
obtaining good representative samples of the ground and field and laboratory test
data, if the information is not reported accurately and in a logical fashion so
that the design or project engineer can readily extract the information he
requires.

The first step in reporting is made by the rig operator who should produce a
daily borehole record (Fig. 11.9) in which full factual information is recorded
concerning the boring. This includes details of samples and cores taken,
groundwater levels, depth and size of casing used, strata descriptions, details
of in situ testing, any chiselling required to penetrate obstructions or bedrock,
any instrumentation inserted in the completed borehole such as standpipes or
piezometers, and any other observations that may be of geotechnical interest.
The next step is for a qualified geotechnical engineer/engineering geologist to
produce a finished borehole record for report presentation (Fig. 11.10) by
marrying the operators field record with the results of the various field and
laboratory tests and a full description of all the samples taken.

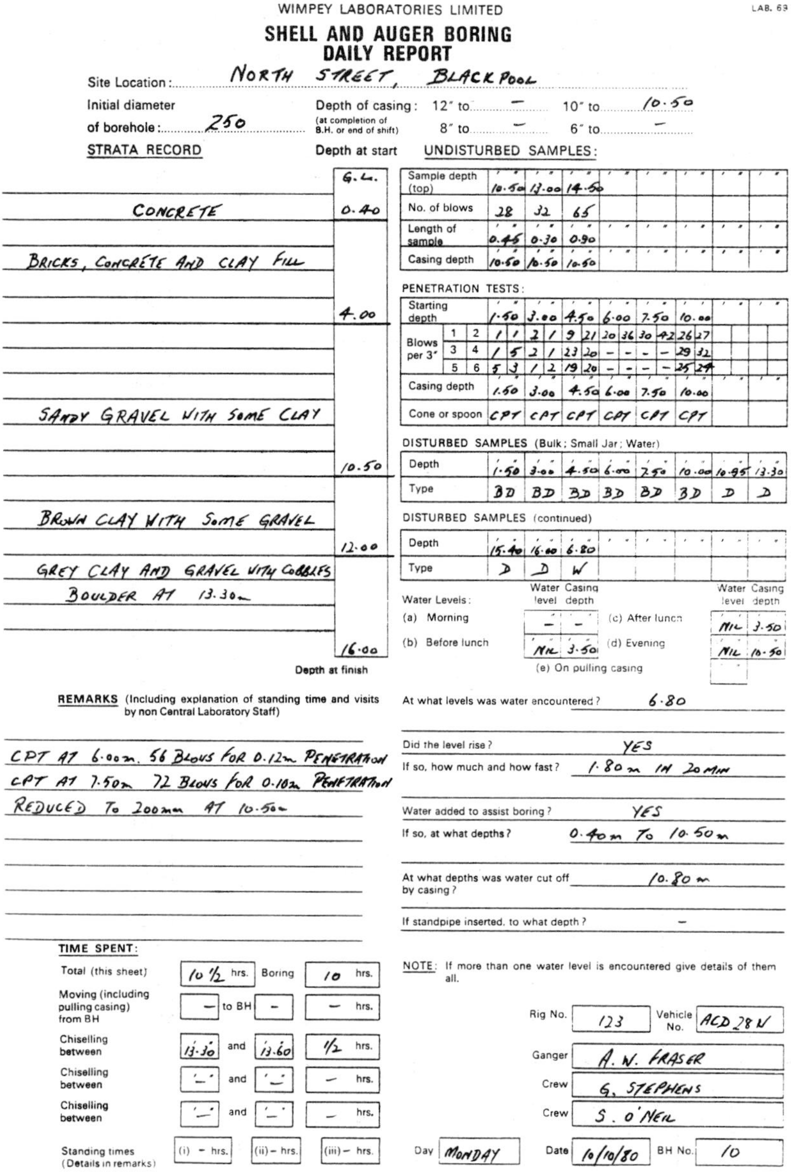

WIMPEY LABORATORIES LIMITED

SHELL AND AUGER BORING
DAILY REPORT

LAB. 63

Site Location: _NORTH STREET, BLACKPOOL_

Initial diameter of borehole: _250_

Depth of casing: 12″ to _—_   10″ to _10.50_
(at completion of B.H. or end of shift)
8″ to _—_   6″ to _—_

STRATA RECORD   Depth at start   UNDISTURBED SAMPLES:

STRATA RECORD	Depth
	G.L.
CONCRETE	0.40
BRICKS, CONCRETE AND CLAY FILL	
	4.00
SANDY GRAVEL WITH SOME CLAY	
	10.50
BROWN CLAY WITH SOME GRAVEL	
	12.00
GREY CLAY AND GRAVEL WITH COBBLES	
BOULDER AT 13.30m	
	16.00

Depth at finish

UNDISTURBED SAMPLES:

Sample depth (top)	10.50	13.00	14.50							
No. of blows	28	32	65							
Length of sample	0.45	0.30	0.90							
Casing depth	10.50	10.50	10.50							

PENETRATION TESTS:

Starting depth		1.50	3.00	4.50	6.00	7.50	10.00			
Blows per 3″	1 2	1 / 1	2 / 1	9 21	20 36	30 42	26 27			
	3 4	1 / 5	2 / 1	23 20	— —	— —	29 31			
	5 6	5 / 3	1 / 2	19 20	— —	— —	25 24			
Casing depth		1.50	3.00	4.50	6.00	7.50	10.00			
Cone or spoon		CPT	CPT	CPT	CPT	CPT	CPT			

DISTURBED SAMPLES (Bulk; Small Jar; Water)

Depth	1.50	3.00	4.50	6.00	7.50	10.00	10.45	13.30
Type	BD	BD	BD	BD	BD	BD	D	D

DISTURBED SAMPLES (continued)

Depth	15.40	16.00	6.80					
Type	D	D	W					

Water Levels:	Water level	Casing depth		Water level	Casing depth
(a) Morning			(c) After lunch	NIL	3.50
(b) Before lunch	NIL	3.50	(d) Evening	NIL	10.50
			(e) On pulling casing		

At what levels was water encountered? _6.80_

REMARKS (Including explanation of standing time and visits by non Central Laboratory Staff)

CPT AT 6.00m. 56 BLOWS FOR 0.12m PENETRATION
CPT AT 7.50m. 72 BLOWS FOR 0.10m PENETRATION
REDUCED TO 200mm AT 10.50m

Did the level rise? _YES_
If so, how much and how fast? _1.80m IN 20 MIN_

Water added to assist boring? _YES_
If so, at what depths? _0.40m TO 10.50m_

At what depths was water cut off by casing? _10.80m_

If standpipe inserted, to what depth? _—_

TIME SPENT:

Total (this sheet)	10 ½ hrs.	Boring 10 hrs.
Moving (including pulling casing) from BH	— to BH —	— hrs.
Chiselling between	13.30 and 13.60	½ hrs.
Chiselling between	— and —	— hrs.
Chiselling between	— and —	— hrs.
Standing times (Details in remarks)	(i) — hrs. (ii) — hrs. (iii) — hrs.	

NOTE: If more than one water level is encountered give details of them all.

Rig No. _123_   Vehicle No. _ACD 28 N_

Ganger _A. W. FRASER_

Crew _G. STEPHENS_

Crew _S. O'NEIL_

Day _MONDAY_   Date _10/10/80_   BH No. _10_

Fig. 11.9. Borehole record (drillers).

Boring method	Shell and Auger						Location SD 310356	Record of BOREHOLE 10	
Boring diameter (mm)	250 to 10.50m; 150 to 20.00m							(sheet of)	
Casing diameter (mm)	250 to 10.50m						Orientation	Ground level (m.O.D.) 24.10	
Boring equipment	Pilcon Wayfarer 20							Date commenced 10.10.60	

Samples and in situ tests		Casing Depth (m)	Water Depth (m)			Date and Depth (m)	DESCRIPTION OF STRATA	O.D. Level (m.O.D.)	Legend
Depth (m)	Type								
			–			10/10 0.40	CONCRETE	23.70	
0.40	BD								
1.50	C14	1.50					FILL (brick fragments, gravel, sandy clay with some organic material)		
1.50	BD								
3.00	C6	3.00							
3.00	BD								
						4.00		20.10	
4.50	C82	4.50							
4.50	BD								
6.00-6.12	C56●	6.00							
6.00	BD								
			6.80*				Very dense sandy fine to coarse GRAVEL with traces of clay and occasional cobbles		
7.50-7.60	C72●	7.50							
7.50	BD								
9.00	BD								
10.00	C110	10.00							
10.00	BD								
10.50	U	10.50	10.30**					13.60	
10.95	D					10.50			
							Stiff fissured brown silty CLAY with scattered gravel		
12.00	D					12.00		12.10	
13.00-13.30	U●	10.50							
13.30	D								
14.50	U	10.50							
15.40	D						Stiff to very stiff fissured grey sandy CLAY of low to medium plasticity with fine to coarse gravel and occasional cobbles and boulders (Lodgement Till)		
16.00	D								
17.00	U	10.50							
17.90	D								
19.00	U	10.50							
19.90	D					20.00		4.10	

END OF BOREHOLE

REMARKS

Ground-water observed at 6.80m below ground level rose to 5.00m in 20min. Water was added to facilitate boring from 0.40m to 10.50m below ground level.
Borehole was advanced by chiselling from 13.30m to 13.90m below ground level (½h).
A standpipe was inserted to 7.20m below ground level.
The borehole was backfilled with natural spoil from 20.00m to 7.20m, pea gravel to 4.00m, bentonite to 3.50m, natural spoil to 0.50m and concreted stop cock box to ground level.

For explanation of symbols and abbreviations see Notes, pages (i) and (ii)   SCALE 1:100

LAB Ref. No. SI 28432	NORTH STREET, BLACKPOOL	Fig. 10

Fig. 11.10. Finished borehole record for report presentation.

The borehole records (and similarly records of any trial pitting or trenching carried out) with a site plan clearly showing their locations, together with the tabulated results of field and laboratory tests and any field observation such as variation in piezometric levels taken during site work, or in situ permeability tests, form the basic factual part of a report. A report may be produced as evidence in a public court or arbitration hearing at some future date and it is absolutely essential that all factual information presented is completely accurate.

Such reporting should be clear, precise and free from ambiguities with discussion given in simple terms reaching sound and logical conclusions from the facts and observations presented in the report. Moreover, the report should be a start to finish account of the whole job containing all the technical facts, good and bad, and where appropriate analyses and recommendations given in a simple logical manner so that the client can see his problem(s) clearly and simply stated in a way he can pass the information on to a specialist or contractor without unnecessary further work.

In general, a site investigation report should be prefaced by standard explanatory notes to avoid repetition throughout the text, test results sheets and borehole records, etc. Details are normally given of symbols and letters used to refer to basic information concerning laboratory and field tests and items such as rock core sizes. The report text should open with an introductory section giving details of the client, brief title and location of the project, reason for the investigation and dates during which the work was carried out. Following sections should deal with site topography and geology, field work including boring, trenching (e.g. Fig. 10.5d) sampling, in situ tests and details of any special instrumentation, etc. following by results of laboratory testing, and the site plan.

This normally completes the factual reporting of the investigation. Where comprehensive reporting is required, sections will follow dealing with, for example, a discussion on the soil, rock and groundwater conditions, foundation design, earthworks, pavement design, slope stability, any special problems likely with temporary works, and chemical aggression of buried concrete and pipework. Detailed information on field and laboratory tests, field observation such as piezometer readings or borehole location control, should be tabulated separately followed by figures showing the finalized borehole and trial pit records, laboratory test results such as Mohrs circles, consolidation tests, grading curves and triaxial test effective stress information, geological sections, and graphical display of field tests such as field permeability draw-down curves.

For the larger factual or comprehensive reports a frequent requirement is to provide a complete colour photographic record of any rock core recovered. Photographs are particularly useful for recording rock outcrops, slope failures, structural cracking, and other conditions that are difficult to convey by the written word.

Foundation Engineering in Glaciated Terrain

W. F. Anderson

INTRODUCTION

The foundation of a structure is that part of the substructure which is in direct contact with the ground and is transmitting the structural loads to the ground.

A well designed foundation will be economical and should be designed so that there are adequate factors of safety against either structural failure within the foundation or shear failure in the soil in the vicinity of the foundation. Also, expected movements of the foundation both during and after construction should be computed to check that they are within acceptable limits and will not cause damage to the structure. Because different types of structure will have different responses to ground movements, the foundation should not be considered in isolation. The foundation and the structure form an integrated unit and act together under the influence of the applied loads.

By making assumptions about applied loads and soil properties, and applying the appropriate soil mechanics theory, it is possible to compute the factor of safety against shear failure in the soil and to compute the expected amount and rate of settlement. It should always be borne in mind that assumptions are being made, and the predictions resulting from these calculations should be critically examined considering the actual ground conditions. This is particularly important when constructing foundations on glaciated terrain where considerable variation in soil type and properties usually occurs both laterally and vertically on any site.

Basic Definitions. Total overburden pressure, p. This is the intensity of total pressure, i.e. soil + water, on any horizontal plane before foundation construction begins.

Effective overburden pressure, P_o. This is the intensity of intergranular pressure on any horizontal plane before foundation construction begins. It is equal to the total overburden pressure minus the pore water pressure, u. Usually the pore water pressure will be equal to the depth of the horizontal plane below groundwater level multiplied by the unit weight of water, γ_w.

$$p_o = p - u = p - \gamma_w \, h \qquad (12.1)$$

Total bearing pressure (gross loading intensity or gross bearing pressure), q_n. This is the total vertical normal stress on the ground at the base of the foundation after the foundation has been fully loaded. It incluudes the gross load of the foundation, structure and backfilled soil, and soil water.

Net bearing pressure (net loading intensity or net foundation pressure), q_n.

This is the increase in pressure at foundation level after the construction of the foundation and full structural loading. It is equal to the total bearing pressure minus the total overburden pressure at the base of the foundation.

$$q_n = q - p \qquad (12.2)$$

Ultimate bearing capacity, q_f. This is the value of bearing pressure at which the ground beneath the foundation fails in shear. Generally it is defined in terms of gross loading intensity, but the British Standard Code of Practice for Foundations (B.S.1, 1972) defines it in terms of net loading intensity. It is less ambiguous to have a further definition for net ultimate bearing capacity.

Net ultimate bearing capacity, q_{nf}. This is the value of the net bearing pressure at which the ground beneath the foundation fails in shear.

$$\text{i.e } q_{nf} = q_f - p \qquad (12.3)$$

Allowable bearing pressure, q_a. This is the maximum allowable net loading intensity at the base of the foundation, taking into account the ultimate bearing capacity, the expected settlement and the ability of the structure to accommodate this settlement.

Presumed bearing value. This is net bearing pressure considered appropriate to the particular type of ground for preliminary design purposes. Values for different rock and soil types are available (B.S.1 1972).

Factor of safety (Load factor),F. This is the ratio of the net ultimate bearing capacity to the allowable bearing pressure.

$$\text{i.e } F = \frac{q_{nf}}{q_a} \qquad (12.4)$$

12.1 Types of Foundation

Foundations can be divided into shallow and deep foundations. Foundations may generally be considered as shallow when the depth below finished ground level is less than three metres. However if the depth to breadth ratio is high, then foundations nearer than three metres to the surface may have to be designed as deep foundations.

A number of different types of shallow foundation may be constructed, and the choice of which type to use normally depends on the magnitude and distribution of the structural loads and on the bearing capacity of the ground. Pad foundations are used for supporting structural columns, and non-load bearing walls between columns may be supported on ground beams spanning between the columns. For minor structures pad foundations may be constructed of unreinforced concrete, but for heavier loads reinforced concrete pads should be used. Strip foundations are normally used as foundations for load bearing walls or for rows of columns which are so closely spaced that construction of separate pad foundations would be uneconomical. If the bearing capacity of the soil is low and the width of strip footing is large in relation to the width of wall, then it may be necessary to put transverse reinforcement in the strip to prevent the foundation cracking. Raft or mat foundations are used where the bearing capacity of the soil is low or where columns or load bearing walls are so close that pad or strip footings cease to be economical. Another situation where a raft foundation may be used is where excessive differential settlement is anticipated if pads or strip footings are used.

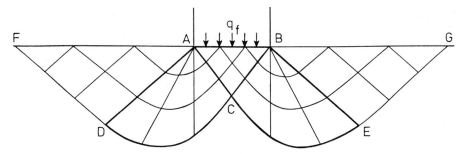

Fig. 12.1. Typical rupture surfaces beneath a shallow strip footing at failure.

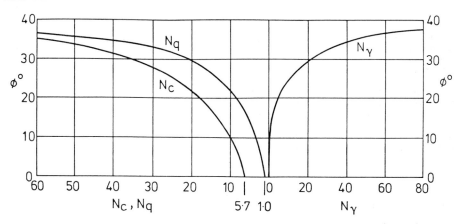

Fig. 12.2. Terzaghi bearing capacity coefficients.

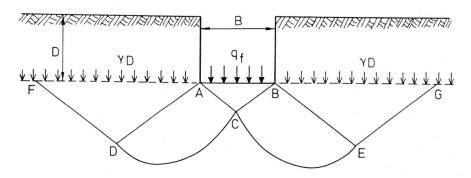

Fig. 12.3. Terzaghi's assumptions for bearing failure of a foundation at depth D.

Deep foundations include basements, piles and diaphragm walls. Basements may be constructed in an open excavation if there is sufficient room. If the excavation goes beneath groundwater level, then it is necessary either to lower the water table so that construction can progress unhindered or to construct a cofferdam which prevents excess water getting into the excavation. One advantage of using a basement, or hollow box, construction is that the net loading intensity at foundation level may be quite small as the structural load intensity due to the basement is likely to be considerably less than the overburden pressure. Basements which perform in this manner are sometimes called buoyancy rafts.

Pile foundations may be used where the soil immediately beneath the structure has insufficient strength to support shallow foundations, or where compressibility of the soil would lead to excessive settlement if shallow foundations were used. Piles may also be used where load variations or soil variation may result in excessive differential settlement of the structure. Piles are also used to resist uplift forces. Piles not only support the structure by end bearing, but also by shaft friction or adhesion. If a pile gets most of its support by resting on a hard layer such as rock or very dense sand or gravel, it is known as an end bearing pile. If, however, it derives a considerable proportion of its carrying capacity from its shaft friction, it is known as a friction pile.

Although there are numerous types of pile and methods of installation, they may be subdivided into two groups, displacement and non-displacement piles. Piles are referred to as displacement piles when due to the method of installation, usually driving or jacking, the ground is displaced. Non displacement piles are constructed by boring a hole in the ground prior to the placement of the pile. The method of installation affects the ultimate bearing capacity of the pile.

Diaphragm walls are occasionally used as extensions to load bearing walls or to closely spaced columns on soil which has insufficient bearing capacity for shallow foundations to be used. These walls are formed in deep trenches excavated by a special grab and supported by bentonite slurry. Reinforcement is lowered into the trench and it is then filled with concrete using a tremie pipe to displace the bentonite slurry from the bottom upwards.

12.1.1 Determination Of Ultimate Bearing Capacity Of A Long Shallow Strip Footing.

A number of approaches have been used to try to predict the ultimate bearing capacity of a foundation. The earliest theories were based on limiting states of equilibrium in the ground, i.e. active and passive failure zones, but because of the number of simplifying assumptions, these early theories were found to be inadequate. The theories which are now in general use are based on plastic equilibrium in the soil beneath the foundation. Because it is much simpler to work in two dimensions, most analyses have considered the case of a long strip footing. A typical pattern of rupture surfaces in the soil at failure is as shown in Figure 12.1. Depending on the roughness of the foundation base, the wedge of soil ABC under the foundation may or may not be in a state of plastic equilibrium. As this wedge of soil punches into the ground, zones of radial shear (ACD and BCE) will be created and the horizontal displacement due to these will give rise to zones of passive failure (ADF and BEG).

All bearing capacity theories based on the above mechanism give an equation of the form

$$q_f = c.N_c + p_o (N_q - 1) + 0.5\gamma.B.N_\gamma + p \qquad (12.5)$$

where c is the soil cohesion

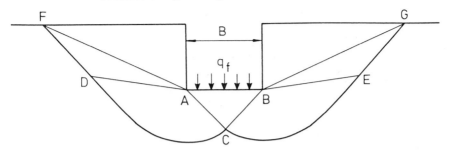

Fig. 12.4. Meyerhof's assumptions for bearing failure of a foundation at depth D.

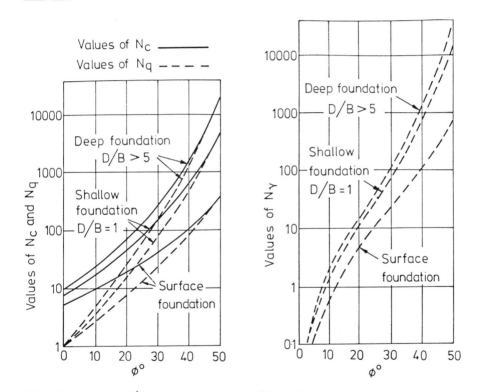

Fig. 12.5. Meyerhof's bearing capacity coefficients.

P_o is the effective overburden pressure

p is the total overburden pressure

γ is the soil unit weight

B is the breadth of the foundation

and N_c , N_q and N_γ are bearing capacity coefficients.

The values of the bearing capacity coefficients depend on ϕ , the angle of shearing resistance of the soil, and on the assumptions made in their derivation. Bearing capacity coefficients may be obtained from Figure 12.2. These values were derived assuming that the foundation had a rough base and was situated on the soil surface. When the foundation was at a depth D beneath the surface, Terzaghi (1943) argued that the soil above the foundation base level could be considered as a surcharge loading as shown in Figure 12.3. This assumption ignores the shearing resistance which may be mobilized in the soil above foundation level. The Terzaghi bearing capacity coefficients may therefore be conservative.

Meyerhof (1951) overcame this problem by assuming failure zones for shallow footings as shown in Figure 12.4. The bearing capacity coefficients which he computed therefore depend on the depth of the foundation base as well as the angle of shearing resistance of the soil. Values of Meyerhof's coefficients are given in Figure 12.5.

12.1.2 Effect Of Foundation Shape On Ultimate Bearing Capacity.

The ultimate bearing capacity of a square, rectangular or circular foundation is difficult to calculate using a rigorous mathematical approach. Semi-empirical methods are therefore used which involve modifying the general bearing capacity equation for a strip footing. Terzaghi has suggested the following.

For a square foundation

$$q_f = 1.3 \ c.N_c + p_o.(N_q-1) + 0.4 \ \gamma .B.N_\gamma + p \qquad (12.6)$$

and for a circular foundation

$$q_f = 1.3 \ c.N_c + p_o.(N_q-1) + 0.3 \gamma .B.N_\gamma + p \qquad (12.7)$$

where $N_c{'}$, N_q and N_γ are the bearing capacity coefficients for a strip footing.

For a shallow rectangular foundation Skempton (1951) has suggested that

$$q_f = (1 + 0.3 \tfrac{B}{L})c.N_c + P_o.(N_q-1) + (1-0.2 \tfrac{B}{L}) \tfrac{B}{2} . \gamma .N_\gamma + p \qquad (12.8)$$

where B and L are the breadth and length of the rectangular footing.

12.1.3 Application Of Bearing Capacity Theory

Excess pore pressures are set up when loads are applied to the ground. The rate of pore pressure dissipation is related to soil permeability. The low permeability of matrix rich tills and clay soils means that little or no dissipation occurs during construction. With time the excess pore pressures dissipate and the soil consolidates with a resulting increase in strength. The factor of safety against a shear failure occurring under the foundation is therefore lowest at the end of construction when the applied stresses have reached their maximum value and no increase in strength due to consolidation has occurred. Foundations on matrix rich tills and clay soils should therefore be examined using a total stress analysis with shear strength parameters obtained

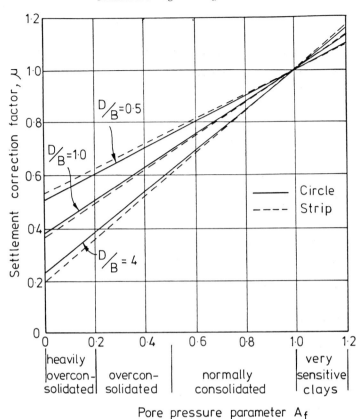

Fig. 12.6. Skempton and Bjerrum correction factors for consolidation settlement.

from undrained triaxial tests on undisturbed samples (Sects. 8.3.2; 11.8).

When samples of a saturated clay soil are tested under undrained conditions in a triaxial test the angle of shearing resistance is found to be zero and the undrained shear strength is equal to the cohesion, c_u. For a $\emptyset$ value of zero the bearing capacity coefficients N_q and $N\gamma$ are 1 and 0 respctively. The equation for the ultimate bearing capacity of a strip footing on clay then becomes

$$q_f = c_u.N_c + p \qquad\qquad (12.9)$$

or the net ultimate bearing capacity

$$q_{nf} = c_u.N_c \qquad\qquad (12.10)$$

where c_u is the undrained cohesion of the saturated clay

N_c is the bearing capacity coefficient

and p is the total overburden pressure.

Since sands and gravels are free draining materials excess pore pressures set up during construction will normally be dissipated during the construction period and the foundation can be designed on the basis of an effective stress analysis i.e. use values of the effective stress parameters c′ and ∅′. Unless there is a cementing agent between the sand grains the value of c′ will be zero. The ultimate bearing capacity of a strip footing therefore becomes

$$q_f = p_o.(N_q-1) + 0.5\gamma.B.N_\gamma + p \qquad\qquad (12.11)$$

where P_o is the effective overburden pressure

p is the total overburden pressure

γ is the unit weight of the soil

B is the breadth of the footing

and N_q and N_γ are bearing capacity coefficients.

Because it is impossible to obtain undisturbed samples of cohesionless soils, the value of ∅′ obtained from laboratory tests may not be exactly the same as that in situ. It is therefore preferable to use the more conservative Terzaghi bearing capacity factors when computing the ultimate bearing capacity of footings on sand or gravel.

12.1.4 Settlement Of Foundations.

Foundations have to be designed so that settlements, both total and differential, are tolerable. The settlement of a foundation can be considered as two parts. During construction elastic deformation of the soil occurs, without any change in water content, as the stresses are increased. This is known as immediate settlement, P_i. The increasing stresses applied to the soil during construction are, in the case of a saturated soil, carried initially by the pore water. The excess pore pressure thus created causes water to drain out of the soil with a corresponding decrease in volume until the excess pore water pressure is zero. This volume change manifests itself as consolidation settlement, P_c. The rate at which consolidation settlement occurs depends on the compressibility and permeability of the soil. Fine-grained soils may take some considerable time to complete settlement, whereas coarse grain soils will have consolidated by the end of construction or soon after completion.

The total settlement, $P_{t'}$, is given by

$$P_{t'} = P_i + P_c . \qquad\qquad (12.12)$$

Consolidation settlement is usually predicted on the basis of one dimensional consolidation tests carried out using an oedometer. Skempton and Bjerrum (1957) have suggested that, particularly in the case of overconsolidated clays, the consolidation settlement, $P_{c'}$, is not equal to that predicted from an oedometer test, P_{od}. They proposed a correction factor, μ.

Values of μ are related to the stress history of the soil which in turn governs the value of the pore pressure parameter A at failure, A_f. The relationship

Influence value = 0·02

Use of Newmark's chart

1. Draw foundation plan to convenient scale
2. Determine length, L, to this scale
3. Place point under which settlement is required
   over the centre of the chart
4. Count the number of influence areas, n, covered by foundation plan
5. Compute influence factor $I_\varrho = 0·02 \times n$
6. Compute immediate settlement $\varrho_i = q \times L \times \dfrac{1-\nu^2}{E} \times I_\varrho$

Fig. 12.7. Influence coefficients for determining immediate settlement (Newmark's chart).

between μ and A_f is shown in Figure 12.6.

The immediate settlement, P_i, may be found from the equation

$$P_i = q \times B \times \frac{1 - \nu^2}{E} \times I_p \qquad\qquad (12.13)$$

where q is the loading intensity on a uniformly loaded area

    B is the breadth of the foundation

    ν is Poisson's ratio for the soil

    E is the modulus of deformation of the soil

and I_p is an influence factor.

Values of ν range from 0.15 for a dense coarse sand to 0.5 for a saturated clay. Values of E may be obtained from laboratory tests, but because of the sensitivity of modulus values to sample disturbance, values obtained from in situ tests, such as the plate bearing test, may be more reliable. Values of the influence factor I_p for regular shaped foundations are given in Table 12.1. These values can be used to find the mean settlement of a rigid foundation or the settlement under the edge or centre of a flexible foundation. For irregular shaped foundations use may be made of influence charts such as the Newmark chart shown in Figure 12.7. This allows elastic settlement under any part of the foundation to be determined.

Table 12.1 Values of the influence factor, I_p, for computation of

        immediate settlement on a semi-infinite soil mass.

Shape of	Flexible foundation			Rigid foundation
Foundation	Settlement of centre	Settlement of corner	Mean settlement	Settlement
Circular	1.00	0.64	0.85	0.79
Square	1.12	0.56	0.95	0.88
Rectangular L/B=2	1.53	0.77	1.30	1.22
L/B=5	2.10	1.05	1.83	1.72
L/B=10	2.53	1.27	2.25	2.12
L/B=100	4.00	2.00	3.69	-

Consolidation settlement may be estimated using the compressibility found from oedometer tests. Since the compressibility in any stratum varies with depth, and also the stress increase due to the foundation loading varies with depth, the compressible layer is usually subdivided into a number of sub layers when computing consolidation settlement. The settlement, ΔH, of each sub layer is calculated by substituting appropriate values in the following equation for the compressibility and effective stress change.

$$\Delta h = m_v \cdot \Delta \sigma_z' \cdot \Delta H \qquad (12.14)$$

where m_v is the average coefficient of volume compressibility of the sub layer,

σ_z' is the average increase in vertical effective stress in the sub-layer due to foundation loading.

H is the thickness of the sub-layer.

The settlements of the sub-layers are then summed to get the total oedometer settlement, P_{od}', for the whole compressible layer.

Since the value of the coefficient of volume compressibility, m_v, is dependant on the stress level, it is important to use the appropriate value for the particular loading range in any sub-layer. Values of m_v for glacial deposits range from less than 0.05 m^2/MN for heavily overconsolidated lodgement till to 0.30 m^2/MN for glaciolacustrine clays.

Values of σ_z' for each sub-layer may be found using a stress influence factor, I_σ. Values of I_σ vary depending on the geometry of the foundation, the depth at which the stress is computed and its position underneath the foundation. For a uniformly loaded rectangular area the vertical stress beneath a corner can be computed using Fadum's chart (Figure 12.8). The increase in vertical effective stress, $\Delta \sigma_z'$, is given by

$$\Delta \sigma_z' = q \cdot I_\sigma \qquad (12.15)$$

where q is the foundation loading intensity.

and I_σ is the influence coefficient.

Since the principle of superposition can be applied when determining vertical stresses due to applied loading, the vertical stress at any depth under any part of a foundation which can be subdivided into rectangles can be found. A number of other charts and tables are available for other shapes of foundations and loading conditions.

It is usually necessary to predict the rate of consolidation settlement. The time, t, for any percentage consolidation settlement to occur is given by the equation

$$t = \frac{T_v \cdot d^2}{c_v} \qquad (12.16)$$

where T_v is a time factor

d is the length of the drainage path.

c_v is the coefficient of consolidation found from oedometer tests.

The values of the time factor T_v for various percentages of consolidation are given in Figure 12.9. The length of the drainage path depends on the materials

above and below the consolidation layer. If both are pervious, and concrete of a
foundation may be considered pervious, then a double drainage condition exists.
Pore water will drain from the centre of the layer both upwards and downwards and
the length of the drainage path, d, will be equal to half of the thickness, H, of
the compressible stratum. If, on the other hand, the compressible layer overlies
an impermeable stratum then a single drainage condition exists and the drainage
path length, d, equals the thickness of the compressible stratum, H.

12.1.5 Bearing Capacity Of Pile Foundations.

The ultimate bearing capacity of a single pile, Q_{ult}, is the sum of the pile base
resistance, Q_b , and the skin friction or adhesion on the pile shaft, Q_f.

$$Q_{ult} = Q_b + Q_f. \qquad\qquad (12.17)$$

Because of the difficulty in finding values of $\emptyset'$ for granular soils, theoretical
methods for finding Q_{ult} are usually confined to cohesive clayey soils.

The base resistance of a pile in a clay soil may be calculated using the general
bearing capacity equation.

$$Q_b = N_c \times c_b \times A_b \qquad\qquad (12.18)$$

where N_c = bearing capacity coefficient

c_b = undrained shear strength of clay immediately below the
            pile base.

A_b = area of the pile base.

For a deep circular footing the value of N_c can be taken as being equal to 9.

The adhesion on the pile shaft in a clay soil may be calculated using the
equation.

$$Q_f = \alpha \times \bar{c} \times A_s \qquad\qquad (12.19)$$

where α = adhesion factor whose value depends on the soil
            strength and the method of installation of the pile.

$\bar{c}$ = average undrained shear strength over the length of
            the pile.

A_s = surface area of the pile shaft.

Piles are rarely used singly so the interaction between adjacent piles in a group
has to be examined to find the ultimate bearing capacity of the group. If the
spacing between piles is too close, then "block failure" may occur in clays where
the soil between the piles moves with the piles (Figure 12.10). This is unlikely
to happen if the pile spacing is greater than three times the pile diameter.   At
close spacings the bearing capacity of the pile group failing by this type of
action should be checked using the equation.

$$Q_u \quad = \quad 2D\,(B + L)\,\bar{c} + 1.3\ c.N_c.BL \qquad\qquad (12.20)$$

where D   is the length of the piles

B   is the breadth of the group

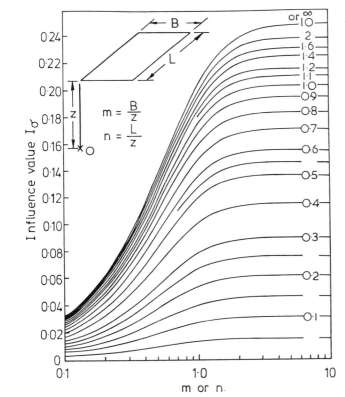

Vertical stress at O = Uniform loading intensity on rectangle x influence factor

$$\sigma_z = q \times I_\sigma$$

<u>Fig. 12.8</u>. Influence coefficients for determining vertical stress below the corner of a uniformly loaded rectangular area (Fadum's chart).

L   is the length of the group

$\bar{c}$ is average undrained shear strength over the length of the piles

c   is the undrained shear strength at depth D

N_c is a bearing capacity coefficient.

Sometimes piles are required to resist uplift forces, such as in foundations to resist overturning movements or beneath basements below groundwater level where the buoyancy effect gives a net uplift force. In these cases the resistance to uplift comes from skin friction. This may be calculated in a similar manner to

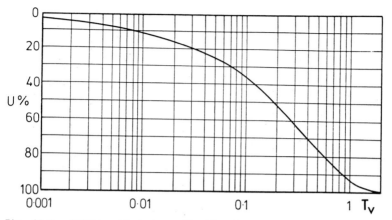

Fig. 12.9. Relationship between the time factor T_V and percentage consolidation assuming a uniform initial pore pressure distribution.

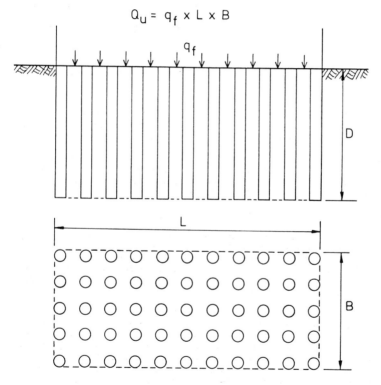

Fig. 12.10. Block failure of a pile group.

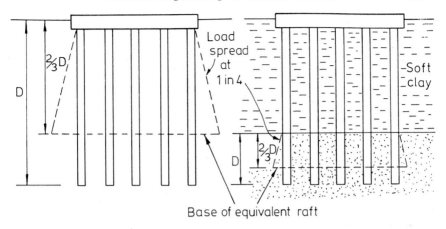

(a) Pile group supported
    by adhesion

(b) Pile group penetrating soft clay
    and supported by adhesion and
    end bearing in dense granular
    soil

Fig. 12.11. Assumed load transfer for determination of settlements of a pile group.

the skin friction on bearing piles, but pile test results have suggested that the mobilized skin friction in tension may be much lower than that in compression. It has been suggested that a 50% reduction should be used for tension piles.

The immediate and consolidation settlement of a pile group have also to be calculated to obtain an allowable bearing pressure. This can be done by assuming transfer of load so that the group is acting as an equivalent raft at a depth determined using the methods shown in Figure 12.11.

12.1.6 Shallow Foundation Design Based On In Situ Tests.

Because of the difficulty in getting undisturbed samples of cohesionless soils, the allowable bearing pressures are usually based on in situ test results. The allowable bearing pressure on most cohesionless soils is governed by settlement criteria rather than shear failure. Shear failure is only likely to be the governing factor if a narrow foundation is bearing on sand or gravel with the watertable above the base of the foundation.

The most common in situ test is the standard penetration test. The number of blows, N, required has been related to the allowable bearing pressure (Fig. 12.1.2) by Terzaghi and Peck (1967). N values should be corrected for overburden pressures particularly at shallow depths and also for fine sands or tills (Section 11.6.1: Fig. 11.6). It should be noted that in Figure 12.12 it is assumed that the water table is at least depth B below the footing and that the values of allowable bearing pressure are for a settlement of 25 mm. If the water table is at foundation level then for a wide shallow foundation the settlement will be doubled.

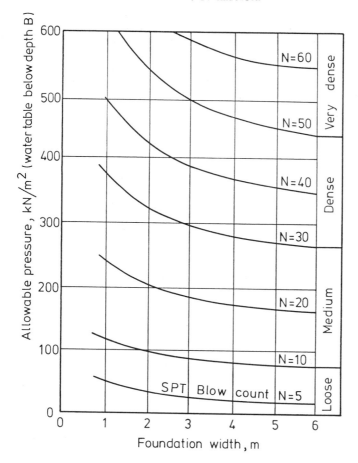

Fig. 12.12. Use of standard penetration test results to predict allowable bearing pressure.

Static cone tests (Dutch cone) are also used in cohesionless soils (Sect. 11.6.2). Meyerhof (1956) has suggested that for square or strip footings less than 1.2 m wide on dry sand the allowable bearing pressure is given by

$$q_a = 3.6 \, c_{kd} \text{ kN/m}^2 \tag{12.21}$$

where c_{kd} = cone resistance in kg/cm^2

For wider footings the equation becomes

$$q_a = 2.1 \, c_{kd} \, (1+\frac{1}{B}) \text{ kN/m}^2. \tag{12.22}$$

where B is the width of the footing.

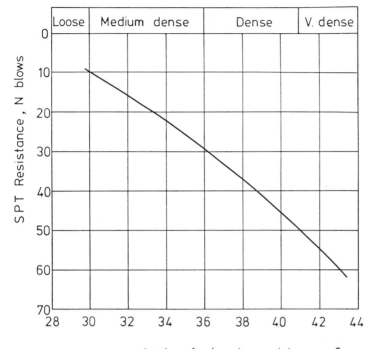

Fig. 12.13. Use of standard penetration test results to estimate density and angle of shearing resistance.

Again if the water table is at the level of the base of the footing the settlement will be doubled.

A plate loading test (Sect. 11.6.3) can be carried out in a trial pit and is used as a miniature foundation. Settlements measured using the plate can be scaled up to predict footing settlements using the relationship

$$\frac{P}{P_o} = \frac{4}{(1 + D_o / D)^2} \tag{12.23}$$

where P = settlement under foundation of diameter D

P_o = settlement under plate of diameter D_o for same loading intensity as foundation.

12.1.7 Pile Design Based On In Situ Tests.

The standard penetration test may be used to estimate a value of the angle of shearing resistance, $\emptyset$, and the relative density of the soil (Figure 12.13). The net base resistance of a pile in sand may be found from

W. F. Anderson

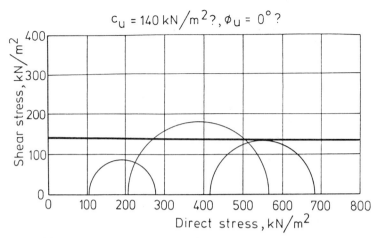

(a) 38mm diameter specimens

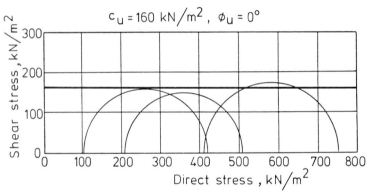

(b) 100mm diameter specimens

Fig. 12.14. Typical sets of triaxial test data for 38 mm and 100 mm diameter specimens of lodgement till.

$$Q_b = A_b \cdot p_o \, (N_q - 1) \qquad\qquad (12.24)$$

where A_b = area of base of pile

p_o = effective overburden pressurre at pile base level

and N_q = bearing capacity coefficient whose value depends on .

The shaft friction may be found using the values of unit friction, f, given in Table 12.2 and the equation

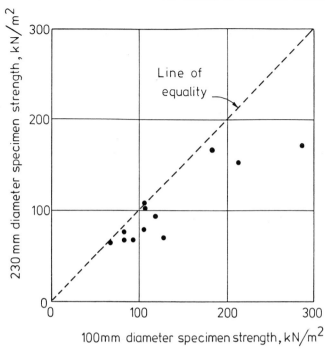

Fig. 12.15. Undrained triaxial tests on 100 mm and 230 mm diameter specimens.

$$Q_f = f \times A_s \qquad\qquad (12.25)$$

where A_s = the surface area of the shaft.

Static cone test results may also be used to predict end bearing resistance and
shaft friction.   End bearing resistance is simply taken as being equal to the
cone end resistance. Since this varies somewhat it is usual to take an average
value of cone resistance over a depth from 3D above the pile base level to D
below the pile base level, where D is the pile diameter.   The shaft friction
component may either be found by measuring shaft friction directly with a
friction jacket cone or by using the empirical relationships given by Meyerhof
(1956).

For circular displacement piles

$$q_{ult\ f}\ kN/m^2 = \frac{\bar{q}_c}{2} \qquad\qquad (12.26)$$

and for H piles

$$q_{ult\ f}\ kN/m^2 = \frac{\bar{q}_c}{4} \qquad\qquad (12.27)$$

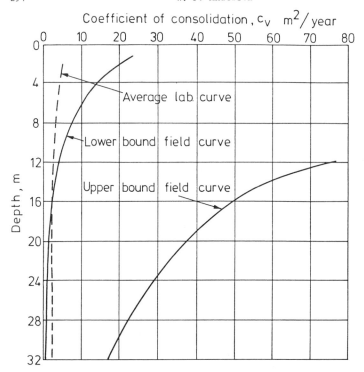

Fig. 12.16. Coefficients of consolidation for glaciolacustrine clays derived by oedometer and in situ permeability tests.

where $q_{ult\ f}$ is the ultimate unit shaft friction and q_c is the average cone resistance in kg/cm² over the length of the pile shaft.

Table 12.2 Average unit skin friction values for straight sided piles in cohesionless soils.

Relative density	Average unit skin friction (kN/m²)
< 0.35 (Loose)	10
0.35 - 0.65 (medium dense)	10- 25
0.65 - 0.85 (dense)	25- 70
>0.85 (very dense)	70-110

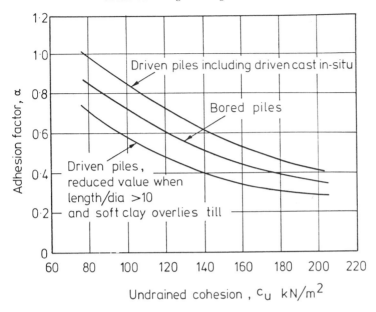

Fig. 12.17. Suggested relationship between the adhesion factor and undrained cohesion in lodgement till.

12.2 Design Of Shallow Foundations For Glacial Soils

With the exception of silts and clays, glacial and glaciofluvial soils may be regarded as good foundation materials (DeJong and Morgenstern, 1973; Milligan 1976). These deposits are however variable and softer or looser soils in what are elsewhere incompressible deposits may lead to differential settlement (see Sects. 2.9.2, 13.4). For this reason pad or strip footings are rarely used on glacial soils unless loadings are very light. It is more common for shallow foundations to be in the form of a mat so that these potential differential settlements can be accommodated. Construction on silts and clays will usually require pile foundations, and piling may also be necessary for heavy structures on till or glaciofluvial sands and gravels.

A major problem in the design of foundations for glacial soils is finding representative values for the design parameters; while this problem is stressed in Chapter 11 it is worth repeating here. Figure 12.14 shows a not untypical set of 38 mm diameter triaxial test results for a site on lodgement till. There is so much variation, as a result of sample disturbance, in the measured strengths at failure that it is impossible to decide where to draw a horizontal failure envelope. Figure 12.15 shows the results of a set of tests carried out on 100 and 230 mm diameter specimens from the same site. Use of these larger specimens has resulted in a considerable reduction in the variability but the fissures which are included in the larger specimens results in lower strengths being obtained. The expense of taking and testing specimens larger than 100 mm diameter is so great that there are extremely few jobs on which this could be justified.

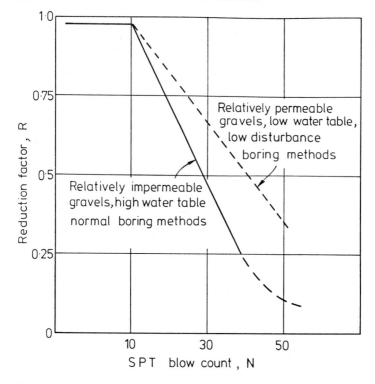

Fig. 12.18. Reduction coefficients for boring disturbance for standard penetration test values in granular soils.

Testing of 100mm diameter specimens of till in the laboratory normally results in considerable scatter in the undrained strength values for any site. This is due to natural variability, fabric effects and sample disturbance; the scatter usually found when a strength-depth plot is drawn makes it difficult to draw a design curve through the points. Generally an average curve is drawn through the points and Tomlinson (1980) states that evidence from field observations indicates that this average curve is likely to be conservative.

Having found values for the undrained shear strength the ultimate bearing capacity can be found using bearing capacity theory (Sect. 12.1.3). A factor of safety of the order of 2.5 to 3 should be applied to the bearing capacity to find the safe bearing pressure. The allowable bearing pressure is equal to or lower than this safe bearing pressure and is such that settlements will be tolerable.

Accurate predictions of both the amount and rate of settlement are difficult in glacial soils. In stiff lodgement tills settlements will generally be small and not be a problem unless loading intensities are very high. However, for complex or sensitive structures several tests are available for assessment of settlement behaviour (Sect. 11.6). Settlement on lodgement tills are usually negligible and usually results from elastic compression.

Estimates of immediate settlement are based on the elastic modulus of the soil and as with strength determinations it may be difficult to get meaningful values from laboratory tests. Modulus values are much more sensitive to sample disturbance than strength values, and it is therefore recommended that in situ tests, in particular plate bearing tests, are used to estimate the modulus when accurate predictions of immediate settlement have to be made (Sect. 8.4, Fig. 8.16). Even with plate tests there will be some disturbance of the soil on which the plate is laid. The effects of this disturbance can be minimized by carefully preparing the surface and using plaster of Paris as a bedding material. Probably the most meaningful value of the modulus will be found by carrying out a loading/unloading cycle before reloading the plate, and then using the slope of the reload portion of the test curve to determine E. Because of the variability of till a reasonable number of plate tests should be carried out to get a measure of the lateral and vertical variations of modulus on any site (Sect. 11.6.3). The low modulus values reported by Soderman et al (1968) appertain to soft ´silty-clay tills´ of glaciolacustrine origin (Table 12.3; Sect. 2.9).

Consolidation settlements of matrix dominant tills are commonly predicted on the basis of conventional oedometer tests in which a specimen 76mm diameter and 19mm thick is subjected to loads of increasing intensity. It will be appreciated that specimens of this size are likely to be disturbed and are also unlikely to be representative in respect of either containing the larger clasts in the till mass or containing the soil macrofabric. Comparative studies using different size specimens both in the U.K. and in Scandinavia have shown that compressibilities measured using large specimens may be considerably lower than those found using standard size specimens. This is probably due to gravel and cobble size particles acting as displacers and bridges with the more compressible matrix being sheltered from the effects of full loading intensity. For this reason settlements of structures on lodgement till are often grossly overestimated.

The rate at which consolidation occurs depends partly on the mass permeability of the soil. In tills fissures and lenses of sands and gravels can act as drainage channels, increasing the mass permeability of the soil by a large factor (Figs. 1.3, 1.4, 1.5, 1.6). Settlements therefore normally occur much faster than predicted on the basis of laboratory consolidation tests.

With fine-grained soils such as laminated clays, consolidation settlements may be considerable but because of silt and fine sand layers present in these materials the mass permeability will be quite high and is unlikely to be estimated accurately by oedometer tests. Tests carried out on larger samples in 150mm or 250mm diameter Rowe consolidation cells will usually give good estimates of settlement rate. Another method is to carry out in situ permeability tests in the strata to find the coefficient of permeability, k, of the soil mass. The coefficient of consolidation, c_v, may then be calculated from the relationship.

$$c_v = \frac{k}{\gamma_w m_v} \qquad\qquad (12.28)$$

where γ_w is the unit weight of water

and m_v is the coefficient of volume compressibility determined at the appropriate pressure from an oedometer test.

Figure 12.16 shows a comparison between c_v values from oedometer tests and those calculated from in situ permeability tests for a thick deposit of glacial lake clays (Threadgold and Weeks, 1975).

When soft silts and clays overlie gravel deposits, not only should the settlements due to elastic compression and consolidation be considered, but also settlements resulting from the soft silts and clays being squeezed into the gravel.

The use of the standard penetration test for determining settlement is discussed in detail in Sect. 11.6.2. Stuart and Graham (1974) monitored the behaviour of a mat foundation, designed using this method, supporting a thirteen storey building on a site in Belfast consisting of fine silty sand underlain by compact gravel and lodgement till. An extensive penetration test programme was carried out to identify pockets of loose sand close to foundation level which were then excavated and replaced with weak concrete. With hindsight it appears that pad footings were a more economical solution, since the possibility of differential settlements was virtually eliminated by the excavation and backfilling of loose pockets.

Pad footings were used for support of two twelve storey towers in Birmingham founded on dense sands and gravels (Levy and Morton, 1974). The structure was very stiff and could only tolerate limited differential settlements. This possible problem was overcome by leaving recesses on the starter blocks of the tower columns so that Freyssinet flat jacks could be inserted and differential settlements rectified. About 80% of the settlement occurred during construction and the measured settlement was less than predicted. Differential settlements were about 30% of the maximum settlement and jacking was not required. Similar examples from North American experience are discussed by Milligan (1976).

On cohesionless soils which are likely to contain occasional pockets of looser material the effects of vibration should be considered. Vibrations from machinery can lead to soil densification and differential settlements which in turn may result in failure of the machinery. In these situations it is prudent to mount the machines on blocks and put pile foundations under the blocks.

Shrinkage and swelling usually does not have to be considered for shallow foundations on tills. However, some may show signs of shrinkage during hot summers. For light loadings it is suggested that on these soils the minimum foundation depth should be 1.2 m instead of the usual 0.9 m. The presence of tree roots close to a foundation may aggravate shrinkage problems particularly in silty-clay 'tills' and other soft clays (Sect. 2.9.2, Ch. 6).

Softer or more compressible layers may underly stiff deposits at or near the surface. This situation can arise when till overlies glaciofluvial and glaciolacustrine silts and clays, or where dessication has given a stiff crust to a soft fluvioglacial silt or clay deposit. Strength changes due to weathering should also be considered (Sect. 8.2.4). It is true to some extent that the more variable the ground conditions the less sophisticated should be the testing procedures (Dowding, 1979).

12.3 Design Of Piled Foundations For Glacial Soils

The bearing capacity of piled foundations is derived both from shaft friction or adhesion and end bearing.

The end bearing component for piles in cohesive tills may be determined using bearing capacity theory and an N value of 9 for a deep circular footing. The contributions of end bearing to total bearing capacity can be increased by underreaming the base giving a larger bearing rate for the pile. Underreaming may have the effect of lowering shaft adhesion immediately above the underream; the shaft resistance for a distance of two times the pile diameter above the top

Table 12.3 Summary of published values of modulus of elasticity of tills (After Fisher, 1983)

Reference	Location	Undrained compressive strength, in kN/m² or penetration resistance N (blows/30 cm)	Method of determination	Modulus of elasticity E_s (MN/m²)
Klohn (1965)	Northern Sask.	1000	Plate load test	324-697, ave.=497
			Excavation rebound	608-1220, ave.=967
			Triaxial UU test	31-152, ave.=76
			Unconfined compr. test	83-283, ave.=145
DeJong and Harris (1971)	Edmonton, Alta.	800 N=111	Foundation settlement	280-490
			Triaxial UU test	10
			Triaxial CU test	97
			Consolidation test	21
Eisenstein and Morrison (1973)	Edmonton, Alta.	300 N=111	Pressure meter	55-245, ave.=173
Radhakrishna and Klym (1974)	Toronto, Ont.	1300-2900 N=200	Plate load test	97-207, ave.=180
			Pressure meter	97-290
			in situ shear box	133-332
			Triaxial CU test	41-242, ave.=138
Soderman et al (1968)	Southern Ont.	5	Plate load test	38-69
Fisher (1983)	Central Maine		Plate load test	103-130, ave.=115

NOTE: UU = unconsolidated, undrained;   CU = consolidated, drained

of the underream is usually ignored.

Shaft friction in cohesive tills is usually estimated using the   factor method.
The method of installation of the pile influences the   value. If bored cast-in
situ piles are used in till, because of the nature of the soil the surface of the
pile will be quite rough and the interlocking this produces with the soil might
be expected to increase the shaft adhesion. However, there is always a delay
    een boring and concreting and although good practice is to keep this to a
minimum, some stress relief and swelling of the soil will inevitably occur. This
will lead to a softening of the till at the interface with the pile, and adhesion
is likely to be less than that found with a driven pile in the same material.
When till is overlain by softer fluvioglacial or postglacial deposits and driven
piles are installed, the softer material may be dragged down during driving
giving a reduction in the adhesion factor, . Based on numerous pile loading
tests, and taking the above factors into account, Weltman and Healy (1978) have
suggested a relationship between   and undrained cohesion, c_u (Figure 12.17).
When using this plot the cohesion values should be from standard commercial site
investigations and undrained tests on 100mm specimens. Because of fissuring in
heavily overconsolidated clays it may be prudent to limit the value of adhesion
to 100 kN/m^2, the maximum value used when estimating shaft friction in fissured
London Clay.

In granular tills and glaciofluvial sands and gravels piles may be designed
either on the basis of driving formulae or using the results of in situ tests
(Sect. 12.1.7). If bored piles are used in granular soils casing is necessary
and considerable loosening of the soil may occur. Bearing capacities based on
the results of standard penetration tests may therefore be overestimated. Figure
11.18 gives a relationship suggested by Weltman and Healy (1978) between a
reduction factor and the standard penetration test blow count.

It is unlikely that glaciofluvial silts and clays would be able to provide
sufficient bearing capacity on their own to support a heavy structure. When
piles are used on sites where these soils are present they are generally driven
to a depth of at least five times the pile diameter into the underlying deposits,
usually fluvioglacial sands and gravels or till. The stiffer sands, gravels and
tills then support the piles in end bearing. Particularly problematical are soft
clay zones occurring within variably consolidated waterlaid tills;   depths to
refusal when pile driving can be underestimated (Milligan, 1976).

Because of the uncertainties involved with the design of piles a comprehensive
pile load test programme should be undertaken at every site. Some piles should
be loaded to failure and others to 1-1/2 times working load.   Precontract pile
testing may be very useful.

12.4 Foundation Construction In Glacial Soils: Some Common Problems

Excavations for shallow footings rarely cause problems even in the stiffest
materials.   Excavations for rafts may be somewhat deeper, and the stability of
the sides of the excavation have to be considered.   Problems arise when the
strata through which the excavation is progressing contain permeable materials.
This may lead to slope instability or increased pore water pressures under the
finished foundation, causing a reduction in the bearing capacity. During
construction, excavations below the water table can be dewatered, and
satisfactory drainage measures and waterproofing should be adopted for the final
structure.

Temporary excavations in stiff tills can usually be cut with a vertical face.
However, because of the occasional presence of cohesionless pockets and possible

fissuring, it is advisable to give some support using open timbering. Where boulders are exposed on the slope surface close timbering should be used to support them. In other soils temporary support to the excavation sides can be provided by driving interlocking steel sheet piles. Although no difficulties are likely to be encountered using this method in glaciofluvial soils, boulders in tills may prevent satisfactory driving of the sheet piles. Diaphragm walling may be the best solution in tills requiring continuous support.

When piles are mainly supported by end bearing in dense sand and gravel deposits, driven piles are used but sands and gravels are often so dense that it is impossible to penetrate into them for a sufficient distance. The only solution in this situation is to prebore. Another problem is where a bearing stratum of gravel or sand is thin and it is difficult to position the pile tips accurately. If the tips puncture these dense stratum then there is no alternative but to continue driving, sometimes for a considerable distance, until sufficient resistance is achieved. Because of driving difficulties due to the presence of boulders, bored and cast in situ piles tend to be used in lodgement tills. If lenses of cohesionless and permeable soil are present the hole has to be supported by either a temporary casing or a bentonite suspension. Large diameter bored piles are often used so that the boreholes may be entered and the obstructions dealt with by hand.

Debris falling from the borehole sides during or after boring can collect at the bottom of the hole, and if this is not cleaned out properly loss of end bearing can result. Overbreak during concreting without casing, or when a temporary casing is being withdrawn, can contaminate the concrete or cause a loss of integrity in the pile. These problems are usually associated with water bearing cohesionless pockets being penetrated in the till, but fissuring of the till may also cause similar problems. If there is any appreciable delay between boring and concreting stress relief may cause an opening of the fissures leading not only to partial collapse of the borehole sides, but also to a loss of adhesion on the pile shaft.

CHAPTER 13

Road Construction in Glaciated Terrain

J. E. Cocksedge

INTRODUCTION

The site occupied by a road is long and relatively narrow and as a result a road project often crosses varying geological deposits and associated topography. Major roads demand high standards of vertical and horizontal alignment which often involves the use of deep cuttings, high embankments, retaining walls, bridges and viaducts and, occasionally, tunnels.

For the purposes of road design, construction and performance, the geotechnical properties and behaviour of the soils, rocks and groundwater lying beneath and adjacent to a proposed road line must be assessed. In appropriate situations these will include their ability to support embankments and the magnitude and rate of any resulting settlements; the steepest angle of construction of cutting slopes consistent with long term stability; the pattern of groundwater movement, its influence on stability and its control where appropriate; the usability of excavated materials, with or without processing, as embankment fills, road sub-base or base and as backfill or filter media in drains; the compacted strength of the materials at road formation level as relevant to the design of the pavement; the parameters of the soils and rock relevant to the design of retaining walls, culverts, underpasses, tunnels and foundations to structures; and the behaviour of and changes to the materials which occur during construction and subsequent use. The aim of this chapter is to review the above in the context of glaciated terrains.

13.1 Soil Classification And Relationships

Soils are generally classified and placed in groups on the basis of their grading and the plasticity of their fine fraction which are either estimated by look and feel (field method) or determined by testing (laboratory method). These parameters, and therefore the resulting groupings, are largely independent of the in situ state of the soil: they are not influenced by structure (bedding, discontinuities etc.), moisture content, in situ density, strength or compressibility. However, classifying soils into broad groups can be a useful preliminary guide to their behaviour, particularly when they are to be used as a construction material in earth works, pavement layers and drains or when forming the natural sub-grade for a road. Such classification has the virtue that it simplifies the handling, presentation and digestion of a large volume of information concerning the soils. While a detailed description of each individual sample will be much more comprehensive, few such descriptions will be identical and with a large number of samples the picture can become very confused.

There are numerous classification systems in existence: the two most commonly encountered are the AASHTO Classification System (1970) and the Unified Soil

302

Classification System (1968). A more recently developed system is the British Soil Classification System for Engineering. However, in the context of roads in glaciated terrain experience has shown that the use of these classification systems is not very helpful to either designer or contractor. Ad hoc classifications tailor made for particular purposes (e.g. usability), are generally more useful.

As outlined in Chapter 8 the search for and development of reltionships, if any, between simple index tests on samples (e.g. Liquid Limit, Plasticity Index, Liquidity Index) or the position of origin of the samples (depth and/or chainage) and those soil parameters which are of direct interest to design and construction can prove very useful when selecting design parameters, assessing quantities of a particular type of material, planning mass haul etc. Relationships developed for similar materials elsewhere can be a useful preliminary guide as well as a check on the results obtained from a specific site. However many sites are too complex in the range of soils they present to the engineer to allow simple relationships to be developed.

13.2 Usability

Material arising from excavations may or may not be suitable for use elsewhere as fill. It is seldom sufficient to base judgement on a comparison of properties of the material in situ and the specified requirement at the point of application. Factors which should be considered include, changes in the material which may occur due to the methods of excavation, transport and placing, its behaviour under plant and vehicles during construction, the use to which it might be put, the need for processing, if any, and its immediate and long term behviour in the works. In practice it is very difficult to allow for some of these factors, even in a broad qualitative sense, without direct personal experience or access to well documented records. A few examples will help to illustrate the problem.

An excavation for a long shallow cutting was made through a wet till into underlying thinly bedded limestone. The surface of the limestone had been shattered and disturbed by ice/or frost action and considerable penetration of till into joints and fissures had taken place. This till was compact in situ but moisture sensitive and, when disturbed during the ripping of the limestone, it was loosened and rapidly absorbed moisture from rain and groundwater seepage. The result was a mixture of rock fragments and wet till which was difficult to handle or traffic. After much delay the cutting was eventually completed with the incorporation of temporary drains and a change from rubber tyred scrapers to a tracked excavator and dump trucks. In the absence of a suitable 'wet' fill in which to place the material, a considerable quantity of otherwise sound rock ended up in a spoil tip.

A particular problem is the frequent failure of site-investigation work to identify the full range of soil conditions represented by a single material. Design criteria established on the basis of a single response are often seen to be unrepresentative where excavation starts and there is greater access to the subsurface. The occurrence of boulder pavements in lodgement tills and the unforeseen increased cost involved in their removal is a case in point (Legget, 1974, 1979).

During earthworks excavation it is very necessary to exclude or rapidly remove water from all sources to prevent the soils becoming wet, otherwise they will lose strength, become more difficult to traffic, handle and compact and may prove unsuitable for use in embankments except, perhaps, in special 'wet' fills. Tills are characteristically well graded from gravel and cobbles to clay as a result of which their permeability is low. In many cases however, the ratio of silt to clay content is high: the deficiency in clay binder and a relatively high silt content results in a material which can take in water fairly easily. Even small changes in moisture content can result in a significant decrease in undrained shear strength: any wet surface being trafficked easily ruts which then holds water and further aggravates the situation. Under these conditions, efforts to keep the earthworks shaped to shed water become very difficult. The vagaries of weather are all too familiar to those engaged in road construction but with an appropriate choice of plant, cuttings can be worked with a vertical or steeply sloping face: by virtue of being better drained, this reduces the problem associated with the ingress of water.

Limits of suitability for earthworks based on a   criteria are discussed   later. A problem arises when pockets or layers of sand and silt are mixed with a clay: the resulting reduction in PL is generally greater than any reduction in mc (if the cohesionless soils are saturated mc may even increase) and this leads to an increase in   which may render the soil unsuitable (Cocksedge and Hight 1975).

This situation is quite common with glacial deposits where excavation and handling leads to mixing of sand and silt pockets, included in cohesive tills, and sand and silt laminations in glacial lake clays.

The granular tills found in shield terrains and glaciated valleys (Figs. 1.3 and 1.6) can have many uses, as dug or after processing, including their use as sub-base, in drainage blankets, as filter media and as aggregate for concrete and asphalt. On one project an attempt was made to form drainage blankets with talus debris derived from limestone, which was quite permeable in situ;   this failed because the material broke down under handling and became relatively impermeable. Simple and cheap aggregate testing would have avoided this problem.

13.3 Groundwater

As the hydrogeology of glacial drift is discussed in detail in Chapter 15 only a few examples of the many groundwater related problems which are encountered during road construction in glaciated terrain are given here.

On a major road project the junction between an upper weathered layer of wetter material and underlying unweathered till was invariably marked by seepage in the face of cuttings and required treatment by drainage.   It would appear that weathering had resulted in a relative increase in permeability and a partial perched water table condition.

Another common experience is the presence of zones of relatively high permeability within lodgement till sequences (Figs. 1.4) which when exposed in the slopes and base of a cutting are often the site of persistent or intermittent seepage, the safe disposal of which generally requires the installation of individual and ad hoc drainage arrangements. Some points of seepage may run for only a limited period and not reoccur while others may not become active until some time after exposure. Possible explanations include long or short term rainfall patterns, the presence of pockets of trapped ´fossil´ water and changes in the pattern of groundwater movement resulting from the formation of the cutting and associated drainage works. Therefore, where possible, it pays to keep occurrences of seepage and areas of potential seepage under observation for a period before embarking on expensive special drainage or applying topsoil. The development of seepage on a slope after it has been covered with a layer of topsoil usually results in the sloughing off of the topsoil layer. Both drainage and retopsoiling are then required: repairs are usually more expensive than the original work particularly where completion of permanent works results in more difficult access and restricted working. Isolated, individual seepages are often dealt with by protecting the point of issue by a small zone of suitable filter media which connects with a pipe leading into the main drainage system. More widespread seepages, both in number and size, may be tapped by means of a herringbone drainage system or a general filter blanket with a piped outlet from its lower end. Unstable till slopes with failures lubricated by persistent ingress of water from permeable layers are a persistent problem with many road cuttings and natural slopes. An inexpensive solution is to regrade the slope to a low angle and install deep drains to prevent the watertable rising in the slope. Unfortunately this requires a greater land ´take´ along the routeway in order to accommodate the increased length of low slope angles.

Surface and buried deposits of permeable granular soil, either in layers or pockets are frequently the source of unwanted water flow into the works. Sand and gravel lenses in lodgement till and strata close to valley sides are most problematical. On one project in a glaciated valley, a weathered scree layer obscured the base of a steep valley side slope of slate at its junction with lodgement till filling in the valley floor. This scree contained a lot of water. A road cutting skirting the base of the hill intersected the lower edge of the scree. In this instance the presence of the scree and its contained water were known in advance from site investigation and discussion with local landowners and a special drainage system was devised. The cutting was first excavated to a level just above that of the water in the scree, following which a subsoil drain was installed through the remaining scree and run-off elsewhere. With the water thus controlled, the cutting was continued but not completely in the dry as had been hoped. The fill produced its share of random seepages, fed perhaps from the water in the overlying scree uphill of the drainage. At one point in the cutting a long pocket of gravel was intersected and produced a strong flow of water. This was controlled by pumping; the flow persisted for many weeks, but eventually ceased.

Particular care should be exercised when excavating in laminated lacustrine clays. The coarser sandy layers may be a direct hydraulic connection with any nearby rivers or lakes and problems of water ingress into the site may only be apparent at certain times of the year (Parsons, 1976; Thompson and Mekechuk, 1982).

A groundwater problem which is frequently encountered in glaciated terrain is that of the confined aquifer which may lead to the development of artesian or sub-artesian water pressures (Fig. 13.1). The phenomenon is encountered where a valley infill mantles and feathers out against a steep valleyside rockslope. Tills are relatively impermeable (10^{-5}m/sec or less is common; Table 15.1) and infiltration in the hill above the till edge is impeded from escaping into the valley bottom except by very slow seepage through the till cover. Storage within the hill builds up and may eventually overflow in the form of springs and seepage lines above the till edge. In this situation the piezometric head in the bedrock below the till would reach the level of the till edge. This produces an artesian condition relative to a falling till surface. When the storage within the hill is below the maximum or there is some loss of head due to seepage, the bedrock piezometric head could be sub-artesian. With the very low permeability of the till the rate of seepage is small and natural evaporation may remove any visible sign at the ground surface or at worst the surface may be slightly marshy which is by no means unusual on many freely-drained hillsides. Under artesian conditions upward seepage gradients would reduce effective pressures within the till and hence the factor of safety of the slope; it would also add to the latent instability of any superficial solifluction deposits that may be present on the surface of the till (Fig. 5.1).

In a widely-met situation the bedrock surface is either fragmented (frost

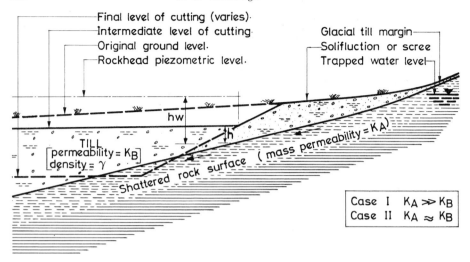

Case I $K_A \gg K_B$
Case II $K_A \approx K_B$

<u>Fig. 13.1</u> . Groundwater conditions below a glaciated valleyside slope.

shattered) and of high permeability relative to the overlying till, or relatively sound and with a permeability of the same order of magnitude as that of the overlying till. In both cases the presence of artesian piezometric pressures at bedrock level can be established by means of simple standpipe piezometers. By reference to Fig. 13.1 it can be seen that before any cutting excavation is started $\gamma w.hw < \gamma.h$. As excavation proceeds a critical depth is reached where $\gamma w.hw = \gamma.h$. Any further excavation results in $\gamma w.hw > \gamma.h$. What then happens varies according to the character of rockhead.

Where the bedrock is shattered $(K_A > K_B)$ the till heaves upward but any drop in h is rapidly re-established by flow through relatively permeable Zone A of the rock. The heave results in fracture of the till accompanied by some outflow of water to the excavation surface but still remains high. Any attempt at further excavation results in slurrying of the disturbed till and the start of progressive slope failure. The situation can be stabilized by relieving the high porewater pressures by means of various drainage measures, the main feature of which might be a drain to bedrock along the cutting slope. This can be installed through the unstable ground by means of a fully supported trench excavation system. Initial heavy water flows in the drainage system may become less over a period of time: the rate of drainage of the aquifer may exceed the rate of recharge. Problems such as described above with reference to puncturing of permeable horzons with artesian water conditions are frequently met where waterlaid tills (Sect. 2.9) and glaciolacustrine clays overlie lodgement till. Penetration through the latter into bedrock results in flooding of the excavation.

In case II $(K_A \approx K_B)$ the sequence of events is different because there is no observable surface expression of seepage nor are there any piezometers down to rock in the critical region of the cutting because, they interfere with the movement of excavation plant; a common occurrence in constricted construction sites. When the cutting surface moves below the critical depth local bedrock pore pressures are relieved by a slight heave of the overlying till and do not re-establish themselves: the rate of seepage through a diminishing thickness of till probably exceeds the rate of inflow through the rock, the flow in both cases being imperceptibly slow. Pore water pressures in the rock both uphill and downhill of the critical zone are not affected and remain high. To guard against the possibility of a slow, long term re-establishment of adverse piezometric heads below critical portions of the till cutting slope, a number of counterfort type slope drains should be installed; these also take care of ad hoc seepages resulting from infrequent high rainfall storms and the possibility of deep winter freezing of the slope generating critical porewater pressures.

Deep excavations for pipes, foundations or other works often encounter problems due to a combination of variable soil and groundwater conditions associated with glacial drift. A number of examples are discussed in Arrowsmith (1978) and Dowding (1979).

13.4 Earthworks

Hight and Green (1978) provide a general review of earthworks in relation to road design and construction. The following concentrates on aspects of earthworks which are influenced by the special nature of the materials encountered in glaciated terrain.

At sites of major excavations for cuttings or foundations, there are several factors to be considered. These include choosing plant for efficiency of excavation, loading and cartage, assessing material qualities and use and destination, separating materials as necessary and preventing their deterioration.

In the rugged topography, wet climate, frequent groundwater seepage and restricted hauls commonly associated with upland glaciated terrain, the operation of rubber tyred scrapers is often not the best choice. Tractor drawn box scrapers are more suited to wet conditions. However, with any type of scraper it is difficult to be selective - pockets of wet or otherwise unsuitable material get mixed in with suitable material. The excavation face is trafficked and the material is more easily rendered unsuitable in the presence of water; drainage is difficult if rutting occurs and this is worse with heavier plant. Excavator-loaders and dumptrucks present many advantages such as better ability to cope with wet conditions and negotiate difficult ground, greater flexibility and selectivity in excavation, and ability to maintain a steep, well drained working face. However, under dry conditions cost and output rate may be unfavourable compared to scrapers.

There are many other considerations which influence the choice of plant, including the topography, length and difficulty of haul, obstructions, cost of tips and borrow areas, likely disruptions to construction by hauling through, type and availability of plant at the time. Operational economics may be in conflict with the most efficient use of the excavated material and broader environmental considerations. Where the latter are considered to be of sufficient importance it is appropriate to specify the plant and methods which would be acceptable for that contract.

In a steep sided narrow glaciated valley a new road often has to vie for space with a river and existing roads and/or railways. This, combined with the higher standards of vertical and horizontal alignment demanded by modern roads, often requires the use of major cuts and fills in which the type and quality of rockhead may be critically important (Sects. 4.6.2, 5.4).

Embankments are generally used to carry a road across valleys or flood plains where ground conditions are often difficult. Conditions resulting from impeded drainage are a common feature of glaciated terrain; many valley bottoms are superficially wet and marshy while others are underlain by substantial deposits of silt, peat or soft laminated glacial lake deposits. An embankment foundation must remain stable under the imposed load, and the long term total and differential settlement of the embankment must be acceptable in terms of the pavement riding quality and integrity as well as the safety and correct functioning of drains. It is not always economical or practical to remove all soft deposits from below embankment foundations. Shallow soft deposits are sometimes built over, starting with a blanket of free draining material. (A drainage blanket is sometimes needed below embankments constructed on firm ground if the latter is subject to seepage). In other cases deposits of silt or laminated clay may be left in place with only overlying peat being removed and replaced with free draining fill below water level. However, difficulties of separation may dictate total removal.

When soft deposits are left in place a number of options are available. One of these is to build the embankment and monitor its settlement until the rate of settlement is acceptable before completing the pavement. In many cases a settlement period not exceeding three months has been found to be satisfactory. Care is needed if failure is to be avoided by too rapid construction; stage loading and varying side slopes with height may be necessary (Threadgold and Weeks, 1975). Surcharging may be used to accelerate settlement and light weight fill used to restrict total loading. Ground treatment by sand drains or dynamic compaction (Menard and Brosse, 1975) may be appropriate. Design, construction control and monitoring is now much easier because of advances in knowledge of ₃oil behaviour, and the results of full scale field studies (Bishop and Green, 1974; Vaughan et al, 1978), better and quicker methods of analysis using computers, and advances in instrumentation.

A suitable fill material needs to be sufficiently incompressible to insure stability of the embankment and to carry construction traffic. In practice most tills are suitable for use as fill provided their moisture contents are not too high: in certain circumstances wetter tills can also be used. Granular moranic deposits are generally suitable with limitations on cobbles and boulders; peat is obviously unsuitable and in most cases silts, laminated glacial lake clays, solifluction deposits etc. would not be used unless relatively dry and mixed in small quantities.

In Britain a Department of Transport Specification (1976) leaves the choice of an upper limit of moisture content (mc), for each type of soil encountered on a road project, to be decided by the Engineer. Logically such limits should be derived from correlations of mc with strength, compressibility and efficiency of compaction. However, because in any natural deposit there are variations in grading and plasticity which influence other properties there are complications to choosing a moisture content limit in practice; none of the approaches to date has proved completely satisfactory (Arrowsmith, 1978). One common practice has been to express mc limits relative to the Plastic Limit (PL) for cohesive soils and relative to the optimum moisture content (omc) for a given standard of compaction for cohesionless soils (Fig. 13.2). For cohesive soils a commonly used limit is 1.2 x PL; this is about the practical limit for the efficient operation of towed and small self propelled scrapers up to 15 m capacity (Sherwood, 1970). At 1.1 x PL medium and large motorised scrapers are able to operate but for most efficient plant use 0.9 x PL is desirable. Embankment settlement increases with increasing $\frac{mc}{PL}$ ratio. One of several disadvantages attaching to the PL criteria is that the measurement of PL is not consistently reproducable (Sherwood, 1970). Some investigators suggest that the use of the

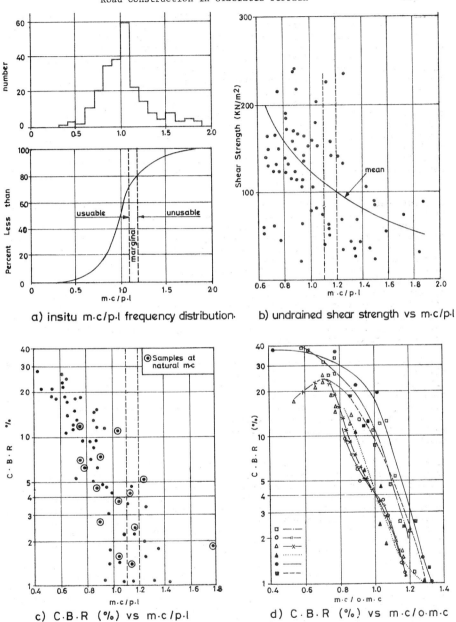

a) insitu m·c/p·l frequency distribution.

b) undrained shear strength vs m·c/p·l

c) C·B·R (%) vs m·c/p·l

d) C·B·R (%) vs m·c/o·m·c

Fig. 13.2. Index properties for lodgement tills with reference to California Bearing Ratio and suitability for fill.

Liquid Limit (LL) would give more consistent results. The problem becomes more involved when dealing with stoney cohesive soils typical of many tills. For a given matrix moisture content increasing stone content would give reducing total moisture contents (see Sects. 8.2.3, 8.3.2). The matrix has a considerable influence on strength and compressibility which the total moisture content will not reflect. Stones above a certain size (a typical practical limit is 5mm) can either be removed before testing or a correction made based on the percentage by weight of such stones present in each specimen. However, the PL and LL are measured on the portion less than 0.425mm, so a problem of compatability still remains. This problem of stone content also effects the measurement of shear strength and compressibility.

Another approach which avoids measurement of moisture content and makes direct use of the soil in its existing state is the Moisture Condition Test (Parsons, 1978, 1981). This is a compaction test carried out in a special apparatus and the Moisture Condition Value of a sample is defined as the number of blows required to fully compact that sample. Samples with stones up to 20mm in size can be used which somewhat eases the stone content problem. The direct measurement of shear strength of the compacted soil in a quick undrained test or unconfined compression test is sometimes used. In many cases usability is dictated by considerations of trafficking rather than stability or compactibility; usability typically falling in the range 35 to 100 kN/m^2 (Kennard et al, 1978) (as a rough guide, 1/5 tyre inflation pressure), with increasingly heavy modern self propelled rubber tyred scrapers being at the upper end of the scale (Farrar, 1971; Farrar and Darley, 1975).

Assessment of the suitability of the materials to be excavated from any cutting is made on samples whose volume is a minute fraction of the volume of material being assessed. Glacial deposits are notoriously variable and extrapolation from a small sample must be complimented by a clear understanding of the geological geometry of the deposit and the processes involved in its formation. Allowance must be made for subsequent changes, discussed previously, that could occur to the materials during excavation, handling, placement and adjustment to a new equilibrium as well as those caused by seasonal climate factors. Incorrect assessment can be very costly: claims may arise whether the volumes of suitable materials fall short of or exceed the stated quantities.

Excavation in tills often produces large volumes of material which are too wet to qualify as suitable in terms of the criteria previously discussed. On projects which have a shortfall of suitable material the possibility of using wet material on embankment fills is therefore attractive. The cost and difficulty of finding suitable borrow areas and spoil tips together with environmental constraints on their siting could prove a further incentive. The first major road project in Britain where wet fills were successfully used was on the M6 motorway in Cumbria. The project involved about 5.5 x 10 m^3 of earthworks and of this about 1.3 x 10 m^3 was estimated to be too wet for normal use. A 6m high instrumentated trial embankment was first constructed to determine the most appropriate embankment configuration, trials of plant operation and methods of field and laboratory testing (McClaren, 1968). The trial embankment was constructed with lodgement till, artificially wetted as necessary. Drainage layers were incorporated at various spacings to speed up the rate of excess pore pressure dissipation thereby improving stability, accelerating settlement, and allowing earlier completion of the road pavement as well as ensuring that subsequent differential settlements would be negligible.

In practice, due to the clast content of the till, the magnitude of embankment settlements was small (25 to 40mm). The wet fill embankments subsequently included in the motorway contract were generally of greater height than the trial

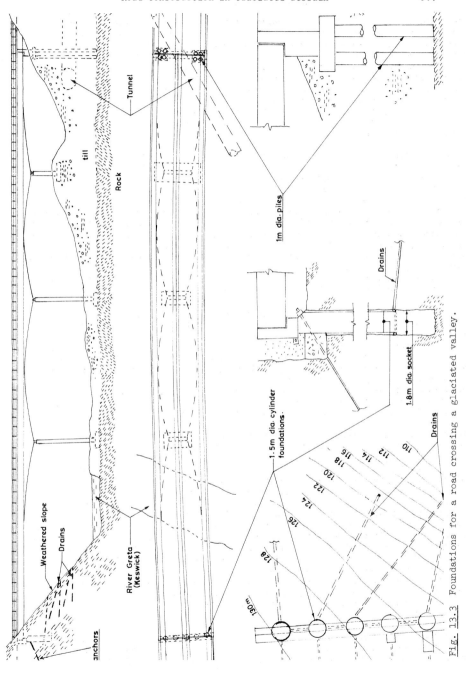

Fig. 13.3 Foundations for a road crossing a glaciated valley.

embankment and involved a much wider range of material type. They were provided with drainage layers at 3m maximum spacing including one at existing ground level and one capping the wet fill. The four month settlement period which had been specified proved to be more than adequate and was occasionally relaxed. During construction wet fill was brought in by dump trucks which were able to run on the stone drainage layers; it was tipped and spread by dozers fitted with special wide tracks. No other compaction was given. All wet fills were topped with about 3m of ordinary suitable material. Consequently this method of construction can only be used for relatively high embankments.

Since then wet fills have been successfully used on other contracts. On the M2 in Northern Ireland till which was only marginally too wet was laid in alternate layers with dry material. Soils having a compacted permeability greater than 10 cm/sec. are considered suitable for successful construction of wet fills. Investigations into the possible use of a wider range of materials in wet fills have since been carried out (Grace and Green, 1978).

Large recoverable deformations are known to occur when compacted fills are subjected to cyclic loading by moving wheels. This resilient behaviour may be sufficiently severe to reduce the effectiveness of compaction of subsequently placed material, especially pavement layers. Rigid pavement layers may be damaged by flexing. It has been argued that these large deformations increase the energy required to propel the moving wheel loads and hence increases operating costs. If high moving wheel loads are kept off for a period, the resilient condition seems to slowly disappear; in some cases it has been known to reoccur under subsequent wheel loading but it appears to be essentially a short term effect. It is commonly believed to be a problem associated with temporary excess pore fluid pressures and is similar to that experienced by cyclic loading transmitted to foundations by marine structures as a result of wave loading (Sect. 11.8). Such behaviour does not occur with wet fills.

The design of pavement thicknesses in Britain is generally based on Road Note 29 and involves an assessment of the California Bearing Ratio (CBR) of the road formation. The CBR of till varies widely with moisture content and dry density and there is usually a wide scatter in the test results (Fig. 13.2). Near the upper limit of moisture content CBR values are low, typically 2 percent or less. Soils at formation are generally partially saturated and their strength (CBR) and stiffness will vary with the porewater suction. Because the relationship between soil suction and moisture content has hysteresis, depending on whether the soil is wetting or drying, subgrades which are wet during construction will have a higher equilibrium moisture content, particularly with more plastic soils. It has been suggested that wet weather working can halve the life of a road. Low strength at the time of construction results in an inadequate platform for effective compaction of granular sub-base and road-base.

In the context of low cost roads morainic soils are not ideal for road construction because they lack binder, require a high compactive effort for good CBR and are susceptible to frost heave. Particular problems occur in shield terrains and glaciated valleys where tills are generally of low clay content and low in place density as a result of their mode of deposition by melt-out and dumping from the ice surface (Figs. 1.3, 1.6) and lack of consolidation by glacier ice. Under wet climatic conditions and high groundwater regimes soil liquefaction caused by high pore pressures as a result of heavy plant movement, leads to disruption of construction. More precise mapping and selection of fill material is required (Woods et al, 1959; Eden, 1976).

Fig. 13.4. High angle reverse faults generated by the collapse of sand units over melting glacier ice at depth

13.5 Slopes

The construction of a road in the vicinity of a natural slope will cause a change, for better or worse, to the stability of that slope. When a cutting is made, slopes are created which are steeper than and therefore generally less stable than the pre-existing ground at that point. A useful review of methods for assessing slope stability is given by Sarma (1973) and Chandler (1977). Methods of stabilization and their effectiveness are reviewed by Hutchinson (1977).

In a glaciated terrain there can be a number of situations where a natural slope has marginal stability with a previous history of instability and may fail if any adverse change is imposed, (e.g. loading by an embankment, removal of support by a cutting or a rise in water table by the blocking of natural drainage and removal of surface vegetation). In glaciated valleys one of the main triggers for previous instability in glacial soils is the loss of buttressing support by the ice, coupled with excessively high disturbed groundwater regimes (Sects. 4.6.2, 4.7, 5.4.4). Erosion due to ice movement often results in a steepening of natural slopes. Another potentially unstable situation is that often associated with solifluction sheets or lobes widely encountered in periglacially modified terrains. During construction of road embankments on sloping ground, large movements occur which can subsequently be shown to be due to the presence of solifluction deposits. Such failures have been extensively studied and reported (Fookes and Higginbottom, 1970; Symons and Booth, 1971; Horton, 1974; Sect. 5.4). Due to earlier movements these deposits contain slip surfaces along which the shear strength was at a low residual value. The solution adopted to improve stability is drainage and reduction of embankment loading by changing to a light weight fill. The metastability of many soils such

Fig. 13.5. Pock marked lodgement till plain formed by subsurface collapse into solution pockets in limestone. Note dense concentration of pocks in the right background.

as loess and 'quickclays' has already been commented on (Sects. 5.6, 6.5.1). Faulting in fine-grained sediments (Fig. 13.3) as a result of the melt of buried periglacial ground ice or glacier ice may necessitate removal of the entire soil mass and backfilling.

Another frequently encountered problem is of anistropic permeability; more permeable bands in lacustrine laminated clays directing groundwater seepage horizontally to the cutting face where erosion and softening causes slips to develop (Sauer, 1975). Some embankment failures after heavy rain have been attributed to the presence of silt and sand layers incorporated in fills.

The parameters that control slope behaviour are the drained and undrained strength, consolidation or swelling and permeability (Vaughan and Walbancke, 1975). Lodgement tills have low rates of permeability and swelling and consolidation so that undrained conditions apply during slope cutting and drained conditions govern long term behaviour. The stress relief of unloading produces pore water suction in cut slopes; the suctions dissipate with time so that stability is most critical with regard to long term pore pressures. In compacted fills, end of construction pore pressures are generally high. In high embankments consolidation under self weight will tend to produce an increase of density and a reduction of permeability with depth. Water infiltration may exceed evaporation and lead to the development of perched water tables in the top of high embankments.

Well graded sandy tills of low plasticity have a relatively high drained strength and are of low sensitivity and brittleness. In cuttings the operational drained strength will therefore not be far short of the strength of intact specimens, even if the till is fissured unless these fissures have a critical orientation

(Sect. 8.3.3). Similar drained strength will apply in fills because remoulding causes little reduction. Therefore relatively steep slopes (1:1) can be adopted. With more plastic tills the limiting stable slope is about 2.5 to 1. Because of the low brittleness of sandy tills, slope failures are often very shallow and the distance moved by slips is small. Tills of higher plasticity and brittleness will behave differently. At a P.I. of about 20 to 25% there is an abrupt transition from low to high brittleness (Sect. 8.3.1, Fig. 8.9). Slope failures as a result of greater cohesion are deep-seated and rotational in type. The majority of lodgement tills are of reduced P.I.'s and are therefore of reduced brittleness.

13.6 Earth Retaining Structures

Problems associated with the construction of retaining structures in glaciated terrain are generally similar to those discussed in other contexts: obstruction due to cobbles and boulders, water problems associated with permeable inclusions, random seepage paths, etc. The presence of cobbles and boulder pavements can seriously handicap the driving of sheet piling, whether for temporary or permanent support. Sheet piling for the support of an excavation for a conventional retaining wall has often been abandoned due to this cause and with walls redesigned and built as a diaphragm wall excavated with bentonite support. Grab and chisel coped successfully with the cobbles and boulders. This technique is commonly employed where space restrictions rule out normal sloping sides to cuttings; particular attention should be paid to identifying any buried sand and gravel horzons in till which may promote sidewall collapse during excavation and loss of bentonite.

13.7 Foundations To Road Structures

This section is confined to aspects of foundations for road structures (i.e. bridges, viaducts, underpasses and culverts) in glaciated terrain. A general overview of other foundation types can be found in Chapter 12.

In the case of many road bridges adequate founding for simple pad foundations can be found within tills or, where the cover is thin, on underlying rock. The main problem is often in assessing likely differential settlement when faced with variable subsoil conditions, particularly in buried drift-filled valleys lying parallel to the modern valley, in addition to the general problems of determining subsoil stress changes and load redistribution due to structural stiffness.

An example which illustrates the variable conditions which sometimes have to be accommodated on a single bridge structure is shown in Fig. 13.4. The east abutment would have been founded on a simple strip footing if it had not been for the presence of the railway tunnel. This was disused and its long term stability could not be guaranteed. It was therefore considered prudent to support the footing below the tunnel and a piled solution was adopted. In the choice of type of pile, consideration was given to the dense and bouldery nature of the till, the load to be carried and the depth to the required founding level. The choice of large diameter bored piles proved successful: a bucket auger and heavy chisel coped adequately with conditions which had put considerable strain on both rotary and percussive site investigation borings. The main concern with the pad foundation to the eastern pier was to limit its settlement relative to the adjacent rock bearing foundations. The west abutment was perched on a small rock bluff which formed the lower end of the western shoulder of a massive multiple landslip, resulting from postglacial valleyside failure. The disturbed, frost shattered and weathered nature of the rock together with its steep angle and inclination of the slate cleavage dictated a foundation which avoided loading the slope: the solution developed consisted of deeply founded cylindrical piers and inclined rock anchors.

The requirement for end bearing piles to rock (see Sect. 12.1), the presence of cobbles and boulders and adverse groundwater conditions led to the adoption of steel H-piles for the foundations of some bridges on the section of the M6 motorway through the Lune Gorge; a classic glaciated valley in the English Lake District where rail and roads compete for space along the confined valley floor. Problems arose due to the deflection during driving of some of the raking piles. The stoney soils also prevented the cutting of sleeved holes around the pile heads for concrete encasement above the water table. Total excavation, concreting and backfilling had to be employed.

In another project a motorway underbridge founded on strip footings resting on compact till was found to have settled by an unexpectedly large amount. Underlying the till was a limestone surface with open joints containing water at sub-artesian pressure. A possible explanation was erosion of the till from below by groundwater flow in the limestone (analagous to incipient swallow hole formation; Fig. 13.5). Although the rate of settlement of the structure had become negligible it was decided to stabilise the situation by grouting the till and upper zone of limestone with a cement-PFA grout.

Culverts invariably occupy low points in the terrain which often possess poor ground conditions. In such situations culverts are subject to large differential settlements and lateral strains which require good articulation if severe cracking is to be avoided. The presence of large buried boulders or sudden changes in bedrock profile are common occurrences in or below glacial deposits but their detailed distribution is impossible to determine during site investigation. Unexpected hard spots or rapid changes in thickness of compressible soil below a culvert founding level can grossly aggravate local differential settlements. Too rapid construction of the embankment can lead to slope instability which may displace the ends of the culvert.

After excavation highly overconsolidated lodgement tills, tend to remain in lumps with high internal suctions due to stress relief. These lumps resist complete breakdown when compacted in a fill and are susceptible to wetting and softening with time. If heavily loaded, unacceptably large settlements may occur.

13.8 Tunnelling

For the tunneler, working in glacial materials can involve every form of soft ground (Bevan and Parkes 1975). About half the soft ground tunnels in the UK are formed in glacial materials. Because of varying soil conditions access to the working face is essential which precludes the use of full face machines. Machines of the road header, hydraulic impactor or back-acter type are preferable. The use of compressed air tools and explosives are sometimes necessary in the harder clays. For safety a tunnel shield should be used throughout even though it may only be needed for 20 to 50 percent of the length. Linings should normally be placed immediately behind excavation. Circular linings of precast concrete or cast iron are commonly used, bolted and grouted for positive support. In addition to comprehensive investigation from the surface, by borings and geophysical techniques, probing ahead during construction is highly desirable.

Two major problems which are often encountered during tunnel construction in glacial deposits are, firstly, unpredictable lateral and vertical variation of material type and, secondly, troublesome groundwater flow, sometimes under considerable pressure, from included beds or lenses of silt, sand and gravel. The glacial succession of buried valleys, from weathered rock head upwards, commonly consists of lodgement till overlain by fluvial valley infill overlain by thick clay sequences resulting from glacial lake ponding along the old valley. The most troublesome deposits are the laminated clays, strong massive clays, silts, sands and gravels. Beds of silt with groundwater under sub-artesian pressure are frequently encountered in laminated clay requiring use of low pressure compressed air for the main drive. Complicated depositional history makes the correlation of successions difficult even with closely spaced boreholes and shield driving is advisable. The frequent presence of silt and clay in water bearing granular deposits often makes their treatment by grouting impossible such that full scale trials are a necessary preliminary to tunnelling work. The importance of very careful groundwater observations during site investigation is stressed.

In complete contrast, consultation of the construction record for the disused tunnel shown in Fig. 13.3 during the desk top stage of site investigation indicated that, because of the very stoney and dense nature of the till through which it penetrated, blasting had to be employed to aid excavation.

CHAPTER 14

Dam and Reservoir Construction in Glaciated Valleys

M. S. Money

INTRODUCTION

Design and construction of dams and reservoirs places a great responsibility on engineering geologists and geotechnical engineers. To the problems posed by extensive and often heavily loaded foundations are added those of seepage control, materials and of slope stability both inside and outside the reservoir. The aim of this chapter is to provide a summary guide to the foundation conditions and designs encountered in British practice within glaciated valleys with brief accounts of case histories which illustrate not only varied geology but a variety of approaches to deal with difficult foundations. Relevant reviews of engineering experience in other high mountain valleys are provided by S.A.W.E. (1970).

14.1. Previous Research

In Britain, the majority of reservoirs have been sited in upland glaciated areas and, from about 1820 to 1950, most were earth embankments with central clay cores, the so-called "Pennine Dam", a design which was successfully but perhaps often unnecessarily applied to a wide range of foundation geology. Since World War II howver, increasing numbers of concrete dams have been constructed, in particular for hydroelectric schemes in Scotland and Wales, and designers of earth dams have adopted alternative methods of seepage control. Several reviews of British dam geology have been published, of which the account by Lapworth (1911) of cut-off trenches with its detailed (but sometimes obscure) drawings is essential reading. Later reviews by Walters (1971), Morton (1973), and Knill (1974) each cover a wide range of case histories and provide useful references. In recent years it has become common in British practice to employ an engineering geologist full-time on site during construction of the larger reservoir schemes. Regrettably this has not lead to an increase in the number of published detailed sections and plans from which so much valuable information can be obtained.

14.2 Classification and Legislation

An overview of British dams in glaciated valleys is hampered by the lack of a comprehensive published list or gazeteer. The most useful list is that in the International Commission on Large Dams ´World Register of Dams´ (ICOLD 1979) although the 559 or so dams registered include only about a quarter of the British reservoirs. ´Large Dams´ as defined by ICOLD are those with heights of more than 15 m from the general foundation area to the crest. Dams of heights between 10 and 15 m are also listed if the dam or reservoir also meets one of the following criteria; if the crest length is more than 500 m, if the reservoir capacity is more than 1 million m^3, if flood discharges are more than 2000 m^3/s, if there are particularly difficult foundation problems or an unusual dam design.

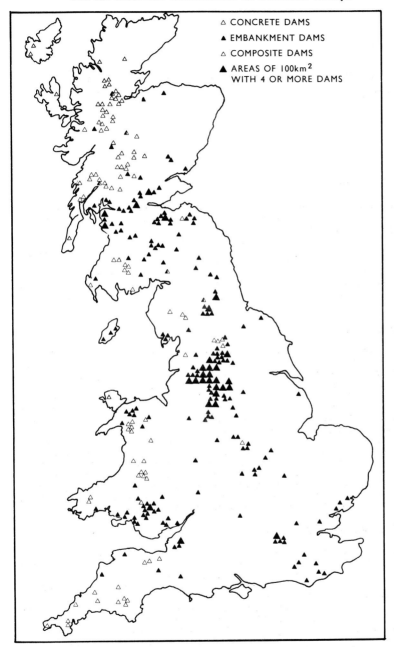

Fig. 14.1. The distribution of Large Dams in mainland Britain based on ICOLD (1979).

The ICOLD list gives the following distribution of Large Dams, the great majority of which are in public ownership

England and Wales	372
Scotland	155
Northern Ireland	22

The list is not completely exhaustive since the criterion of height which qualifies dams for inclusion means that some dams impounding very small reservoirs are listed while some large reservoirs are excluded.

Reservoir construction, maintenance and inspection in the United Kingdom is at present regulated by the Reservoirs (Safety Provisions) Act of 1930 which was introduced after the loss of life resulting from dam failures at Dolgarrog, N. Wales and Skelmorlie, Ayrshire in 1925. A reservoir capacity in excess of 22,700 m (5 million gallons) above ground level is the sole criterion for the application of the Act but there is no central register of reservoirs.

Although the 1930 Act has operated with reasonable success in that there has been no major disaster or loss of life since its introduction, there have been several serious incidents (Moffat 1976). Newer legislation in the form of the Reservoirs Act 1975 provides for a more rigorous system of inspection and will require Local Authorities to maintain a register. Recent, probably incomplete, estimates of the number of reservoirs subject to the 1930 Act have been given in Parliamentary Replies (Hansard 1981).

England and Wales	in public ownership	775
	in private ownership	700
Scotland	in public ownership	430
	in private ownership	95
	Total	2000

14.3 Historical and Geographical Perspective

14.3.1 Distribution Of Dams

The earliest dams built in Great Britain were mainly weirs designed for river diversion to mills and navigation. Significant numbers of impounding dams were also built to provide water power for mining in upland Wales, the Pennines and the Lake District, and for the iron industry of the Weald. In the second half of the eighteenth century numerous dams, some of considerable size, were constructed purely for landscaping (Kennard 1976). The construction of large reservoirs for canal systems began in the last decade of the eighteenth century. As the Industrial Revolution got under way industrial and domestic water demand rose rapidly and between 1847 and 1917 over 230 Large Dams were completed. Further bursts of activity in dam construction occurred in the two post-world war periods, notably from 1950 to 1970 when most of the development of hydroelectricity in the north of Scotland took place.

The great majority of Large Dams are located in upland glaciated areas (Fig. 14.1, 9.1) of enhanced glacial erosion (Fig. 9.4). The distribution of concrete dams is predictable, being concentrated mainly on Precambrian and Lower Palaeozoic rocks. The greatest concentrations of dams are those around the industrial areas of south Wales and northern England and the Midland Valley of

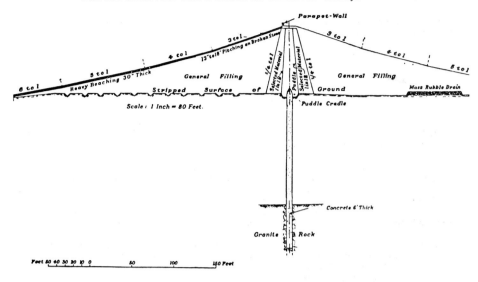

Fig. 14.2. Cross-section of a ´pennine type´ dam. The design follows C19th British practice but was used as late as 1932 (McIldowie, 1934).

Scotland. Typically these are nineteenth century dams, in narrow glaciated valleys cut in sandstones shales and limestones. Most are of a similar design with a narrow central core of puddled clay extended into a deep cut-off trench excavated through drift material to rock (the so-called ´Pennine Dam´).

14.3.2 The Pennine Dam

The Core. The most important feature of the dam is the puddle clay core (Fig. 14.2). This was relied upon as the sole watertight membrane which must not and, if properly constructed, would not leak. Seepage was not expected to reach the downstream shoulder and drains were installed only to dry out the site at foundation level or to remove surface water from the downstream slope.

Clay for the core was usually selected and processed with great care and was often imported from outside the reservoir basin. Long haul roads or extensive networks of light railways were constructed for haulage. Before the widespread introduction of mechanical excavators the borrow pits were generally shallow and it seems likely that weathered shales and soliflucted clays were used. After removal of the larger stones, the clay was often processed to provide a uniform material suitable for puddling and this might involve stockpiling, and hence weathering, the direct addition of water, and reworking, even to the extent of using pug mills. Regardless of the clay type and original moisture content good puddling produced a material with an average undrained shear strength of 10 kN/m . Final placing was achieved by workmen ´heeling´ the clay into place with their boots. If the clay was either too wet or too dry it was unworkable and thus the moisture content was strictly controlled. This laborious method of placing died out in the U.K. in the 1950´s with the introduction of modern compaction plant. Detailed reviews of modern clay core construction are provided by Kennard et al, (1978) and Vaughan et al, (1978).

The principle of graded filters was put forward in 1872 by Latham and successfully employed by Jacob (1894) in Indian dams but did not find favour in Britain. Stone toes were commonly used in clay embankments but neither these, nor drains nor the original river bed were protected from seepage erosion. Engineers did of course add their own individual features. J.F. Bateman (1841) for example, advocated placing a layer of peat 375 mm thick adjacent to the core so that if a leak occurred the fibrous particles would be drawn into it and help to seal it.

The Cut-Off. Graphic descriptions of cut-off trench construction are given by Watts (1897). Excavation, which might well occupy two years, was carried through the glacial drift into rock until a watertight horizon was reached. This was often found at considerable depths, in some cases down to 60 m, and the resulting trench had an irregular stepped profile which tends to promote differential compaction of the core. Substantial quantities of water flowed in the trenches and high piezometric levels led to heave of the floor and difficulties with backfilling. Contemporary accounts also speak of spring flows; Steer and Binnie (1951) describe the selective erosion by springs of puddle clay placed on rock resulting in the removal of fines.

Although cut-off trenches were timbered some relaxation of the trench margins and hence increase in permeability must have occurred. Watts (1906) described cracking of a culvert at Underbank by major joints in bedrock which he ascribed to opening of the cut-off trench 24 m away.

Towards the end of the 19th century concrete began to be used at the base of the trench and later as a substitute for puddle clay. The quality of this early concrete may be suspect. Watts (1904) for example, advocated that it should be placed as wet as possible without compaction.

General Fill. Till, properly compacted at a suitable moisture content, can make excellent fill but its use is not without problems. Excavation, even with modern plant is impeded by boulders and the high undrained strength when dry, while wetting promotes rapid softening. Glacial deposits are often highly variable in composition and it is therefore likely that 19th Century fill was often extremely heterogeneous. It was common practice to zone the shoulders by placing material with the highest clay content nearest to the core with the more granular material on the slopes. Success in practice depended on the materials available and the quality of supervision. Leslie's specification for earthwork published by Conybeare (1859) shows that some Engineers took great care with their fill, while others were criticised for making dams like railway embankments.

Wave protection for the upstream slope was normally provided by laying masonry pitching on a bed of gravel. Downstream slopes are generally laid to grass, often kept immaculately trimmed.

14.4 General Features of Glaciated Valleys

14.4.1 Valley Profiles

The storage characteristics of reservoirs and the choice of the most appropriate dam type are largely controlled by the topographic and geological profiles of valleys (Sect. 4.1, Fig. 9.4). An apparent U-shaped cross-profile in many valeys is produced by a series of slope facets, the highest and steepest being on bare rock, the intermediate slopes formed by screes and the base of the valley occupied by till sequences, outwash or solifluxion debris. In this account valleys are broadly classified on the basis of longitudinal and transverse topographic profiles as follows:

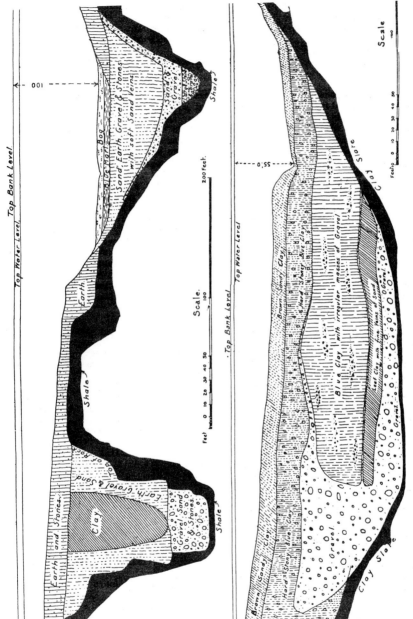

Fig. 14.3 Dam sites showing variability of rockhead profile and valley infill (from Lapworth, 1911).
a) Turners Embankment, Yarrow Reservoir. b) Llanefydd Reservoir.

1. Longitudinal profiles

   (i)   Normal valleys, which slope downstream throughout, although
            not necessarily at an even or steadily changing gradient.

   (ii)   Dammed valleys, in which a natural rock dam or bar (reigel) obstructs
            the valley

2. Transverse profiles

   (i)   Symmetrical valleys, in which the surface profile and rock head
            are broadly symmetrical about the centre.

   (ii)   Asymmetrical valleys in which these profiles are not symmetrical
            and, commonly, the modern river course bears no rlation to
            the infilled valley.

   (iii) Divided valleys in which a rock ridge divides the profile.

3. Corries, which may be the steep-sided termination of a glaciated
   valley but can also be independent features of mountain summits.

There are no firm boundaries between the various types, but the classificatin
forms a useful framework for the discussion of engineering problems and is used
here to group the case histories quoted. The nature of sediments along glaciated
valleys and appropriate techniques for subsurface investigation and sampling have
been covered in Chapters 4,5,6,7 and 11. Figure 14.3 shows two of Lapworth's
1911) sections of dam sites and illustrates the variability of subsurface
conditions.

14.4.2 Rock Head

Proving rock head is an essential objective of the dam site investigations.
Difficulties tend to result either from the nature of the soil/rock contact or
from the presence of large boulders or rafts of rock in the overlying till.
Rockhead contacts can vary within a single site and over a distance of tens of
metres from a knife-sharp junction to a zone several metres thick of boulders, or
shattered or glacitectonized rock (see Sect. 2 for discussion). Every
opportunity should be taken of examining natural outcrops or trial pits to assess
the likely conditions before deciding on a strategy for drilling. The dangers of
mistaking boulders for bedrock are well known but often ignored. It is usually
recommended that rock should be proved with a minimum of 3 m of rotary coring but
this figure may be inadequate. It must always be increased if there is any
likelihood of large boulders or rock rafts being encountered and if core recovery
is poor. Three metres of coring is also insufficient for representative
permeability tests to be carried out in the rock, particularly when drill casing
has to be set. Minor structural features of the rock recovered as core should be
checked to ensure that they confirm to local or regional orientations.
Bedding/cleavage intersections and other lineations are useful indicators even
without sophisticated core orientation techniques.

Rock head profiles must be interpreted with care and with an eye to anomalies
caused by glacitectonic deformations, deep-seated structures such as valley
bulges and cambers and buried landslides (Figs. 2.19, 5.8, 14.4).

14.4.3 Rock Conditions

Conditions below rock head may be just as variable as those above it. The

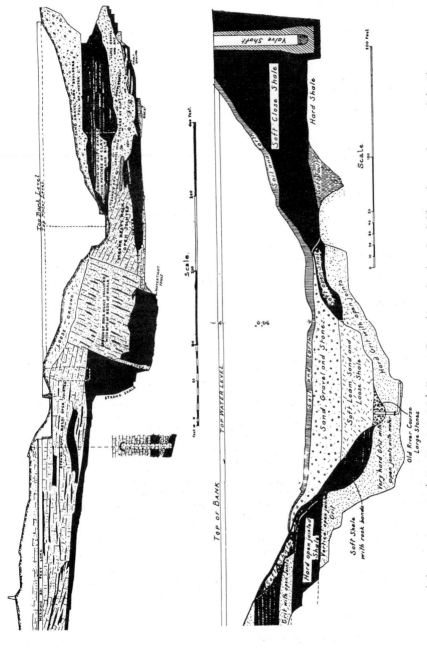

Fig. 14.4 Sections of a) Woodland and b) Yarrow Dam sites showing landslipped ground exposed in the cut-off trenches (from Lapworth, 1911).

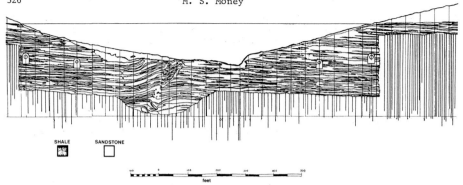

SHALE   SANDSTONE

feet

Fig. 14.5. Central section of the cut-off trench and grout curtain of the
Ladybower Dam from Hill (1949) showing valley bulging (Figs. 5.1, 5.8). Vertical
lines are boreholes.

effects of glacitectonic thrusting and periglacial ground freezing may extend
many metres below rock head (Sects. 2.8.2, 5.3, 5.4, Figs.2.19, 5.1, 5.8). Knill
(1968) has described foundation conditions of several sites for concrete dams at
which discontinuities infilled with fine sediment were found to occur to
substantial depths below rock head. Joints were sub-parallel to the ground
surface and had a curvi-planar form dividing the rock into lenticular masses.
These discontinuities were found generally within 15 m of the surface and were
infilled with clay-silt or silt seams. Knill suggested that the joints were
initiated as shear fractures due to ice movement and that they were then opened
by permafrost, eventually allowing fine sediment to be washed in and deposited.
The discovery of these discontinuities lead to extra excavation of the dam
foundations and, in several cases, to slope instability.

Glacial shear zones sub-parallel to bedding planes in sedimentary rock sequences
are common although rarely documented. They may be similar to and sometimes
exploit pre-existing intraformational ´bedding plane´ shears between shales and
limestones or sandstones resulting from regional tectonism. Ice movement may
also open up vertical or steeply dipping joints and fill them with till.

Discontinuities in the rock are not of course all filled with sediment and the
rock in many cases is shattered by periglacial processes to provide local or
general zones of high permeability (Sect. 13.3). The opening of joints is
particularly associated with cambering of valley sides and valley bulging
(Fig. 15.8, 14.5) At many Pennine dam sites in cambered terrain wing trenches
were constructed on the dam abutments connecting with the dam core to prevent
seepage through the valley side slopes. Walters (1971) has discussed the
implications of valley bulging for grouted cut-offs.

14.4.4 Groundwater

It is important to understand the groundwater regimes both at the dam site and in
the reservoir rim in order to assess the potential for leakage and possible
remedial measures, not least the design of the cut-off for the dam. Although
19th century and later engineers insisted on extending cut-offs to rock it is
clear from modern permeability measurements that the permeability of rock masses,
particularly near rock head, may be one or more orders of magnitude greater than
that of the overlying till. Spring lines on the till/rock contact and artesian

flows from site investigation holes often confirm the presence of a confined aquifer below the drift (Fig. 13.1). Within the drift itself sand or silt horizons may constitute individual aquifers that may or may not be interconnected to each other or to the rock.

Site investigations should include careful observation of water levels during drilling, permeability tests in rock and soil and the installation of piezometers at geologically controlled positions. Interconnection of aquifers and their hydraulic properties may need to be determined by pumping tests (Sects. 11.6.5, 15.2). The experience gained at the Backwater, Cow Green and Derwent sites, all described briefly below, illustrates the importance of these techniques (Sects. 14.5.2, 14.5.3).

14.4.5 Construction Materials

The materials required for embankment dams include soil or rock fill, and material suitable for filters, drains, rip-rap and concrete aggregates. Although in the past materials were often imported to sites, current practice is to win as much material as possible from within the reservoir basin, not only to reduce haulage costs but to minimise damage to the natural environment by the opening of quarries and by construction traffic. Such policies may impose considerable restraints on the dam design. Material requirements for concrete dams are less varied but equally demanding.

The majority of concrete dams in Britain are situated in the Scottish Highlands, where schist bedrock predominates and in the Southern Uplands, the Lake District and Wales where the rocks are predominantly argillites at a low grade of metamorpism. Aggregates for these dams have generally been quarried locally from igneous intrusions, or are imported. Glaciofluvial deposits have in general been found to be unsuitable for concrete because of their heterogeneity.

The properties of tills and other glacial deposits as fills for earthworks have been thoroughly discussed in Section 13.4. Most dams are sited in areas of high rainfall and potential fill material may be wasted in borrow areas especially if they are poorly drained or worked by scrapers. Material that appeared satisfactory at the investigation stage may prove to be unsuitable when excavated (Peters and McKeown, 1976; Pepler and Mackenzie, 1976; McKeown and Matheson, 1979) and it is advisable to locate and prove several times the quantity of materials actually required.

Another consideration is that when siting borrow pits it is important to ensure that drift materials providing a low permeability seal to the reservoir are not breached.

14.5 Case Histories

14.5.1 General

The case histories given here have been chosen to illustrate the range of ground conditions encountered and a variety of approaches to dam design, in particular the choices of dam type and of measures to control seepage. Some of the examples have appeared in print many times but Walters (1971) in his well known book 'Dam Geology' wrote

"... bad mistakes are made from borehole information; a classic case was that at the Silent Valley dam for the water supply of Belfast when a granite boulder at a depth of 50 ft below the surface was assumed to be rock which proved to be nearly 200 ft below the surface. This mistake is so well known that it is not

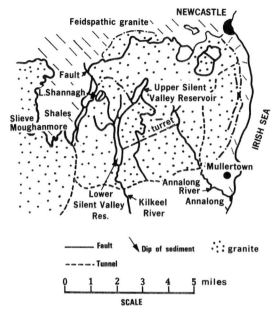

Fig. 14.6 Reservoirs and geology in the Mourne Mountains.

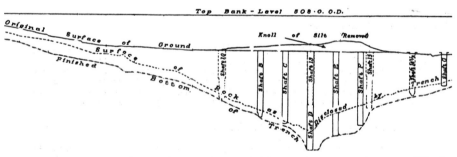

Fig. 14.7. Section of the cut-off trench and dewatering shaft at Lower Silent Valley Dam (After McIldowie, 1934; Fig. 14.6).

likely to occur again!"

Regrettably this mistake is still being made and no apology is needed for including such cases, whose lesson must in any event be learnt by each generation of engineers and geologists. The schemes to be described have been grouped according to valley profile as previously explained:-

Normal valleys

Symmetrical:

> Silent Valley Dam
> Backwater Dam
> Derwent Dam

Asymmetrical:

> Cow Green Dam
> Brenig Dam
> Cold Beck Dam

Divided:

> Thirlmere Dam
> Glen Shira Dams
> Orrin Dam

Dammed Valleys:

> Vrynwy Dam
> Haweswater Dam
> Loch Sloy Dam

Corries:

> Cruachan Dam
> Stwlan Dam
> Lough Shannagh
> Marchlyn

14.5.2 Normal Symmetrical Valleys

The Silent Valley dam illustrates not only the problems of site investigation but the difficulties of forming a completely effective cut-off in thick permeable glacial deposits. These difficulties proved too much for the original engineer and the completion of the cut off was only achieved by extraordinary measures. The other two dams, Backwater and Derwent, illustrate better investigations and also alternative forms of cut-off tailored to the geological conditions.

Lower Silent Valley, Co. Down, Northern Ireland

The Silent Valley is a broad glaciated valley running north-south through the centre of the granite Mourne Mountains (Fig. 14.6). By 1901 the area had already been exploited for water supplies for Belfast by the construction of river intakes and a 35 mile-long conduit. In 1923 a contract was let for the construction of an embankment dam 460 m long with a maximum height of 27 m with a puddle core and a concrete cut-off to rock (McIldowie, 1934). By 1926 the preliminary works were complete and about half of the length of the cut-off trench had been excavated from the two ends and had been bottomed out in rock. The materials met with were "sand, gravel and boulders well bound together" and "capable of standing vertical for long periods without support". In attempts to excavate in the centre of the valley, however, the Contractor had met with water-bearing silts which had 'boiled' into the pit and "in addition to the liveliness of the material, boulders interfered with pile driving". The contractor sank additional boreholes on the trench line and these indicated that rock levels in the centre of the valley reached a maximum depth of 60m, more than three times the depth assumed from the original investigation, in which, as is now well-known, boulders had been mistaken for bedrock.

Ten shafts were sunk upstream and along the trench (Fig. 14.7) and were used for pumping to dewater the ground. Previously the river was diverted and measures taken to drain and exclude water from the trench area, following which the shafts were sunk successfully using compressed air. Pumping from the shafts and the open trenches resulted in a groundwater lowering of 27 m, sufficient to permit opening out of the trench between the shafts using strutted cast-iron segments and heavy timbering. The central section of the trench was sunk through more than 6 m of "laminated silts", which had proved so troublesome to excavate when saturated, and these were underlain by silty, sandy material containing numerous large boulders. Concreting of the trench was completed in 1930, the embankment itself in 1932. There was no clay in the valley suitable for the puddle core and material was imported by light railway. Most of the general fill for the dam was also won from glacial deposits outside the reservoir.

A second dam, the Ben Crom dam was later constructed further upstream in the same valley to impound the Upper Silent Valley reservoir (Anon, 1955). Foundation conditions at the two sites could hardly have been more different. At the upper site granite bedrock was covered with little sediment cover and the foundations of the concrete gravity dam required only minimal dental treatment.

Backwater, Tayside Region

The Backwater Dam which impounds water for the supply of Dundee is situated in the Grampian Highlands on schists and phyllites with a foliation dipping at 20 -30° and striking at roughly right angles to the dam axis. The schist was found to be weathered and extensively faulted and jointed, the discontinuities frequently being iron-stained and clay-filled. (Geddes and Pradoura, 1967; Geddes, Rocke and Scrimgeour 1972).

The valley cross-section is similar in scale to the Silent Valley, although with a length of 549 m and a height of 43 m the finished dam is considerably larger. As at the Silent Valley the rock head profile is an irregular flat 'V' but with a maximum thickness of infilling of 50 m, at the base of which was some 6 m of clast-rich till overlain by 6 m of laminated silts and sands. These are followed by a complex sequence of sands and gravels up to 24 m in thickness and containing some highly permeable horizons and finally a mantle of soliflucted till (Fig. 14.8).

It was found at an early stage of the investigations that the hydrogeology of the foundations was complex. The permeability of the sand and gravel complex ranged between 5×10^{-4} and 1×10^{-6} with an average of about 1×10^{-5} m/sec whilst artesian pressures were found 3 m above ground level in the drift and 12 m above ground level in the rock. Comprehensive instrumentation revealed that water levels in both rock and drift responded rapidly to rainfall and that there were seasonal variations.

It was decided to form the cut-off by grouting with tubes-a-manchette and a full scale trial was carried out. In situ tests and inspection from a trial shaft showed that grouting was effective and reduced the permeability of the sand and gravel to 5×10^{-7} m/sec, comparable to the permeability of the till. At both ends of the dam where the drift was relatively thin a full concrete cut-off was formed entrenched a minimum of 6 m into reasonably sound rock, with grouting in a single row of holes to a minimum depth of 15 m. The remainder of the cut-off was formed by one, three or five rows of holes depending on depth and soil type (Fig. 14.8). Three types of grout were used (a) bentonite cement, (b) deflocculated bentonite (c) silicate based. The rock head zone proved difficult

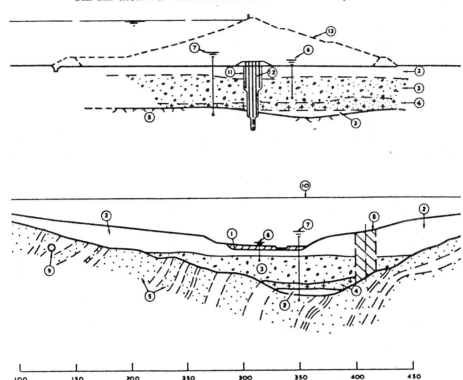

Geological Sections.

1. Recent Alluvium.
2. Mountain Till.
3. Sand and Gravel.
4. Lake Deposits.
5. Rock - Schistose Grits and Phyllites.
6. Artesian Water Level — Overburden.
7. Artesian Water Level — Rock.

8. Test Section.
9. River Diversion Tunnel.
10. Crest of Dam.
11. Grout Curtain.
12. Rows of Grout Tubes.
13. Profile of Dam.

Fig. 14.8. Sections at the Backwater Dam site, from Geddes and Pradoura (1967).

both to define in drill holes and to grout successfully. The use of bentonite
cement failed to achieve the required level of permeability and the zone was
reinjected with silicate aluminate grouts which gave satisfactory results.
Observation of piezometric levels both before and after impounding have confirmed
the effectiveness of the cutoff and although, as predicted, there is significant
seepage this is not lost to supply since it is impounded further downstream.

Till excavated from the vicinity of the dam was used to construct the embankment
with sand and gravel being used for drainage blankets.

a

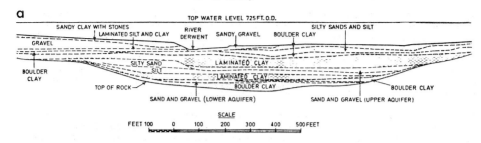

b

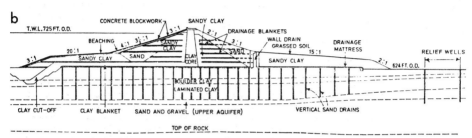

Fig. 14.9. Longitudinal (a) and cross-section (b) of the Derwent Dam. After Ruffle (1965).

Derwent Dam. Northumberland & Durham.

The Derwent Dam was completed in 1966. Its design construction and performance are comprehensively described by Ruffle (1965), Ruffle (1970), Buchanan (1970) and Rowe (1970). Bedrock in the Derwent Valley consists of essentially flat-lying sandstones, shales and thin limestones. A broad U-shaped valley in these beds has been filled with a varied sequence of glacial deposits (Fig. 14.9) with a maximum thickness of 55 m. The dam, 915 m long and 36 m high was originally intended to have a conventional concrete cut-off extending some 6 m into unweathered rock and tenders were invited on this basis. It was recognised that construction of such a cut-off would present difficulties, not least because a layer of silts, sands and gravels (the upper aquifer, Fig. 14.9) contained artesian water. The design also provided for a large number (ultimately 4475) of vertical sand drains to dissipate excess pore pressures in the laminated clay.

Construction of the cut-off wall commenced on both sides of the valley but meanwhile the Contractor had undertaken an additional investigation in the centre of the valley to establish the number and spacing of the wells required to de-water the trench excavation. This investigation established that the drift deposits were more predictable than had been assumed from the original investigation and lead to the recognition of three aquifers, ´upper´ and ´lower´, both of which were sands and gravels, and ´rock´. Pumping tests and piezometric observations showed that while the lower aquifer was connected to the rock, albeit not very effectively, the upper aquifer was not so connected.

These results indicated that the trench could be dewatered using about 22 wells pumping 14,000 m³/day instead of the 36 wells pumping up to 55,000 m³/day originally envisaged.

At this stage an alternative design was adopted in which a partial cut-off connected the core via a clay blanket to the till overlying the laminated clay. This cut-off was constructed along the upstream toe so that seepage could not by-pass it via the sand drains. The cut-off was tied in to the previously constructed core walls and in addition, relief wells were constructed at the downstream toe to avoid uplift pressures resulting from artesian water in the upper aquifer. Relief wells were also constructed in rock in one abutment since the rock was thinly mantled and exposed near top water level.

The clay blanket and core were excavated from borrow pits close to the dam, substantially increased quantities being required for the altered design. One borrow pit accidentally exposed rock in several areas and these were blanketed by 1.2 to 1.5 m of clay to prevent leakage. The shoulders of the dam were formed of sand and sandy clay, the sand proving particularly hard to find in suitable quantities. Sandy gravel used to form drainage blankets was composed of soft sandstone and broke-down badly under compaction.

14.5.3 Normal Asymmetrical Valleys

Dam sites may be classed as asymmetrical when either the present topographical cross-section of the valley is marked asymmetric, or where the axis of a buried valley is offset from that of the modern valley. It is common to find that the modern river has cut down on one side of the valley in rock whilst a deeper buried channel lies adjacent. Such rock exposures can easily mislead the engineer into thinking that those exposures represent the lowest rock head levels. These sites pose problems of foundation conditions which may vary from sound rock to deep drift and the difficulty of establishing the position, profile and contents of the buried valley. The tranditional and conservative approach has been to design the whole dam as an earth embankment. Three examples of such valleys are given here: Cod Beck, in which the traditional aproach was followed but difficulties occurred due to seepage through the buried valley; Cow Green where a composite earth and concrete dam was constructed, and Brenig where a rock-fill embankment was constructed.

Cod Beck Dam, North Yorkshire.

Cod Beck is a tributary of the River Swale and its valley forms a small embankment on the western side of the Cleveland Hills opening into the Vale of York.

Both valley sides at the dam site are steep and landslipped. The right bank and the lower part of the left bank are cut in shales but most of the left bank is composed of superficial materials underlaying a broad terrace. The level of this terrace effectively limited the top water level since a very much longer dam would have been needed to impound to a higher level (Fig. 14.10). As at the other sites discussed above the buried valley is both wider and deeper than the present valley. The infill in this case however appears to be mainly fluvioglacial materials (Anon, 1954). The Cleveland Hills stood above the Devensian ice (Fig. 9.1) and there is no evidence in the valley of glaciation apart from what is shown in the cross-section as 'boulder clay' but which is probably a solifluction sheet. Most of the valley infill results from glacial damming of the river's outlet.

The 30 m high embankment is of traditional design with a central puddle clay core, the material for which was imported from old brickworks some 16 km from the site. The shoulders are zoned, the inner zones consist of shales and clays from required excavation and the outer zones being waste material from a disused sandstone quarry close to the site. The clay core rests on a concrete cut-off.

Wing trenches were constructed on both sides of the valley and boreholes were
sunk on the left bank to find the shortest route across the buried valley. This
investigation showed the presence of sand horizons which had not been identified
in the original site investigation. There are two major sand layers (Fig. 14.10)
and after trials had shown that the sands could be grouted, the concrete cut-off
was bottomed-out in the intervening clay but not extended to the side of the
buried valley. Grouting was continued into bedrock on both sides of the valley.

During impounding in 1953 a slip was reactivated in the glaciofluvial deposits on
the left bank downstream of the dam, and slope movements in the form of mudflows,
continued for two years. Horizontal drains and vertical pumped wells were
installed and the slope was cut back and stablilised. Over the next five years
however, the flow from the horizontal drains diminished and a further major
failure occurred in 1961 (Knill 1974). Investigation of the failure showed that
the slipping had brought the shear strength in the laminated clays from a peak
value of $\emptyset' = 30°$ to a residual value of the order of 16 . There was also
evidence that the clays were dispersing in situ and so material which was
effective as a drain for the sand horizons did not act as a filter for the clays.
Further drains were installed using porous sintered polythene tubing as the main
element and the slope was regraded. No further movement has occurred.

In this case the extent and continuity of the sand layers was not established in
the original site investigation and although the partial cut-off was successful
in preventing large quantities of seepage, it was not successful in controlling
seepage pressures downstream of the dam.

Cow Green Dam. Cumbria/Durham.

Cow Green is a river regulating reservoir situated in Upper Teesdale and
completed in 1970. There was considerable opposition to the scheme on both
technical and environmental grounds; much of the reservoir area is underlain by
limestones while metalliferous mine workings are also present below top water
level and there were considerable doubts concerning the watertightness of the
reservoir (Kennard and Knill, 1969; Kennard and Reader, 1975; James and
Kirkpatrick, 1980).

In addition the area also contains a unique assemblage of plants believed to be
relict tundra vegetation and special measures were required to minimise damage
during and after construction.

The choice of the dam site was limited by the topography to a location some 250 m
upstream of the major waterfall of Cauldron Snout, formed by the Great Whin Sill,
a 75 m thick quartz-dolerite intrusion. The sill crops out at the dam site,
forming the river bed and left abutment but has been eroded to a lower level on
the right bank where the resulting channel has been filled with till
(Fig. 14.11). A composite concrete/earth embankment was judged to be the
cheapest and was adopted because of cost and also because the spillway and
diversion works could be incorporated in the concrete section. A longer
construction season would also be available. The combination of a high altitude
(465 m) and high annual rainfall (1570 mm) restricted the earthmoving season to 5
to 6 months.

No problems were experienced with the rock foundation and grout takes in the
single-row shallow curtain were small.

Till at the site was described as mainly 'stiff, dark brown, poorly-sorted,
unstratified, silty, sandy, clay of medium plasticity containing sub-angular to
rounded gravel, cobbles and boulders ...'. Boulders ranged up to 2 m mean

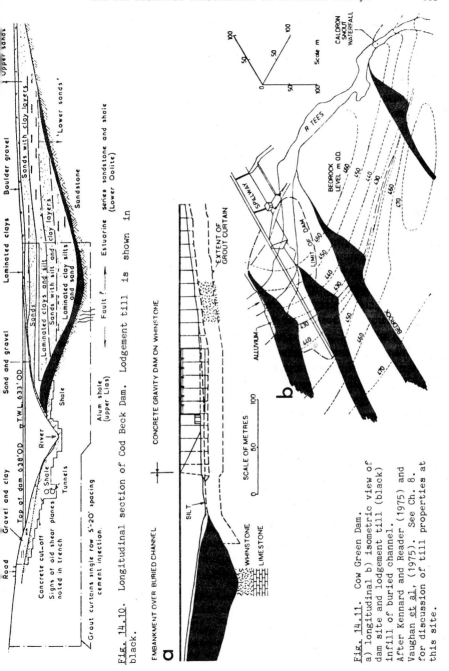

Fig. 14.10. Longitudinal section of Cod Beck Dam. Lodgement till is shown in black.

Fig. 14.11. Cow Green Dam. a) longitudinal b) isometric view of dam site and lodgement till (black) infill of buried channel. After Kennard and Reader (1975) and Vaughan et al, (1975). See Ch. 8. for discussion of till properties at this site.

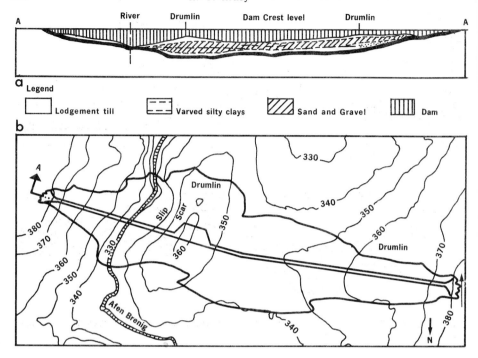

Fig. 14.12 Section (a) and plan (b) of Brenig Dam site from Colback et al, (1975)
Scale 1:5000.

diameter.   Various methods of sampling the clay were tried in the pre-contract
and construction site investigations but rotary drilling with impregnated diamond
bits producing 113 mm diameter cores, with mud flush was found to be most
satisfactory. The till exhibits wide variations in boulder content, plasticity
and strength and it may have been reworked by periglacial solifluction (Vaughan
et al, 1975;   Sect. 5.4).

Stability of the embankment was strongly dependent on the pore pressure generated
in the foundations and in order to accelerate consolidation nearly 1100 sand
drains were installed to maximum depths of 24 m, except where sand layers were
known to be already present. Monitoring of the performance of the sand drains
during construction showed that they were less effective than predicted and that
the values of ·the coefficient of consolidation of the till were extremely low.
There were also significant differences between the field behaviour of the clay
and the performance deduced from laboratory tests (see Sects. 8.3.3, 11.9.9 and
12.2), although it should be noted that the latter had to be performed on
remoulded compacted samples from which the stones had been removed. It was
estimated that the factor of safety at the end of construction fell to about 1.0,
but that due to the non-brittle behaviour of ·the clay this represented a
condition in which significant displacements were occuring, rather than a state
of imminent failure.

Some concern was felt for the stability of the 1-in-3 43 m high till slope

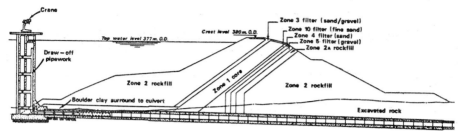

Fig. 14.13. Cross section of Brenig Dam at culvert. Outlet structures of culvert to right, not shown.

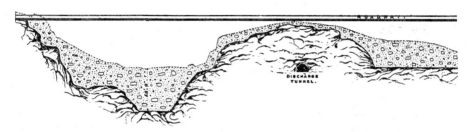

Fig. 14.14. Partial longitudinal section of Thirlmere Dam site from Hill (1896). Scale 1:500.

adjacent to Cauldron Snout where the in-filled channel daylights (Fig. 14.12). The slope sows signs of shallow soil creep and to avoid instability following impounding, relief drains were sunk at the toe of the embankment and other drains installed at the toe of the slope.

During construction of Cow Green Dam, several problems became apparent with internal erosion of the clay core at Balderhead Dam on which the design at Cow Green was based. Internal erosion is normally prevented by a filter placed downstream of the core but the Balderhead filter was proving ineffective. Vaughan and Soares (1982) describe in detail the redesign of the filter during the construction phase at Cow Green.

Brenig Dam. Clwyd.

The Llyn Brenig reservoir was constructed to regulate the River Dee in North Wales and was completed in 1976. Aspects of the design and constructin of the scheme are discussed by Ford et al, (1978) and Colbeck et al, (1975). At the dam site the Brenig valley was partially blocked by two large drumlins overlying mudstones (Fig. 14.12). The most economical position for an embankment dam was such that the maximum volume of the drumlins was incorporated into the structure (Crann, 1978).

Although foundation conditions at Brenig are broadly similar to those at Cow Green and a composite structure might have been feasible, an embankment section was chosen for the whole dam. There was only a limited quantity of till available within the reservoir basin, its winning and placing would have been affected by the weather and the cost would have been greater than using rockfill

in the shoulders. The dam was therefore constructed with a sloping till core
protected by filters and shoulders of mudstone compacted wet. Where the
embankment overlies drift the rockfill slopes are flattened with berms.

The two drumlins which form the central portion of the dam (Fig. 14.12) were
found in the subsurface to be not unlike Figs. 1.4 and 2.10b containing three
principal types of material; lodgement till, lenses of fluvioglacial sands and
gravels and laminated silt and clay which pass laterally into sand and in places
to gravel. The tills contained discontinuities in the form of fissures dipping
at both low and high angles and striking parallel to the inferred direction of
ice movement (Sect. 8.3.3). A slip scar was evident on the right bank of the
river on the flank of one drumlin and it was found that the slip movement had
occured along a silt horizon at the base and a fissure at the back.

Detailed investigations were carried out into the extent and importance of the
laminated silts and clays. These involved conventional soft-ground and rotary
drilling and sampling but recovery was often unsatisfactory and there was
considerable sample disturbance. Test trenches and construction excavations were
therefore used to obtain hand-cut orientated block samples and also permitted
observation of the discontinuity pattern in the till. Following tests on the
silts and clays, the design strength of the upstream shoulder was substantially
reduced and the calculated end-of-construction factor of safety fell from 1.5 to
1.2. Running analyses of stability in 1975 using construction pore pressures
indicated factors of safety less than 1.2 and the upstream berm was raised 5 m to
improve stability. Later, in 1976 a programme of careful coring with duplication
and in some cases triplication of holes was undertaken to investigate the extent
of the laminated silt and clay layers. This cost some £90,000 for about 250 m of
hole length, or about £360 per metre. This work confirmed that the layers were
either too deep to affect stability or were confined to a zone along the upstream
side of the drumlin area.

The core was keyed into a shallow trench into both rock and drift (Fig. 14.13),
the rock being covered with at least 50 mm of gunite before the clay was placed.
Contact grouting was carried out where the core rested on rock or shallow drift.
Grouting was also carried out in the drift using tubes-a-manchette and in rock
where the drift cover was less than some 30 m, but the takes in both materials
were very low.

14.5.4 Normal Divided Valleys

Divided valleys are those in which a rock ridge splits the valley into two parts.
The minimum volume of dam is achieved by using the rock ridge as a central
abutment.

Thirlmere Dam. Cumbria.

Thirlmere was a natural lake with a maximum depth of some 27 m lying in one of
the glaciated valleys radiating from the central Lake District. The lake level
was raised 15 m by the completion of a concrete gravity dam (Hill, 1896).

The valley at the dam site is divided by a small hill of rock which is almost as
high as the dam (Fig. 14.14). Advantage was taken of this hill as the location
of a tunnel for the scour and compensation water pipes. The foundations for the
dam were sunk into rock and as at Vrynwy the depth of excavation exceeded that
normally regarded as suitable for a gravity section.

Glen Shira. Strathclyde Region.

There are two reservoirs in the Glen Shira hydro-electric scheme, the main or upper reservoir which provides most of the storage and a lower reservoir from which water is drawn to the main generating station at Clachan, close to the head of Loch Fyne. A second generating station with a motor generator is located between the two reservoirs which have a head difference between the top water levels of 42 m. The upper reservoir is impounded by a round-head buttress dam with a maximum height of 40 m. The two end sections of the dam, where the height is less than 17 m, were constructed as gravity sections (Paton 1956).

Variable foundation conditions are presented by quartzites, metalimestones, schists and metadolerites sills with a rather irregular rock head profile, the lower dam is sited at the lowest practical point downstream, immediately above a 12 m high waterfall. In this river section a curved gravity dam 18m high has been constructed whilst a drift filled valley 3 to 6 m thick to the west was filled with an embankment 180 m long and 16 m high (Fig. 14.15). Site investigation showed that there was insufficient material of low permeability for the adoption of an earthfill core and thin vertical articulated concrete corewall was constructed. Local granular till was used for the shoulders with both upstream and downstream zones of rock fill.

Glen Orrin Dam. Highland Region.

A similar solution was adopted in Glen Orrin (Gowers 1963) where a gently curved gravity dam was constructed in the main valley and an embankment dam sited in a subsidiary valley. Apart from a mineralised fault and a zone of alteration the granulite in the foundations of the concrete dam was sound and required minimal stripping and grouting.

14.5.5 Rock-Dammed Valleys

Rock-dammed valleys are elongated ice scoured basins in which a rock bar or dam occurs. These natural barriers form obvious sites for dams but experience has shown that the quality of rock is poorer than might be expected. Substantial quantities of excavation may be required to achieve a sound foundation. The examples cited are Vrynwy and Haweswater and Loch Sloy where buttress dams were used to raise natural lakes.

Vrynwy Dam. Powys.

Vrynwy Dam was one of the first large gravity dams to be completed in Great Britain (1892) and impounds water for the supply of Liverpool to a depth of 27 m although the maximum height of the dam above the rock foundation is 49 m. Vrynwy is well known as an early example of comprehensive and successful site investigation (Deacon, 1896, Lapworth, 1911; Walters 1971).

Deacon, the Liverpool Borough Engineer, recognised that the valley had been glaciated, had formerly been occupied by a lake and deduced that there was a concealed rock bar lying at the downstream end of the alluvial plain that infilled the valley. 177 borings and probings and 13 shafts were sunk and a detailed contour plan and model of rock head were constructed.

To accommodate the base of the gravity masonry dam (Fig. 14.16) an excavation up to 39 m wide in rock was required, the side slopes of which were 1 on 1. There is no detailed description of the overburden, it is described by Deacon as being covered with 'impermeable glacial clay' through which a trench was excavated to rock and backfilled with puddle clay.

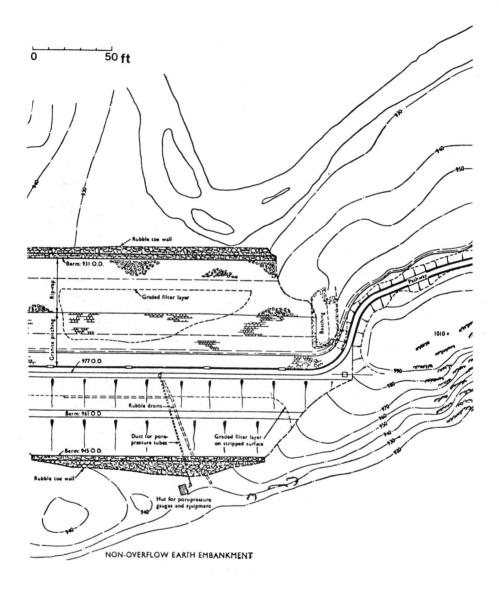

Fig. 14.15. Plan of Lower Shira Dams from Paton (1956).

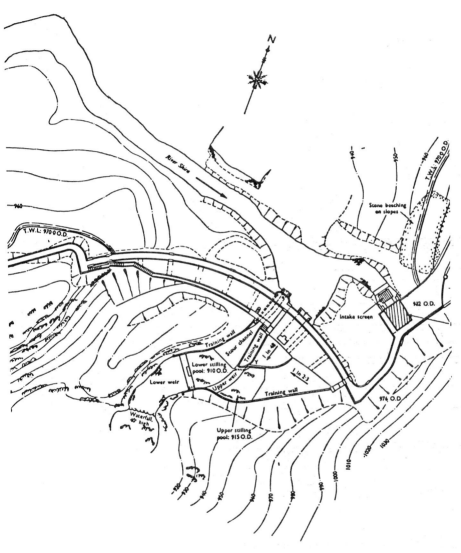

CONCRETE GRAVITY SPILLWAY DAM

In the main excavation, the lower levels consisted largely of 'enormous boulders'. Deacon observed that the rock dipped to the west and was heavily glacitectonised; large "masses weighing several hundreds of tons ... broken from their beds and moved some distance down the valley or only just detached" were met with. The rock below these masses was described as 'dislocated' to an average depth of 2 m and this also was removed, together with sharp irregularities of the rock profile, before being 'rendered scrupulously clean with wire brushes and jets of water'.

Haweswater Dam

Haweswater is situated in the western Lake District and the natural lake, which was impounded by a rock bar was raised by 29 m in 1934 to augment the water supply of Manchester. Taylor (1951) observed that "the site of the dam was to a large extent self selecting, no major movement being possible of the valley in both directions". Bedrock at the site consists of andesites which provided a sound foundation for the buttress design adopted (Walters, 1971). As at Vrynwy, the rock surface was very uneven and much investigation was necessary to establish the foundation level. On the north abutment the rock was covered by up to 9 m of drift whilst the overburden on the south abutment was described as mainly a mass of boulders (e.g. Fig. 4.5). In spite of the apparently good foundation conditions nearly 28,000 m^3 of overburden had to be removed and a further 33,000 m^3 of rock required excavation by blasting. The maximum depth of excavation in rock was 13.5 m in a zone of crushed rock but generally the rock was tight and required minimal grouting (Fig. 14.17). Haweswater was the first buttress dam to be built in the United Kingdom and is unusual in that the downstream faces of the buttresses are linked so that externally the dam appears to be a gravity section.

Loch Sloy. Strathclyde Region.

The original Loch Sloy was a shallow lake impounded by a rock bar in a steep-sided valley cut in schists. The lake level was raised by 45 m by the construction of a concrete dam, mainly of buttress design completed in 1950. (Stevenson, 1952). The valley floor at the site is irregular in section with numerous large boulders of schist, some of cottage size adding to the roughness of the terrain.

27,500 m^3 of peat and sandy material and 43,000 m^3 of rock were removed in the bulk excavation of the foundations. The rock was found to be extensively fractured and contained sandy-filled 'seams' (open joints). These became evident only during the final clean-up and washing of the foundations and had to be dug out. In two areas shafts were used to excavate the cut-off in the areas of major faults and here an additional 20,000 m^3 of excavation was required to a depth of 23 m below original ground level (Fig. 14.18). A buried waterfall was also uncovered, complete with pot hole and pebbles and was bottomed out at 26 m below original ground level.

14.5.6 Corries

Corries can be regarded as a special form of rock-dammed valley but corrie basins do not in general make good reservoir sites. Situated at high elevations with small catchments they generally require long and hence expensive dams for rather limited storage. Corrie sites have however been used for hydroelectric schemes and are of particular value in pumped-storage schemes where high head differences, typically 300 m, are required (Swiger et al, 1980). Investigations and construction has often revealed a variety of conditions of these sites which do not conform to the textbook model of a rock-lipped basin. Major failures of corrie dams have occurred at two sites, Dolgarrog in 1925 (Walsh and Evans 1973) already referred to, and at Keppel Cove in 1927 and the early 1930's where two dams associated with the Greenside Mine failed. In all cases failure appears to have been due to poor design and construction rather than any features of the sites. Examples are given of four corrie dam sites, Stwlan, Cruachan, Lough Shannagh and Marchlyn, all of which have been used or investigated for use as pumped storage reservoirs.

Stwlan Dam. Gwynedd.

The Stwlan Dam impounds the upper reservoir of the Ffestiniog Pumped Storage Scheme, North Wales. The buttress dam raises the level of Llyn Stwlan Geological aspects of the scheme have been described by Anderson (1969, 1971) which occupies a basin with a rock lip below which there is a steep slope to the main valley. The dam was sited on this lip, most of which is composed of rhyolites (Fig. 14.19). It had been thought that these rocks would provide satisfactory foundations but in the central section of the dam large blocks of rhyolite which appeared to be in situ were separated from each other and the underlying rock by clay-filled discontinuities. Over a 150m length of the cut-off trench these effects extended to 12 m below rock head. Discovery of these adverse conditions lead to additional grouting and excavation of the cut-off trench and to re-design of the buttress foundations. These corrie lips are simply large examples of roche moutonnees (Figs. 2.4, 2.13) in which lee-side joints formed by rock expansion in subglacial cavities are important (Sect. 2.3.1).

Cruachan Dam. Strathclyde Region.

The Cruachan pumped storage hydroelectric scheme in Strathclyde constructed on the Etive granite. The lower reservoir is Loch Awe, a natural loch whose drainage has been reversed by a glacial dam, while the upper reservoir has been constructed in a corrie of Ben Cruachan, giving an effective head of 365 m. Here there was little overburden at the dam site which was judged suitable for a buttress dam. Underlying quartz-diorite were well jointed and, as the joints tended to open as a result of blasting, it was expected that the rock might accept a good deal of grout but in the event the diorite proved to tight (Young and Falkiner, 1966).

Lough Shannagh. Co. Down. Northern Ireland

Investigations were carried out in 1965 into the feasibility of a pumped storage scheme in the Silent Valley in the Mourne Mountains, Northern Ireland. This scheme would have utilised the existing Lower Silent Valley Reservoir as the lower pond and a corrie on the flanks of Slieve Moughanmore as the upper pond (Fig. 14.6).

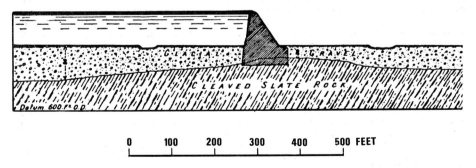

0   100   200   300   400   500 FEET

Fig. 14.16. Cross section of the Vrynwy Dam from Deacon (1896).

The corrie contains a shallow glacial lake, Lough Shannagh which is impounded by a moraine ridge consisting of compact silty sand gravel, granite cobbles and boulders resting on a smoothed granite surface 5-8 m below the lake level.

Numerous granite boulders at the surface led the site investigation contracter to use bored piling equipment with 430 mm diameter casing to penetrate the drift. The use of heavy chisels and balers however reduced the soil to a slurry and attempts to carry out falling head permeability tests produced anomalous low values probably due to clogging of the holes by fines and collapse of the test section. Representative samples were collected from 5 m deep hand-dug trial pits and consistent values of permeability of the order of 2×10^{-6} m/sec were obtained by carrying out well permeameter tests in the pits. Packer tests showed that in most of the drillholes the permeability of the jointed granite immediately below rockhead was higher than that of the drift and artesian water was struck in one hole (e.g. Fig. 3.1).

The circumstances of the Dolgarrog disaster have been described by Walsh and Evans (1973). The Eigiau Dam, a gravity section was built of very poor quality concrete and founded at a very shallow depth on till. The dam was breached, either as a result of seepage erosion of the till or enlargement of a void in the concrete, and the resulting flood flowed into the Coedty reservoir downstream, overtopping and breaching its dam which was an earth embankment with a concrete core wall.

Moffat (1976) lists 4 other failures of British Dams which resulted in loss of life and a total of 77 'incidents', that is 'failure or an event resulting in damage and calling for immediate remedial action or restriction on TWL (top water level) to avert possible failure'. Some dams appear in the list more than once.

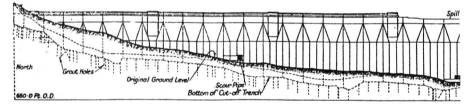

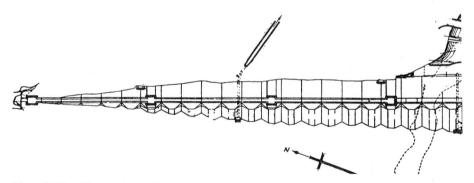

Fig. 14.17. Plan and section of the Haweswater Dam from Taylor (1951).

The Lough Shannagh scheme was abandoned for a variety of reasons, including doubts on the stability of the Lower Silent Valley Dam when subjected to rapid drawdown. If a dam had been constructed at Lough Shannagh it is likely that it would have been an embankment with an upstream impervious membrane. The upstream slope would have been continued in excavation to bedrock with the membrane tied to a grout curtain in the granite.

Marchlyn Dam, Gwynedd.

The Marchlyn reservoir forms the upper pond of the Dinorwic pumped storage hydroelectric project in North Wales. Here advantage has been taken of the 500 m head between the moraine dammed corrie lake of Marchlyn Mawr and the natural lake Llyn Peris on the opposite side of the Elidir Mountains (Carlyle and Buchanan, 1977). The damming moraine ridge was composed in the main of sand and gravel with boulders up to several metres diameter together with some lake deposits. The rock bar or lip in this case was only some 10 m above the lowest point of the basin. The moraine was investigated by rotary drilling with triple tube core barrels which gave a satisfactory recovery.

Provision of the required capacity for the scheme entailed raising the existing water level 33 m and consideration was given to various types of embankment dam. The principal decision to be made was the choice of the form of the impermeable membrane. The provision of a central core would have involved a very large deep excavation for a core trench since a slurry trench cut-off would have been difficult to install because of the large boulders. The solution adopted was similar to that proposed for Lough Shannagh. A rock-fill embankment, composed of

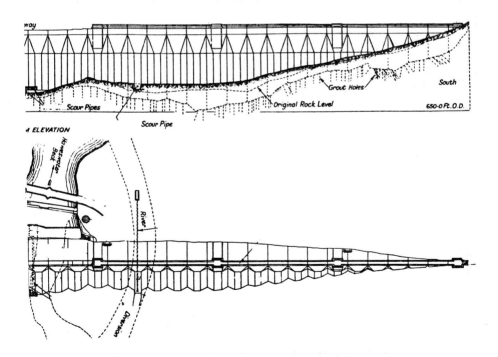

slate waste from a disused quarry, was constructed with an upstream slope of 1 on 2 and this slope was continued in excavation to rock level. An asphaltic membrane was laid on this slope and connected to a toe gallery and hence to an upstream grout curtain (Fig. 14.20). A further advantage of this design is that it is particularly suited to the operating conditions of the scheme which require the reservoir to be filled in 6 hours and drawn down in 5 hours.

14.6 Problems With Old Dams

There have been three major dam disasters in Britain in which there have been serious casualties: the Bilberry (or Homfirth) Dam failed in 1852 with the loss of 81 lives, the Dale Dyke (or Bradfield) Dam which failed in 1864 with a death toll of approximately 250 and the Dolgarrog disaster of 1925 in which 16 were lost. The Bilberry and Dale Dyke dams were both ´Pennine Dams´. In each case the standard of construction was poor and a major spring was encountered in the cut-off trench which was not properly treated and probably caused erosion of the core (Binnie, 1978, 1981). At Dale Dyke the situations was made worse by the stepped profile of the case of the trench which probably lead to differential compaction and cracking (Cavoundis and Hoeg, 1977).

Over half of these incidents have occurred since Dolgarrog and the introduction of the Reservoirs (Safety Provisions) Act, and although no fatalities have occurred in this period there is growing concern of the condition of some of the older embankment dams. Inspections and investigations have revealed numerous defects (Kennard 1972; BNCOLD 1976) and a serious emergency developed at Lluest Wen (Gamblin and Little 1970; Twort 1977).

Some of these old dams have spillway and draw-off arrangements that are considered inadequate by modern standards but equally serious are the defects of embankments themselves. These include settlement, resulting in reduction of freeboard, leakage, internal erosion creating cavities and sinkholes, the use of defective materials such as peat, inadequate cores, some indistinguishable from shoulder fill, hydraulic fracture of a core, shallow slipping of the downstream slope sometimes initiated by spray blown over the dam crest, and damage by tree roots (Hjeldnes and Lavania, 1980).

Internal investigation and instrumentation of such dams is becoming more common and involves the application and instrumentation of geological techniques of exploration and interpretation to a man-made structure. Procedures for inspecting old embankment dams are suggested by Clarke and Le Masurier (1976) and Szalay (1980) and methods of carrying out internal investigations are reviewed by Money (1976) and Simmons (1982).

14.7 Future Trends

In recent years the pace of reservoir construction in the UK has slackened: consumption of water, particularly by industry, has fallen short of the estimates made in the 1970´s and few large schemes are under construction or consideration. Many of the sites most suitable on technical grounds have been used and pressure from local interests and environmentalists further restricts the choice of new sites. There is however considerable interest in the re-use of sites to provide reservoirs of increased capacity where the present storage is insufficient to achieve maximum yields from catchments. Such schemes have the advantages of requiring little additional land, no significant change of land use and, usually, are easily integrated into existing distribution networks. Two options are open to the engineer: raising an existing dam or constructing a new dam usually downstream of the old one. Raising a concrete dam, as has been proposed for

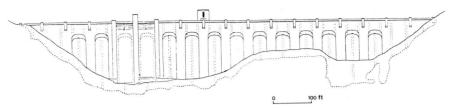

Fig. 14.18. Longitudinal section of the Sloy Dam, after Stevenson (1952), showing irregular rock head surface (dotted line).

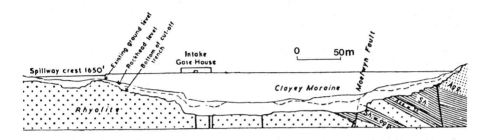

Fig. 14.19. Longitudinal section of the Stwlan Dam site after Anderson (1969). Agg; agglomerate,sh; shale.

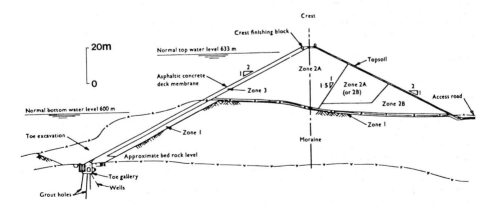

Fig. 14.20. Cross section of Marchlyn Dam after Carlyle and Buchanan (1977).

Haweswater, can be straightforward, but raising embankments, particularly Pennine Dams with their central cores, is more difficult and it may prove more economical to construct a new dam. Opinions differ as to whether any part of the existing dam should be incorporated in the old. At the Lower Lliw reservoir in S. Wales, where the existing dam had given trouble for many years it proved more economical to build an entirely new dam rather than carry out the necessary remedial works on the old dam. At Grimwith in Yorkshire the existing dam forms part of the upstream shoulder of the new dam but its clay core which investigations showed to be ill-defined (Kennard 1972) is not being used and an entirely new core is being constructed downstream. Some Water Authorities are working in the spirit of the 1975 Reservoirs Act and this is likely to result in more investigation, refurbishing and reconstruction of old dams and their associated works.

CHAPTER 15

Hydrogeology in Glaciated Terrains

J. W. Lloyd

INTRODUCTION

Throughout much of the mid-latitudes, glacial and periglacial materials of variable lithology obscure bedrock surfaces and have considerable hydrogeological importance. However because both substantial surface water resources are often present and groundwater can be abstracted in abundance from certain pre-Pleistocene bedrock aquifers, the groundwater resources of surficial geological deposits and their hydrogeology have been inadequately investigated. It is ironic that compilations of well-hole logs frequently provide the first data on subsurface glacial sequences in many areas. With respect to groundwater resources, surficial deposits have been considered almost exclusively in the context of their relationship to major aquifers. The main features that have proved important are the effects of surficial covers on recharge to the main aquifers and the extent to which the groundwaters in these aquifers are confined by surficial materials. Certain very limited hydrogeological information about surficial materials is available from site investigations of civil engineering works. Unfortunately such data are usually for small areas and are not necessarily obtained using consistent techniques; the data therefore provide little real insight into drift hydrogeology. With progress in landfill site technology, data concerning drift permeability and certain chemical characteristics of the materials are being obtained. Again the information is only for a few small and widely distributed areas.

15.1 Aquifer Definition

As glacial sediments range from boulder to clay grades, drift materials cover the whole gambit of aquifer conditions. Terms such as aquifer, aquiclude etc. are relative and in many glacial sequences they are difficult to apply.

In many cases the only glacial drift deposits which transmit significant quantities of water and which may be considered as ´aquifers´ are sand and gravels. These tend to be relatively thin deposits normally less than 10 m in thickness and irregular in geometry. They frequently occur at the base of drift sequences immediately overlying pre-Quaternary materials but can also occur throughout a drift sequence (Figs. 1.3, 1.4, 1.5, 1.6). Many reports of groundwater resources in glaciated terrain emphasize the importance of sands and gravels in bedrock valleys buried and infilled by glacial sequences. On Figure 15.1 an isopachyte map of sand and gravel deposits in northern Lincolnshire is shown. These deposits lie beneath lodgement tills (referred to locally as boulder clays), of last glaciation age, deposited on an eroded Chalk surface and typically exhibit an irregular geometry and limited thickness. In Lincolnshire these sands and gravels form important aquifer materials. The subsurface mapping of buried sand and gravel deposits of this type is particularly difficult and extensive use is often made of surface resistivity techniques with borehole

349

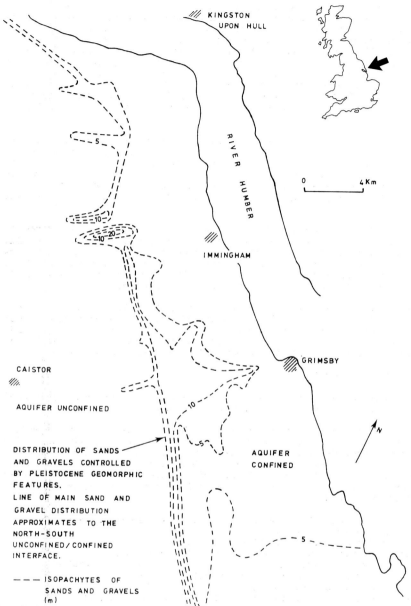

Fig. 15.1. Geometry of glacial sand and gravels, Lincolnshire. The aquifer is confined by lodgement till. The gravels wedge against a buried cliff-line from the previous interglacial (Fig. 15.13) and rest on an extensive wave-cut platform cut across chalk bedrock and till from the previous glaciation. The cliff-line continues to the north and is exposed on the Yorkshire coast at Sewerby (Catt and Penny, 1966).

control.   Unfortunately the interpretation of resistivity data can be difficult
due to resistivity supression where thin sands and gravels occur beneath thick
lodgement tills so that the former cannot easily be recognized. In such
lithological sequences there is no substitute for drilling information in
determining lithological distributions.   In many parts of Britain aggregate
assessments have been carried out in drift materials by the Institute of
Geological Sciences;   their assessments include extensive drilling records and
provide invaluable data concerning the extent of potential aquifer materials in
the deposits.   Similar data are available from State and Provincial Geological
Surveys and Environment Ministries in North America, though the quality of
well-hole logs varies enormously.

In many cases in glaciated terrain clearly delineated lithological units of
coarser grade material with a well defined geometry, are not present in
geological sequences and limiting hydraulic boundaries are diffuse.   To return to
the Lincolnshire example (Fig. 15.1) it is certain that in the east beneath the
boulder clays, more randomly distributed sand and gravel deposits are present
some of which are probably in tenuous but direct hydraulic continuity with the
main body.

While hydraulic boundaries delimiting aquifers in drift materials are difficult
to determine, it is also clear that within the coarser grade aquifer materials,
bodies of finer grained material are frequently present.   Such lithology
variability gives rise to internal groundwater flow barrier conditions in the
drift aquifers which make the assessment of aquifer characteristics particularly
difficult and hinder reliable groundwater resources assessment.

Groundwater movement implying ´aquifer´ conditions in tills is also reported
(e.g. Fig. 2.11, Bonell, 1972). The nature of such flow, however, is difficult
to assess though it can be very important with respect particularly to civil
engineering works that require large excavations. Much of the permeability is
secondary in nature being related to weathering and fissuring (see below).

15.2 Aquifer Characteristics

Values of transmissivity or permeability and storage in surficial geological
deposits are as variable as the materials themselves.

Permeability assessment poses the biggest problem but is the most important
parameter in that the ability of drift materials to transmit water has
significant hydrogeological implications in the area of leachate control in
landfill sites and well yield assessment.   Storage characteristics on the other
hand are only of significance in those units which are important in water
resources terms.   In considering permeability, the drift deposits may be
subdivided as follows:

    (i)   Matrix rich tills (e.g. lodgement till and also various waterlain
          diamicts (Sect. 2.9).
   (ii)   Weathered matrix rich tills (Sect. 8.2.4).
  (iii)   Fractured or fissured matrix rich tills (Sect. 8.2.4)
  (iv)   Tills of variable lithology (e.g. flowed tills, coarse-grained
          supraglacial diamicts, melt-out tills: Chapters 3,4),
          excluding significant sands and gravels
   (v)   Reasonably well defined sand and gravel deposits (Ch. 7).

Permeability determinations in unweathered tills are usually carried out by
testing of undisturbed samples in the laboratory. Mostly, data are available
from civil engineering and landfill site investigation, and the permeability

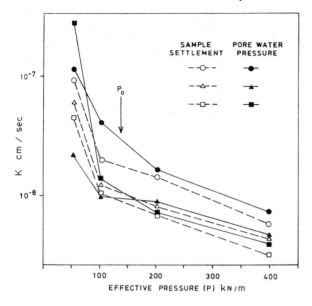

Fig. 15.2. Variation in coefficient
of permeability with applied
effective stress in fissured
lodgement tills. (After McGown and
Radman 1975).

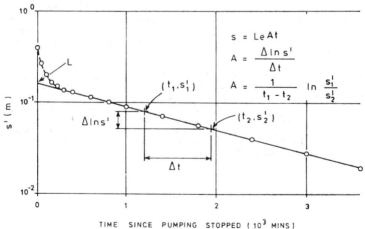

Fig. 15.3.
Permeability
determination
from pumping
test data in
Swedish tills
(After Eng-
quist et al.
1978).

$$s = Le^{At}$$

$$A = \frac{\Delta \ln s'}{\Delta t}$$

$$A = \frac{1}{t_1 - t_2} \ln \frac{s_1'}{s_2'}$$

TIME SINCE PUMPING STOPPED (10^3 MINS)

values are extremely low. Unweathered matrix rich till has the lowest value of
permeability coefficient ranging from 10 to 10 m/day (Table 15.1). Clearly
such material forms an effective aquiclude in groundwater resources terms and may

provide excellent containing ground for landfill sites. The presence of included sand and gravel layers in many till sequences is critical; their general prediction can often be made following identification of the depositional environment from surface landforms (Figs. 1.3, 1.4, 1.5, 1.6). In engineering works the unweathered matrix rich tills provide reasonably consistent hydraulic conditions that can be readily analysed.

TABLE 15.1. Representative coefficients of permeability (K) for glacial sediments (m/day). Variations in measurement techniques and lithologies allow only a very broad representation.

Matrix rich tills[1]	$10^{-3} - 10^{-6}$
Weathered matrix rich tills[1]	$10^{-2} - 10^{-6}$
Fractured or fissured matrix rich tills[2]	$10^{-3} - 10^{-5}$
Variable lithology till and solifluction debris[2]	$10^{-1} - 10^{-5}$
Sands and gravels[2]	$10^{2} - 10^{-1}$

Notes: 1 Laboratory values, predominantly vertical permeability

         2 Laboratory and field values

In many areas of glaciated terrain, till profiles show marked weathering zones in the upper levels, distinguished often by slight red colouration. The weathering, which is the result of periglacial to present day effects, has made these tills friable to variable depths often down to 8 m. Permeabilities in the four weathering zones that can be identified in lodgement till (Sect. 8.2.4) are in general higher than than in the unweathered tills as shown in Table 15.1, but have a large range reflecting the weathering variability from zone to zone. These weathered zones pose problems in civil engineering works where marked pore-water pressure variations can occur spatially over a slope and pressure heads can change rapidly in response to recharge. Weathered zones also cause leachate problems in landfill sites.

Where leachate problems have arisen, simple numerical calculations have normally proved adequate to determine flow. In North America, however, some hydraulic modelling of fractured tills has been carried out (Grisak et al, 1976). McGown and Radwan (1975) report on similar materials in the British Isles and detail sample permeabilities under differing effective pressures (Fig. 15.2). In the North American work, equations given by Snow (1969) have been used to determine permeability and may form the basis for studies elsewhere. The coefficient of permeability (K) can be determined as follows:

$$k = \frac{2 \Sigma b^3}{3 \Delta N}$$

where Δ = average fracture spacing

      b = half-aperture width

and N = total number of fractures in measured area. If Δ is

          not known then the permeability may be determined from:

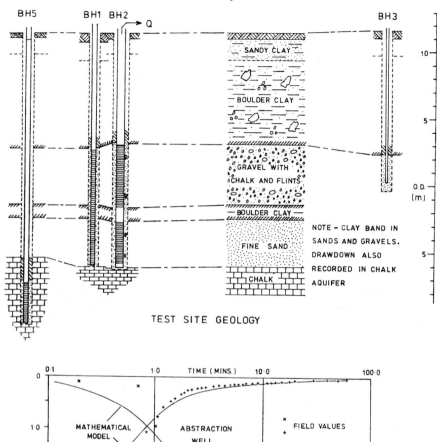

TEST SITE GEOLOGY

NOTE – CLAY BAND IN
SANDS AND GRAVELS.
DRAWDOWN ALSO
RECORDED IN CHALK
AQUIFER

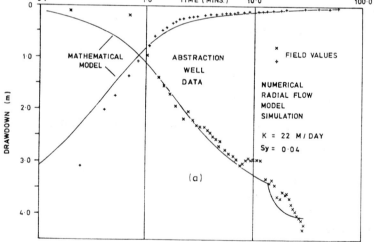

Fig. 15.4. Pumping test details of glacial sand and gravel aquifer, Lincolnshire. (Fig. 15.1).

$$K = \frac{2 \sum b^3}{3W}$$

where W = the discharging width

Grisak (1975) records aperture values of the order of 4.10^{-4} cm and fracture spacings of 4 cm.

Variable lithology till materials (excluding significant sands and gravels), are a common feature of drift deposits in glaciated terrain particularly in glaciated valleys and within supraglacial morainic complexes overlying lodgement till plains (Chapters 3,4). Their hydrogeology is not well known so that information as to aquifer characteristics has to be obtained from elsewhere. Permeability determinations in these types of deposits may normally be carried out using field techniques such as conventional pumping-tests, falling head and slug tests etc. (Sect. 11.6.5, Fig. 11.8). Small well yields can be obtained from these types of deposits so that they are frequently used for farm and individual house supplies. Representative examples of hydrogeological investigations in thicker sequences filling buried valleys are reported by Norris and White (1961) and Stephenson (1967).

An example of the type of permeability study possible in these varied deposits comes from Sweden (Knutsson, 1971) where a large number of small groundwater supplies are obtained from coarse-grained tills on shield terrains. In this Swedish work the permeability equation derived has been established through simple well pumping tests on supply wells (Engquist et al, 1978). Recovery data have been analysed (Fig. 15.3) using the equation for permeability (K) as follows:

$$K = \frac{r^2}{2 \, s_{max} t} \quad \ln \frac{s'}{L}$$

where r = radius of well

s' = residual drawdown

s_{max} = maximum drawdown

t = time since pumping stopped

L = initial drawdown (Fig. 15.4)

To allow for well loss effects in single well tests the following correlation has been used:

$$C + T^{1.25} = 1$$

where C = a well loss constant such that $s_L = CQ^2$

T = transmissivity

Q = discharge of well

and s_L = drawdown due to well loss.

UNCONFINED AQUIFER

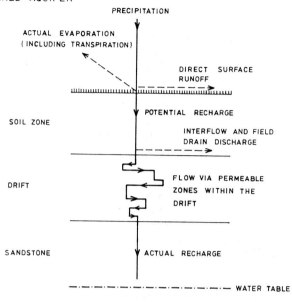

ACTUAL RECHARGE = RECHARGE FACTOR
                  X POTENTIAL RECHARGE

CONFINED AQUIFER

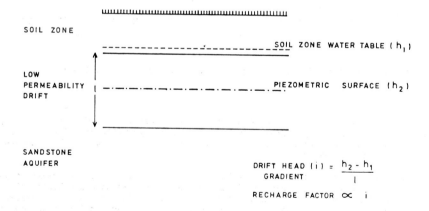

Fig. 15.5. Recharge models for drift deposits covering a major aquifer.

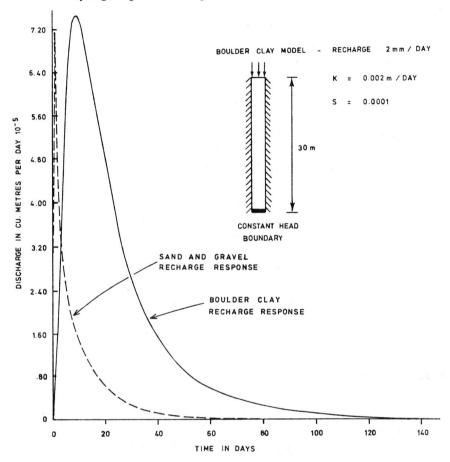

Fig. 15.6. Response delay in spring flows to recharge in glacial soils.

In the Swedish work the reported well losses are normally less than 1% and rarely reach 5%.   Permeability values for variable lithology tills are given in Table 15.1. The values are larger than for matrix rich tills as would be expected while the range is considerable.

Permeabilities for the sands and gravels are variable and can be extremely high (Table 15.1) but no comprehensive data are available. Rushton and Booth (1976) report a coefficient of permeability of 40 m/day for sands and gravels in eastern Wales   and   similar values have been determined using radial flow model interpretation in Lincolnshire (Figure 15.4). In the latter case the aquifer characteristic testing was carried out solely because of the juxtaposition of the sand and gravel deposits to the Chalk aquifer emphasising that drift deposits in Britain are mainly of importance for their influence upon major bedrock aquifers.

Storage values in sands and gravels can be very large and specific yield is an important factor in unconfined circumstances. The test shown on Figure 15.4 was designed especially to determine specific yield. The analysed value of 0.04 is large in comparison to many aquifer materials, but was found to be under-estimated when aquifer modelling was carried out which indicates a specific yield of 0.15 (Fig. 15.13). The example illustrates a typical hydrogeological problem encountered in sands and gravels in that classical pumping-test results tend not to indicate realistic specific yield values. The reason for this is attributed to the presence of small bodies of finer grained materials in the sands and gravels which cause complex internal hydraulic boundary conditions and disturb the classical type of gravity drainage response seen in many other aquifer materials. Thus the presence of 2% illite in otherwise clean sand may lower the permeability two orders of magnitude; 2% montmorillonite may reduce permeability by up to four orders (Haefeli, 1972).

15.3 Recharge

Lodgement tills form the dominant lithology by far in Britain (Chapter 8, Fig. 9.1). As has been discussed above they have extremely low permeabilities so that where they occur at the surface or are high in a geological succession they substantially restrict or totally stop recharge reaching either drift or other aquifers.

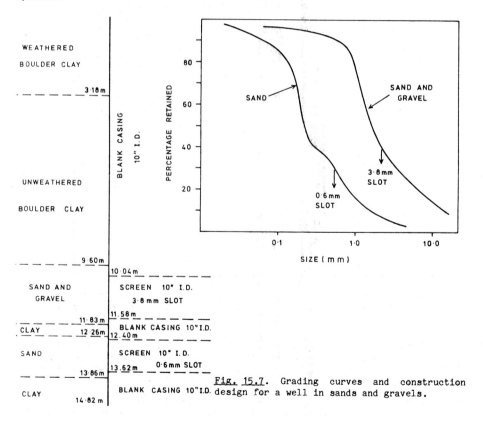

Fig. 15.7. Grading curves and construction design for a well in sands and gravels.

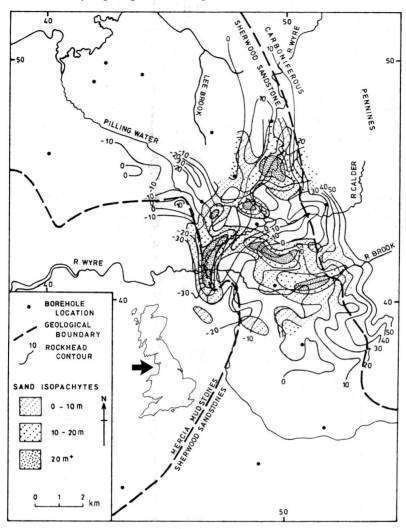

Fig. 15.8. Structure contour map of the top of the Sherwood Sandstone with isopachytes for overlying glaciofluvial sands in the Fylde area of Lancashire.

While direct recharge through the coarser drift deposits is clearly possible the mechanisms of any recharge through the finer grades are difficult to model. A representative natural tritium profile for till pore-waters from eastern England is shown in Table 15.2. In the section, the permeability coefficients for the unweathered till range from 10^{-4} to 10^{-6} m/day. Such low permeabilities will not easily permit recharge yet the tritium concentrations indicate modern influences. Discounting piston-flow the tritium presence is believed to be due to lateral

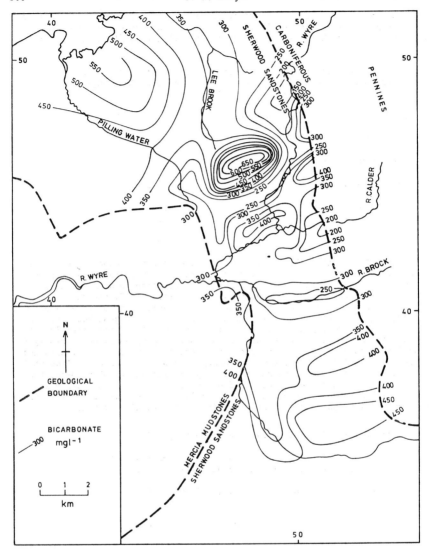

Fig. 15.9. Bicarbonate distributions in groundwaters of the Sherwood Sandstone (Fig. 15.8).

groundwater movement through very thin zones of material with permeabilities higher than the main till body. Interestingly, the groundwater in the sand and gravel aquifer underlying the till sequence has a much lower tritium value though is still modern.

Considering the style of lithological distribution, the mechanism of lateral recharge through permeable zones in the drift is perhaps an obvious conclusion, the quantification of such recharge, however, is virtually impossible except using empirical methods.

TABLE 15.2. Sample profile of Tritium values (T.U.) in porewaters and groundwater from glacial sediments in southern East Anglia. (data from Anglian Water Authority).

Lithology	Sample Depth (m)	T.U.
0-4m Weathered matrix rich till	1.0	77.5
4-31m Unweathered matrix rich till	5.9	8.0
31-33m Sands and gravels	7.7	13.5
Chalk fragment zones (0.1m thick at 3, 8 and 11m depth)	11.7	107.0
	20.3	50.0
	30.0	57.0
	32.0	28.0

In most studies of recharge calculation through the drift, recharge has been considered with respect to underlying aquifers. Drift recharge is therefore normally quoted as a percentage of the recharge to the unconfined main aquifer. The style of recharge model than can be adopted is shown in Figure 15.5. Table 15.3 gives an indication of the importance of variation in surface geology (i.e. recharge surface as defined by Penman (1950)) and shows just how important variation in drift lithology and its distribution can be. For example the significance of recharge through the till is demonstrated by deriving percentages of the total estimated 5-year mean recharge for each soil type. By increasing the till factor from 0.1 to 0.2 the mean recharge estimate is increased from 58.9 to 76 mm. Clearly therefore recharge can be very sensitive to the percentage attributed to the drift deposits. Unfortunately, such values can only be determined on a trial and error basis by correlation with groundwater heads and flows in a properly compounded groundwater model of the aquifer system.

TABLE 15.3. An example of recharge calculations for glacial sediments.

	Sandstone Outcrop	Glacial Sand and Gravel	Recent Alluvium	Peat	Matrix rich till
Total Area, km	4.07	15.82	13.51	9.65	56.95
Infiltration Factor	1.0	0.5	0.1	0.1	0.2
5 year Recharge as mm, weighted	12.1	23.8	2.4	3.3	17.3

Total
Recharge =   58.9 mm

(Lower Mersey Basin, Lancashire, England)

The recharge factors for the lower permeability materials shown in Table 15.3 are

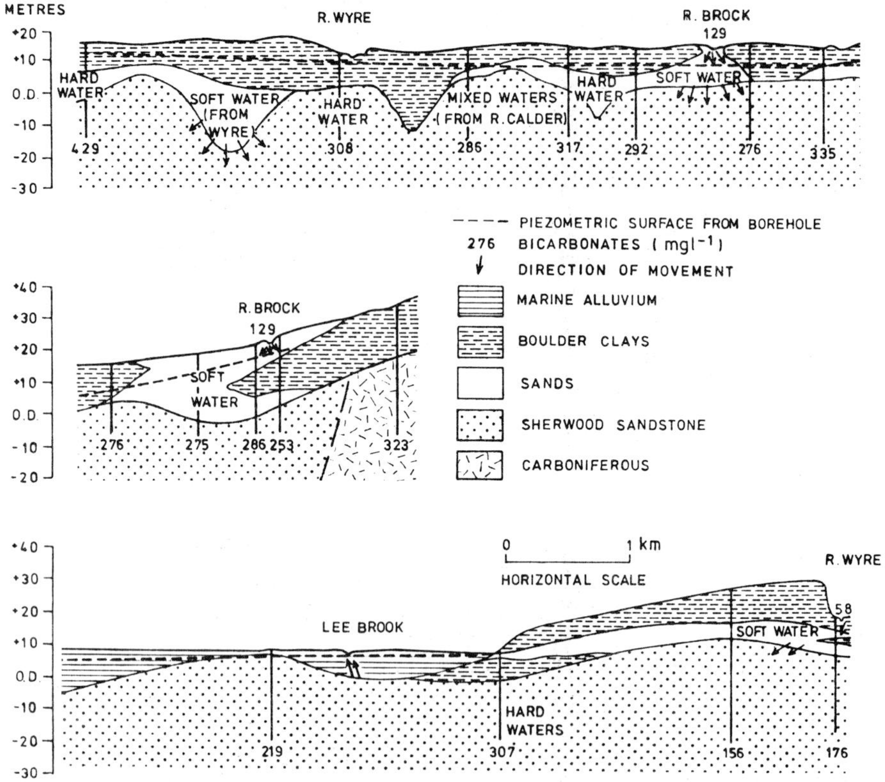

Fig. 15.10. Indirect recharge through glacial sands, Fylde area (Fig. 15.8, 15.9).

maximum values and refer specifically to the condition where the aquifer is unconfined. When the aquifer is confined the recharge factors will vary linearly from their maximum to zero according to the head gradient developed across the confining layer (Fig. 15.5). The importance of this feature is illustrated by a study of the Lower Mersey Basin in Britain where its effect is to provide a gradual increase in annual recharge from about 35 mm in 1850, when the aquifer was predominantly confined, to about 60 mm in 1975 when unconfined conditions prevailed.

Bonell (1972) notes the importance of lateral recharge in drift deposits and also differing seasonal groundwater flow directions. The latter may be attributed to the complex flow paths and consequent flow delays that are a feature of fine-grained deposits. To simulate drift delay effects on recharge entering the Chalk aquifer in Essex, vertical flow models have been used as shown in Figure 15.6 in which it has been necessary to use quite high permeabilities due to the mixed lithologies present. Even with a high matrix rich till permeability of 2.10^{-3} m/day, a 10 day delay in maximum recharge response occurs, while effects continue for some 100 days. Clearly for low permeabilities the responses will be

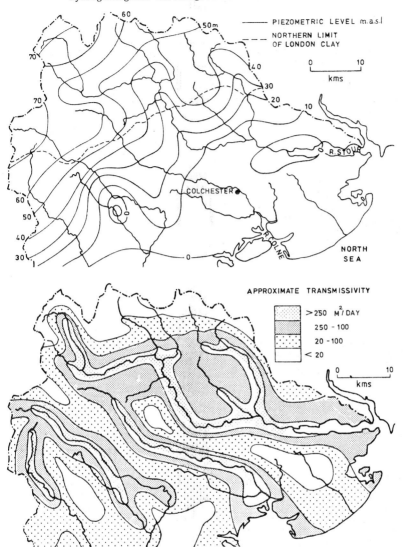

Fig. 15.11. Hydrogeological aspects of the chalk aquifer in southern East Anglia. A: Piezometric surface; B: Transmissivity. See Fig. 15.12 for location.

much longer and account for much of the complexity found in groundwater and surface water hydrographs in drift areas. For comparison note the very limited recharge delay response through sand and gravels (Fig. 15.6).

The pattern of groundwater flow in thick glaciolacustrine sequences deposited in the large lake basins formed during the retreat of ice sheets (Fig. 2.21) is not known in any detail. In the Great Lakes area of Canada, Ostry (1979) has assessed the inflow of water to the present Lake Ontario through a thick (80m) sequence of lake sediments now raised above the present lake level in the Toronto area. The geometry of these basin-infill sequences is relatively simple and comprises layer cake alternations of glaciolacustrine clays, waterlaid ´silty clay tills´ and shallow water lacustrine and deltaic sands (Sect. 2.9). The latter are the most productive units in terms of groundwater resources. Recent interest in the fine-grained ´tills´ stems from concern over the location of toxic waste storage sites (see below). Grisak and Cherry (1975) emphasize the hydrogeological importance of fracture systems in lacustrine clays of the interior plains of North America. The dissipation of large postglacial lakes in Western Canada (Ch. 6, Fig. 6.6, resulted in the generation of modern ´underfit´ streams draining large valleys which were cut during the previous hydrological regime of enhanced meltwater run-off during deglaciation and lake drainage. Isostatic recovery has also tended to reduce the gradient of flow along many rivers.

15.4 Groundwater Abstraction

Groundwater abstraction from glacial sediments in Britain is very localized and poorly documented. Well construction techniques have been extremely variable and range from partially lined dug wells to screen designed structures. In many cases dug well construction has been carried out using retaining concrete rings through clay materials, with an open hole section at the base of the well where sands and gravels are present. In drilled wells most constructions consist of artificially gravel packed, slotted pipe placed against sand and gravel sections. Data of the efficiencies of such wells are not available.

Glacial sand and gravels often provide an ideal environment for naturally developed gravel pack wells. The poorly sorted nature of the deposits allow efficient well development procedures to be operated while the often angular and poorly sorted nature of the grains permit the selection of large slot sizes (Fig. 15.7).

As with construction data, representative well yield data from drift deposits are virtually unobtainable. The yields of many of the wells are generally low and present such a wide range that no sensible characterisation can be made. For an indication of drift well yields reference is made to a study by Olsson (1974) carried out in Sweden where such wells are of much more importance than is the case in Britain. The data may be representative of similar terrains in North America.

15.5 Hydrochemistry

Aspects of glacial drift hydrochemistry have been referred to by a number of authors. Spears and Reeves (1975) working in Britain in the Vale of York, have emphasized the importance of SO_4^{2-} in lodgement till porewaters. The sulphate is derived from the oxidation of pyrite which occurs commonly throughout the drift as a result of incorporation of shales (Madgett and Catt, 1978; Eyles and Sladen, 1981). Spears and Reeves (1975) noted that ionic concentrations in the drift are related to both depth and permeability. Although pyrite oxidation dominates, carbonate dissolution may also be important but the breakdown of feldspars and the leaching of clay minerals play only a minor role. Groundwaters emanating from drift deposits can therefore be distinctly calcium sulphate in character and may be of reasonably high salinity (i.e. up to about 10,000 mgl^{-1} total dissolved solids) after long residence in the drift. Chemical composition however, can be very variable (Table 15.4).

TABLE 15.4. Representative chemical analyses of groundwaters in glacial
             sediments (mg l^{-1})

Location	Lithology	Ca	Mg	K	Na	HCO_3	SO_4	Cl
Vale of York	lodgement till	315	215	16	34	-	1655	34
	Sand	56	11	4	15	-	81	27
N.W. Lancashire	lodgement till	74	20	2	49	37	200	88
	Sand	140	4	1	16	372	53	23
S. East Anglia	lodgement till	160	56	10	70	275	280	59
	Sand	130	12	23	66	310	75	82

Sulphate rich waters can normally be readily recognized where they recharge other
systems. Under natural groundwater flows, they are not necessarily undesirable
but as Spears and Reeves point out groundwater gradients caused by pumping may
induce abnormal flows thereby inducing higher salinity water into a main aquifer.

Groundwaters in glacial sequences are not necessarily always calcium sulphate in
character.   Chemical characterisation is a function of lithology, permeability
and residence, and as a result hydrochemistry may provide an indirect means of
understanding the complex flow distributions within the drift. Differences in
groundwater chemistry coupled with hydrochemical characteristics of underlying
aquifers can also provide invaluable information about the hydrogeological
relationship between the drift and major aquifers which as mentioned above is the
most important hydrogeological feature of drift in many areas with respect to
groundwater resources (Walton, 1970; Charron, 1974).

Considerable hydrogeological interest has been shown lately in the relatively
dense clay-rich tills of the Great Lakes Basin that have been deposited
subaqueously below glaciers standing or floating in large lakes (Sect. 2.8.2).
These units do not yield water and therefore have been little studied from a
hydrogeological viewpoint but they are of interest as watertight seals between
toxic waste disposal and storage sites and underlying bedrock aquifers. Very low
groundwater velocities have been measured from these soils (ranging between 1m in
10^3 yrs. to 10^4 yrs. and very old porewater dating from the early postglacial
exists at depth; molecular diffusion is likely to be the major control on
isotope and ion distribution (Desaulniers et al., 1981). Oxygen isotope studies
in these old waters are described in Section 6.5.6.

15.6 Drift Relationships with Major Aquifers

The concern with drift hydrochemistry in the Vale of York study by Spears and
Reeves (1975) was initiated by a groundwater resources study of the Triassic
Sandstone aquifer in the area which showed high indirect recharge to the
sandstones from glacial sands and gravels amounting to 20 Ml/d out of a total of
70 Ml/d for the whole sandstone aquifer. Under modelled abstraction from the
sandstones it is considered that the drift could provide a possible induced yield
of 80 Ml/d. In many areas the transfer of waters into underlying bedrock
aquifers could affect the long-term viability of abstracting from such aquifers.

The importance of the glacial sediment cover on bedrock aquifer resources can
also be seen in another example from north-west Lancashire. Using hydrochemical

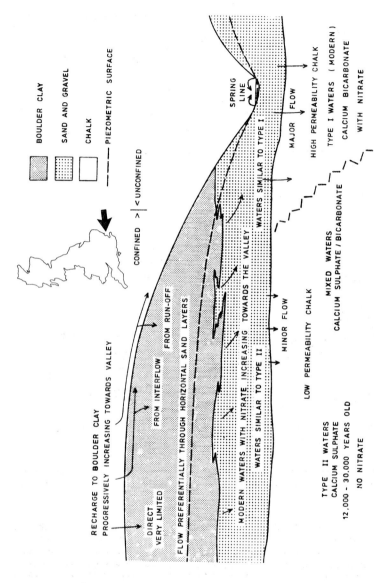

Fig. 15.12 Conceptual model of recharge and groundwater movement in the glacial soils and Chalk aquifer in southern East Anglia.

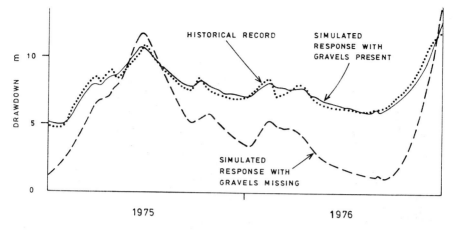

OBSERVATION WELL IN CONFINED REGION

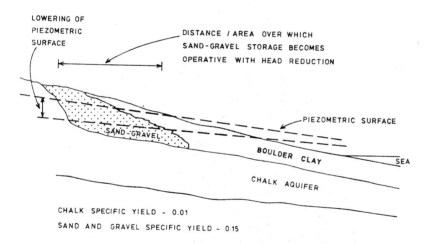

CHALK SPECIFIC YIELD - 0.01

SAND AND GRAVEL SPECIFIC YIELD - 0.15

Fig. 15.13. Significance of storage in glacial sand and gravels in continuity with major aquifer. Note the buried Ipswichian cliff-line and wave-cut platform (Fig. 15.1).

techniques, Sage and Lloyd (1978) have demonstrated indirect recharge through glacial sands and gravels within matrix rich tills overlying the Sherwood Sandstone. Low bicarbonate groundwaters in the sandstones are seen to correlate with stringer deposits of glacial sands which rest upon the sandstones beneath the tills, and which are locally in hydraulic continuity with the rivers, particularly at the foot of the Pennines (Figure 15.8). The bicarbonate distributions and recharge mechanisms are shown on Figures 15.9 and 15.10. It should be noted that the lobes of low bicarbonate modern recharge water retain

their integrity for significant distances across the sandstone aquifer from the
eastern boundary. In part this integrity is due to the conformable alignment of
rivers and sands, but it would also appear that the dominance of sand
permeability over sandstone permeability also plays an important part; their
presence may have implications upon well yields.

The most important British aquifer, the Chalk, is also affected by drift
hydrogeology (Lloyd et al., 1981) and a good example comes from the area of
southern East Anglia (Fig. 15.11). Here the highest transmissivities in the
Chalk occur in the river valleys and correlate with the modern calcium
bicarbonate recharge waters. By way of contrast in till capped interfluve areas,
groundwaters in the Chalk are calcium sulphate in character and extremely old as
a result of the influence of overlying till. Base flow analyses of rivers in the
area indicate that at low flows a substantial amount of water is being derived
from storage in glacial sands and gravels immediately overlying the Chalk. The
drift presence is believed to have severely restricted the development of the
Chalk as an aquifer away from the river valleys while concentrating modern
recharge in the valley areas. A conceptual model for recharge and flow
mechanisms in the area is shown in Figure 15.12.

Sand and gravel groundwater storage effects on the Chalk aquifer can also be
demonstrated in the Lincolnshire example (Fig. 15.1). In an initial study of
groundwater resources the importance of the glacial sand and gravel deposits was
not realized. However initial aquifer modelling indicated insufficient
groundwater in the Chalk aquifer particularly under drought conditions, to
satisfy known historical records of groundwater fluctuation. Subsequent
inclusion of the sand and gravel body into the aquifer system provided the extra
groundwater storage required to obtain satisfactory model simulation and
indicated the considerable importance of unconfined drift storage even when only
relatively small sand and gravel bodies occur. On Figure 15.13 the groundwater
head simulation reflecting the importance of the sand and gravel specific yield
is shown, together with the storage release mechanism invoked.

15.7 Conclusions

The foregoing chapter shows that the hydrogeology of glacial drift deposits is
reasonably understood in concept but is very poorly understood in detail. The
importance of drift hydrogeology in civil engineering and landfill studies is
recognized and conditions are locally well documented but in the broader field of
groundwater resources the importance of drift deposits is perhaps less
appreciated (Lloyd, 1980). While it is accepted that a comprehensive
understanding is virtually impossible and prohibitively costly, it is likely that
in the future the most important aspects of drift hydrogeology will be examined
and determined in a groundwater resources context in relation to major bedrock
aquifers and with regard to the widespread use of glacial soils for toxic liquid
and solid waste storage disposal.

Appendix I

Grain size scales

	British standard test sieves	U.S. Sieve Series No.	Millimeters	Phi Scale	Wentworth Scale	
					Boulder	
			256	−8		
	63		64	−6	Cobble	
	20		16	−4		COARSE
mm	6.3	5	4	−2	Pebble	GRAVEL·MEDIUM
	3.35	6	3.36	−1.75		FINE
		7	2.83	−1.5	Granule	
		8	2.38	−1.25		
	2	10	2.00	−1.0		
		12	1.68	−0.75		
		14	1.41	−0.5	Very Coarse Sand	
	1.18	16	1.19	−0.25		
		18	1.00	0.0		
		20	0.84	0.25		
	600	25	0.71	0.5	Coarse Sand	
		30	0.59	0.75		
	425	35	0.50	1.0		
		40	0.42	1.25		SAND
		45	0.35	1.5	Medium Sand	
	300	50	0.30	1.75		
μm	212	60	0.25	2.0		
		70	0.210	2.25		
		80	0.177	2.50	Fine Sand	
	150	100	0.149	2.75		
		120	0.125	3.0		
		140	0.105	3.25		
		170	0.088	3.5	Very Fine Sand	
		200	0.074	3.75		
	63	230	0.0625	4.0		
		270	0.053	4.25		
		325	0.044	4.5	Coarse Silt	
		400	0.037	4.75		
			0.031	5.0		
			0.0156	6.0	Medium Silt	MUD
			0.0078	7.0	Fine Silt	
			0.0039	8.0	Very Fine Silt	
			0.0020	9.0		
			0.00098	10.0		
			0.00049	11.0	Clay	
			0.00024	12.0		
			0.00012	13.0		
			0.00006	14.0		

SIEVES — mm / μm

PIPETTE, HYDROMETER, COULTER COUNTER, FALLING TUBE SEDIGRAPH ETC

The Wentworth scale uses logarithmic divisions with each successive size separation twice as large as the previous smaller grade limit. The Phi (ϕ) scale is preferred by geologists and is increasingly used by engineers. Note the British terminology for gravel, shown on right, is different from that of the Wentworth scale.

Quantity	Existing unit	S.I. Units & selected decimal multiples & sub-multiples	Unit symbol	Conversion factor
Length	foot inch micron	metre millimetre micrometre	m mm μm	1 ft = 0.3048 m* 1 in = 25.4 mm* 1 micron = 1μm = 10^{-3}mm*
Area	square foot square inch	square metre square millimetre	m^2 mm^2	1 ft^2= 0.0929 m^2 1 in^2= 645.2 mm^2
Volume	cubic yard } cubic foot }	cubic metre	m^3	{ 1 yd^3 = 0.7646 m^3 { 1 ft^3 = 0.028 32 m^3
	gallon } cubic foot }	litre	litre	{ 1 gallon = 4.546 litre } { 1 ft^3 = 28.32 litre }
	cubic centimetre cubic inch	millilitre cubic centimetre	ml cm^3	1 cc = 1 ml = $1cm^3$* 1 in^3 = 16.39 cm^3
Mass	ton	megagramme (tonne)	Mg (tonne)	1 ton = 1.016 Mg (= 1.016 tonne)
	pound hundredweight } gramme	kilogramme	kg	{1 lb = 0.4536 kg {1 cwt = 50.80 kg
Density	pound per cubic foot	megagramme per cubic metre	Mg/m^3	100 lb/ft^3 = 1.602 Mg/m
	gramme per cubic centimetre	(tonne per cubic metre)	(tonne/m^3)	1 gm/cc = 1 Mg/m^3* = (1 tonne/m^3*)
Force	pound force } kilogramme force }	newton	N	1 lb f = 4.448 N 1 kg f = 9.807 N
	pound force } ton force }	kilonewton	kN	{ 1 lb f = 0.004 448 kN { 1 ton f =9.964 kN
	ton force	meganewton	Mn	100 ton f = 0.9964 MN
Pressure and stress	pound force per square inch			1 lb f/in^2 = 6.895 kN/m^2
	pound force per square foot	kilonewton per square metre	kN/m^2	1 lb f/ft^2 = 0.047 88 kN/m^2
	ton force per square foot			1 ton f/ft^2 = 107.3 kN/m^2 1 kg f/cm^2 = 98.07 kN/m^2
	ton force per square foot	meganewton per square metre	MN/m^2	1 ton f/ft^2 = 0.1073 MN/m^2 1 bar = 100 kN/m^2*
Moduli of elasticity (E, G, K)	pound force per square foot	meganewton per square metre (= newton per square millimetre)	MN/m^2 (=N/mm^2)	{1000 lb f/ft^2 = 0.047 88 MN/m^2 {1 ton f/in^2 = 15.44 MN/m^2
	ton force per square inch			
Coefficient of volume conpressibility (mv)	square foot per ton force	square metre per meganewton	m^2/MN	1 ft^2/ton f = 9.324 m^2/MN
Coefficient of consolidation (Cv) or swelling	square foot per year square centimetre per second	square metre per year	m^2/year	1 ft^2/year = 0.0929 m^2/year 1 cm^2/s = 3154 m^2/year
Coefficient of permeability (k)	centimetre per second foot per year	metre per second metre per year	m/s m/year	1 cm/s = 0.01 m/s* 1 ft/year = 0.3048 m/year* = 0.9651 x 10^{-8} m/s

*Exact conversion factor

Appendix III Notation for chapter 8.

Symbol	Represents
A	Absorption coefficient. ie average moisture content of intact clast material.
C	Constant.
Cc	Compression index.
C_s	Swelling index.
Cu	Undrained shear strength.
E_V^1	Youngs Modulus in the vertical axis in terms of effective stress.
E_H^1	Youngs Modulus in the horizontal axis in terms of effective stress.
Gs	Specific gravity of soil particles.
Ko	Coefficient of effective earth pressure at rest.
Knc	Coefficient of effective earth pressure at rest for normal consolidation.
Kp	Coefficient of effective passive earth pressure.
LI	Liquidity Index.
LIo	Liquidity Index in-situ.
LI1	Liquidity Index on the virgin compression curve at unit pressure.
LIm	Liquidity Index on the virgin compression line at the maximum past consolidation pressure.
LL	Liquid Limit.
M	Moisture content of whole sample.
Mm	Matrix moisture content.
Mv	Coefficient of volume compressibility.
NMC	Natural moisture content.
OCR	Over consolidation ratio.
PI	Plasticity Index.
PL	Plastic Limit.
P_V^1	Vertical effective pressure.
P_{Vo}^1	Vertical effective pressure in-situ.
P_{Vm}^1	Maximum past vertical effective pressure.
S_f	Shear strength in terms of effective stress.
c	Proportion of sample by dry weight greater than sand size.
c^1	Cohesion in terms of effective stress.
e	Voids ratio.
eo	Voids ratio prior to loading.
e1	Voids ratio on the virgin compression curve at unit pressure.
e2	Voids ratio prior to unloading.
m	Moisture content.
n	Ratio of the vertical to the horizontal Youngs Modulus in terms of effective stress.
μ	Pore water pressure
Δ	is change eg. ΔP_V^1
υ^1	Poisson's Ratio in terms of effective stress.
υ_1^1	Poisson's Ratio effect of vertical strain on horizontal strain.
υ_3^1	Poisson's Ratio effect of horizontal strain on complementary horizontal strain.
σ_H^1	Horizontal effective stress.
ϕ^1	Peak angle of internal friction in terms of effective stress.
ϕ_r^1	Residual angle of internal friction in terms of effective stress.

References

Aario, R. 1977. Classification and terminology of morainic landforms in Finland. Boreas, 6, 87-100

AASTO 1970. American Association of State Highway and Transportation Officials' Classification System for Soils. AASHTO Designation M 145

Allen, J.R.L. 1971. Transverse erosional marks of mud and rock; their physical basis and geologic significance. Sed. Geol., 5, 167-385

Allen, J.R.L. 1982. Sedimentary Structures. Elsevier Scientific

Allen, V.T. 1959. Gumbotil and Interglacial clays. Geol. Soc. Am. Bull., 70, 1483-86

Alley, D.W. and Slatt, R.M. 1976. Drift prospecting and glacial geology in the Sheffield Lake-Indian Pond Area, north central Newfoundland. In Glacial Till. Roy. Soc. Can. Spec. Pub. 12, 249-66

Al-Saada, R. and Brooks, M. 1973. A geophysical study of Pleistocene based valleys in the lower Swansea Valley, Vale of North and Swansea Bay. Proc. Geol. Ass., 84, 135-153

Al-Shaikh-ali, M.M.H. 1978. The behaviour of Cheshire Basin lodgement till in motorway construction. In Clay Fills, Inst. Civ. Eng. London, 15-23

Andersland, O.B. and D.W. Anderson, Eds. 1978. Geotechnical Engineering for Cold Regions. McGraw-Hill

Anderson, J.B., Kurtz, D.D., Domack, E.W., Balshaw, K.M. 1980. Glacial and glacialmarine sediments of the Antarctic continental shelf. J. Geol., 88, 399-414

Anderson, J.G.C. 1969. Geological factors of the Ffestiniog Pumped Storage Scheme, Merioneth, Wales. Q. J. Eng. Geol., 2, 183-194

Anderson, J.G.C. 1971. Reservoirs for pumped storage. Q. Jnl. Engng. Geol., 4, 370-371

Anderson, J.G.C. 1974. The buried channels, rock floors and rock basins, and overlying deposits of the Smith Wales valleys from Wye to Neath. Proc. Sth. Wales Inst. Eng., 88, 3-17

Anderson, J.G.C. and Trigg, C.F. 1970. Geotechnical factors in the redevelopment of the South Wales Valleys, Proc. Conf. Inst. Civ. Eng., 637-49

Anderson, J.L. and Sollid, J.L. 1971. Glacial chronology and glacial geomorphology in the marginal zones of the glaciers, Midtdalsbreen and Nigardsbreen. Norsk. Geogr. Tidsskr., 25, 1-38

Anderson, M.G. and K.S. Richards. 1981. Geomorphological aspects of slopes in mudrocks in the United Kingdom. Q. J. Eng. Geol., 14, 363-372

Andrews, J.T. 1972. Post glacial rebound. The National Atlas of Canada. Canada Department of Energy, Mines and Resources, 35-36

Andrews, J.T. 1972a. Glacier power, mass balances, velocities and erosion potential. Zeit. Geomorph., 13, 1-17

Andrews, J.T. 1975. Glacial Systems: an approach to glaciers and their environments, Duxbury Press, Massachusetts

Andrews, J.T. 1982. Comment on "New evidence from beneath the Western North Atlantic for the depth of glacial erosion in Greenland and North America" by E.P. Laine. Quat. Res., 17, 123-124

Andrews, J.T. and Peltier, W.R. 1976. Collapse of the Hudson Bay ice centre and glacio-isostatic rebound. Geology, 4, 73-75

Andrews, J.T., W.W. Shilts and G.H. Miller. 1983. Multiple deglaciations of the Hudson Bay Lowlands, Canada, since deposition of the Missinaibi (Last Interglacial) Formation. Quat. Res., 19, 18-37

Andrews, J.T. and Matsch, C.L. 1983. Glacial marine sediments and sedimentation: GeoAbstracts (in press)

Anon. 1954. The Cod Beck Reservoir and Works of the Northallerton and District Water Board. Water and Water Engng., 58, 99-105

Anon. 1955. The upper Silent Valley Works of Belfast and district. Water and Water Engng., 59, 191-196

Anon. 1972. The preparation of maps and plans in terms of engineering geology. Q. Jl. Engng. Geol., 5, 293-381

Anon. 1977. The description of rock masses for engineering purposes. Q. J. Eng. Geol., 10, 355-388

Anon. 1981. New drill gets clean cores from unconsolidated formations. Western Miner, June

Anon. 1982. Recommended symbols for engineering geological maps. Report on the I.A.E.G. Commission on Engineering Geological Maps. Bull. Int. Assoc. Engng. Geol. 24

Anson, W.W. and J.I. Sharp. 1960. Surface and rockhead relief features in the northern part of the Northumberland coalfield. Dept. of Geography. Univ. Newcastle-Upon-Tyne Research Series No. 2, 23

Armstrong, J.E. 1981. Post Vashon Glaciation. Fraser Lowland, British Columbia. Geol. Surv. Can. Bull. 322

Armstrong, J.E. and Brown, W.L. 1954. Late Wisconsin marine drift and associated sediments of the lower Fraser Valley. British Columbia. Bull. Geol. Soc. Am., 65, 749-64

Arrowsmith, E.J. 1978. Roadwork fills - a material engineers' viewpoint. In Clay fills. Inst. Civ. Eng. 25-36

Ashley, G.M. 1975. Rhythmic sedimentation in glacial Lake Hitchcock, Massachusetts-Connecticut. In Glaciofluvial and glaciolacustrine sedimentation. Edited by A.V. Jopling and B.C. McDonald. Soc. Econ. Paleont. Mineral. Spec. Pub. No. 23, 304-320

Atkinson, J.H. 1975. Anisotropic elastic deformations in laboratory tests on undisturbed London Clay. Geotechnique, 25, 357-375

Atkinson, J.H. and Bransby, P.L. 1978. The Mechanics of Soils - An Introduction to Critical State Soil Mechanics. McGraw-Hill Book Company

Attewell, P.B. and I.W. Farmer. 1975. Principles of Engineering Geology. Chapman and Hall

Baguelin, F., Jezequel, J.F. and Shields, D.H. 1978. The Pressuremeter and Foundation Engineering. Trans. Tech. Publications

Baker, C.A. 1976. Late Devensian periglacial phenomena in the Upper Cam valley. Proc. Geol. Ass., 87, 285-306

Banerjee, I. and McDonald, B.C. 1975. Nature of esker sedimentation. In A.V. Jopling and B.C. McDonald, eds. Glaciofluvial and glaciolacustrine sedimentation. Soc. Econ. Paleont. Mineral. Spec. Pub. 28, 132-154

Banham, P.H. 1975. Glacitectonic structures: a general discussion with particular reference to the contorted drift of Norfolk. In Wright, A.E. and Moseley, F. eds. In Ice Ages: Ancient and Modern. Seel House Press, 69-94

Baracos, A. 1977. Compositional and structural anisotropy of Winnipeg soils - a study based on scanning electron microscopy and X-ray diffraction analyses. Can. Geot. J., 14, 125-137

Barnett, D.M. and Holdsworth, G. 1974. Origin, morphology and chronology of sublacustrine moraines, Generator Lake, Baffin Island, NWT Canada. Can. J. Earth Sci., 11, 380-408

Bateman, J.F. 1841. Description of the Bann Reservoirs. County Down, Ireland. Proc. Instn. Civ. Engrs., 1, 168-170

Bayrock, L.A. 1969. Till-like glacial lacustrine sediments in Alberta, Canada: Abstracts. VII Congress. INQUA, Paris, 268

Bazley, R.A.B. 1971. A map of Belfast for the engineering geologist. Q. Jl. Engng. Geol., 4, 313-314

Benedict, J.B. 1976. Frost creep and gelifluction features: a review. Quaternary Research, 6, 55-76

Bentley, S.P. and Smalley, I.J. 1978a. Inter-particle cementation in Canadian post-glacial clays and the problem of high sensitivity ($S_t > 50$). Sedimentology, 25, 297-302

Bentley, S.P. and Smalley, I.J. 1978b. Mineralogy of sensitive clays from Quebec. Canadian Mineralogist, 16, 103-112

Berry, F.S. 1979. Late Quaternary scour-hollows and related features in central
    London. Q.J.E.G., 12, 9-29
Bevan, O.M. and Parkes, D.B. 1975. Tunnelling in glacial materials in the
    British Isles. In The Engineering Behaviour of Glacial Materials, Geo.
    Abstracts, 212-220
Binnie, G.M. 1978. The collapse of the Dale Dyke dam in retrospect. Q. Jnl.
    Eng. Geol., 11, 305-324
Binnie, G.M. 1981. Early Victorian Water Engineers. Thomas Telford. London
Bishop, A.W. 1958. Test requirements for measuring the coefficient of earth
    pressure at rest. Int. Conf. on Earth Pressure Problems. Brussels, 2-14
Bishop, A.W. 1967. Progressive failure with special reference to the mechanism
    causing it. Proc. Geotech. Conf. Oslo, 2, 142-150
Bishop, A.W. and Henkel, D.J. 1957. The measurement of soil properties in the
    Triaxial Test. London. Arnold
Bishop, A.W. and Green, P.A. 1974. The development and use of trial embankments.
    Proceedings of the Conference on Field Instrumentation in Geotechnical
    Engineering, Butterworths, London
Bishop, A.W., Green, G.E., Garga, V.K., Andersen, A. and Brown, J.D. 1971. A
    new ring shear apparatus and its application to the measurement of residual
    strength. Geotechnique, 21, 273-328
Bjerrum, L. 1973. Geotechnical problems involved in foundations of structures in
    the North Sea. Geotechnique, 23, 319-358
Bjerrum, L. and Jørstad, F.A. 1968. Stability of rock slopes in Norway. Norwegian
    Geotech. Inst. Publ. 79
Black, R.F. 1969. Slopes in southern Wisconsin, U.S.A.; periglacial or temperate?
    Biuletyn Peryglacjalny, 18, 69-82
Black, R.F. 1976. Periglacial features indicative of permafrost: Ice and Soil
    Wedges. Quat. Res., 6, 3-26
Blodgett, R.H. and Stanley, K.O. 1980. Stratification, bedforms and discharge
    relations of the Platte braided river system, Nebraska. J. Sediment. Petrol.,
    50, 139-148
Bluck, B.J. 1974. Structure and directional properties of some valley sand
    deposits in southern Iceland. Sedimentology, 21, 533-554
Bluck, B.J. 1979. Structure of coarse grained braided stream alluvium. Trans.
    Roy. Soc. Edinburgh, 70, 181-221
BNCOLD/Univ. Symposium. 1976. British National Committee on Large Dams and
    University of Newcastle-Upon-Tyne. Inspection operation and improvement of
    existing dams. Proceedings of Symposium, Newcastle-Upon-Tyne, 24-26
    September 1975
Boardman, J. 1978. Grezes Litees, Keswick, Cumbria. Biuletyn Peryglacjalny,
    27, 23-54
Boellstorff, J. 1978. Chronology of some Late Cenozoic deposits from the central
    United States and the Ice Ages. Trans. Neb. Acad. Sci., VI, 35-49
Bolton, M. 1979. A guide to soil mechanics. MacMillan
Bolviken, B. and Gleeson, C.F. 1979. Focus on the use of soils for geotechnical
    exploration in glaciated terrane. Geol. Surv. Can. Ec. Geol. Rep. 31,
    295-326
Bonell, M. 1972. As assessment of possible factors contributing to well level
    fluctuations in Holderness boulder clay, East Yorkshire. Jl. Hydrol., 16,
    361-368
Bonell, M. 1972a. The application of the auger hole method in Holderness glacial
    drift. J. Hydrol., 16, 125-146
Boothroyd, J.C. and Ashley, G.M. 1975. Process, bar morphology and sedimentary
    structures on braided outwash fans, northeastern Gulf of Alaska. In A.V.
    Jopling and B.C. McDonald, eds. Glaciofluvial and glaciolacustrine sediment-
    ation. Soc. Econ. Paleont. Mineral. Spec. Pub. 23, 193-222

Boothroyd, J.C. and Nummedal, D. 1978. Proglacial braided outwash: a model for humid alluvial-fan deposits. In A.D. Miall, ed. Fluvial Sedimentology. Can. Soc. Petrol. Geol. Mem., 5, 641-668

Boulton, G.S. 1970. On deposition of subglacial and melt-out tills at the margins of certain Svalbard glaciers. J. Glac., 9, 231-246

Boulton, G.S. 1970a. On the origin and transport of englacial debris in Svalbard Glaciers. J. Glac., 19, 213-229

Boulton, G.S. 1971. Till genesis and fabric in Svalbard, Spitsbergen. In R.P. Goldthwait, ed. Till: A symposium. University of Ohio Press, 41-72

Boulton, G.S. 1972. Modern Arctic glaciers as depositional models for former ice sheets. J. Geol. Soc. Lond., 128, 361-393

Boulton, G.S. 1972a. The role of thermal regime in glacial sedimentation. Inst. Brit. Geog. Spec. Pub. 4, 1-20

Boulton, G.S. 1974. Processes and patterns of glacial erosion. In Glacial Geomorphology. Pub. in Geomorphology. State Univ. New York. 41-87

Boulton, G.S. 1975. Processes and patterns of subglacial sedimentation, a theoretic approach. In Ice Ages: Ancient and Modern. A.E. Wright and Moseley, F. eds. Seel House Liverpool, 7-42

Boulton, G.S. 1976. The origin of glacially fluted surfaces - observations and theory. J. Glac., 17, 287-301

Boulton, G.S. 1977. A multiple till sequence by a late Devensian Welsh Ice Cap. Cambria, 4, 10-31

Boulton, G.S. 1978. Boulder shapes and grain-size distributions of debris as indicators of transport paths through a glacier and till genesis. Sedimentology, 25, 773-799

Boulton, G.S. 1979. Processes of glacier erosion on different substrata. J. Glaciol., 23, 15-38

Boulton, G.S. 1979a. General Discussion. J. Glac., 23, 385

Boulton, G.S., Dent, D.L. and Morris, E.M. 1974. Subglacial shearing and crushing and the role of water pressures in tills from south-east Iceland, Geografiska Annaler, 56, 135-45

Boulton, G.S. and M.A. Paul. 1976. The influence of genetic processes on some geotechnical properties of glacial tills. Q. J. Eng. Geol., 9, 159-194

Boulton, G.S., Jones, A.S., Clayton, K.M., Kenning, M.J. 1977. A British ice-sheet model and patterns of glacial erosion and deposition in Britain. In F.W. Shotton, ed. British Quaternary Studies, 231-246, Oxford

Boulton, G.S. and Eyles, N. 1979. Sedimentation by valley glaciers: a model and genetic classification. In Moraines and Varves. Ed. C. Schluchter. Balkema Press, Rotterdam, 11-24

Boulton, G.S. and Jones, A.S. 1979. Stability of temperate ice caps and ice sheets resting on beds of deformable sediment. J. Glac., 24, 29-44

Boulton, G.S., Morris, E.M., Armstrong, A.A. and Thomas, A. 1979. Direct measurement of stress at the base of a glacier. J. Glac., 22, 3-24

Boulton, G.S., Baldwin, C.T., Peacock, J.D., McCabe, A.M., Miller, G., Jarvis, J., Horsfield, B., Worsley, P., Eyles, N., Chroston, P.N., Day, T.E., Gibbard, P., Hare, P.E. and Von Brunn, V. 1982. A glacioisostatic facies model applied to Spitsbergen and Weichselian glacioclimatic events in the Atlantic Sector of the Northern Hemisphere. Nature, 298, 437-441

Bowen, D.Q. 1978. Quaternary Geology. Pergamon Press

Bradshaw, P.M. 1975. Conceptual models. In Exploration Geochemistry Developments in Economic Geology, No. 2. Elsevier Scientific

Brand, E.W. and Brenner, R.P. 1981. Soft clay engineering. Elsevier Sc. Pub. Co. Dev. In Geotech. Eng., 20

Bretz, H. 1969. The Lake Missoula floods and the channeled scabland. J. Geol., 77, 504-543

British Standards Institution. 1972. Code of Practice for Foundations

British Standard 812. 1967. Methods for sampling and testing of mineral aggregates, sands and tills

British Standard 812. 1975. Methods for sampling and testing of mineral aggregates, sands and fillers. Pt 2. Physical Properties

British Standard 1377. 1975. Methods of test for soil for civil engineering
purposes

British Standard 5930. 1981. (formerly CP 2001) Code of Practice for Site
Investigation

Brodzikowski, K. and A.J. Van Loon. 1980. Sedimentary deformations in Saalian
glaciolimnic deposits near Wlostow. Geol. En . Mijn., 59, 251-272

Brooker, E.W. and Ireland, H.O. 1965. Earth pressures at rest related to stress
history. Can. Geot. J., 11, 1-15

Broscoe, A.J. and Thompson, S. 1969. Observations on an alpine mudflow, Steele
Creek, Yukon. Can. J. Earth Sci., 6, 219-229

Broster, B.E. 1977. Magnetic, physical and lithologic properties and age of till
exposed along the east coast of Lake Huron, Ontaro: Discussion. Can. Jour.
Earth Sci., 14, 2169-2171

Brunsden, D. and Jones, D.K.C. 1972. The morphology of degraded landslide slopes
in south-west Dorest. Q. J. Eng. Geol., 5, 1-18

Brunsden, D., Doarnkamp, J.C., Fookes, P.G., Jones, D.K.C. and Kelly, J.M.H.
1975. Large scale geomorphological mapping and highway engineering design.
Q. Jl. Engng. Geol., 8, 227-253

Buchanan. 1970. Derwent Dam-construction. Proc. Instn. Civ. Engnrs., 45, 401-422

Buck, B.W. and Parry, W.T. 1976. Physico-chemical properties of sensitive soils
in the lower Jordan Valley, Utah. Proceedings. 14th Symp. Eng. Geol. and
Soils Eng. Boise, Ill., 53-70

Budd, W.F. 1979. The importance of ice sheets in long term changes of climate
and sea level. In Sea level, ice and climatic change. I.A.H.S. Publication
131, 441-471

Bull, C. and Marangunic, C. 1968. Glaciological effects of debris slide on Sherman
Glacier. In The Great Alaska Earthquake of 1964. Hydrology Volume. N.A.S.
Pub. 1603. Washington, U.S., 309-317

Bull, P. 1981. Environmental reconstruction by electron microscopy. Progress,
Physical Geography, 5, 368-397

Burrin, P.J. 1981. Loess in the Weald. Proc. Geol. Ass., 92, 87-92

Burnett, A.D. and Fookes, P.G. 1974. A regional engineering geological study
of the London Clay. Q. Jnl. Eng. Geol., 7, 257-295

Cabrera, J.G. and Smalley, I.J. 1973. Quick clays as products of glacial action:
A new approach to their marine geology, distribution and geotechnical
properties. Eng. Geol., 7, 115-133

Calkin, P.E., Muller, E.H. and Barnes, J.H. 1982. The Gowanda Hospital Interstad-
ial Site, New York. Am. Jour. Sci., 282, 1110-1142

Carlyle, W.J. and R.W. Buchanan. 1977. The dams for the Dinorwig project. BNCOLD
News and Views, 18, 21-22

Carnerro, F.B., Ferrante, A.J. and R.C. Batista. eds. 1982. Offshore Engineering.
Pentech. Press, Plymouth, U.K.

Carruthers, R.G. 1939. On northern glacial drifts: some peculiarities and their
significance. Quat. J. Geol. Soc. London, 95, 299-333

Carruthers, R.G. 1947. The secret of the glacial drifts. Proceedings Yorkshire
Geological Society, 27, 43-57, 129-172

Carruthers, R.G. 1953. Glacial drifts and the undermelt theory: Newcastle-Upon-
Tyne, Harold Hill & Son, 42

Casagrande, A. 1936. The determination of the preconsolidation load and its
practical significance. 1st Int. Conf. Soil Mech. and Foundn. Engng., 3,
60-64

Carter, D.J. and Hart, M.B. 1977. Micropalaeontological investigations for the
site of the Thames Barrier, London. Q. J. Eng. Geol., 10, 321-338

Catt, J.A. and Penny, L.F. 1966. The Pleistocene deposits of Holderness, East Yorkshire. Proc. Yorks Geol. Soc., 35, 375-420

Catt, J.A. 1977. Loess and Coversands. In British Quaternary Studies (ed. F.W. Shotton). 221-230. Oxford

Cavoundis, S. and Hoey, K. 1977. Consolidation during construction of earth dams. Jnl. Geot. Eng. Div. Am. Soc. Civ. Eng., 103, 1055-1068

Chan, H.T. and Kenney, T.C. 1973. Laboratory investigation of permeability ratio of New Liskeard varved soil. Can. Geot. J., 10, 453-472

Chandler, R.J. 1969. The effect of weathering on the shear strength properties of Keuper Marl. Geotechnique, 19, 321-334

Chandler, R.J. 1970. The degradation of the Lias Clay slopes in the east Midlands. Q. J. Eng. Geol., 2, 161-181

Chandler, R.J. 1970a. Solifluction on low angled slopes in Northamptonshire. Q. J. Eng. Geol., 3, 65-69

Chandler, R.J. 1972a. Periglacial mudslides in Vestspitsbergen and their bearing on the origin of fossil 'solifluction' shears in low-angled clay slopes. Q. J. Eng. Geol., 5, 223-241

Chandler, R.J. 1972b. Lias Clay: weathering processes and their effect on shear strength. Geotechnique, 22, 403-431

Chapman, L.J. and D.F. Putnam, 1973. The physiography of Southern Ontario. 2nd Ed. Univ. Toronto Press

Charlesworth, J.K. 1957. The Quaternary Era. 2 vols. Arnold

Charron, J.E. 1974. A study of groundwater flow in Russell County, Ontario. Env. Canada. Inland Waters Directorate. Sci. Ser. No. 40

Cheel, R.J. and Rust, B.R. 1982. Coarse-grained facies of glacio-marine deposits near Ottawa, Canada. In Proc. 6th Guelph Symp. Research in Glacial Systems. Geo. Books, Norwich, 279-295

Christiansen, E.A. 1971. Tills in Southern Saskatchewan, Canada. In Goldthwait, R.P. ed. Till a Symposium. Ohio State Univ. Press, 167-183

Christiansen, E.A. 1979. The Wisconsin deglaciation of southern Saskatchewan and adjacent areas. Can. J. Earth Sci., 16, 913-938

Christiansen, E.A. and Whitaker, S.H. 1976. Glacial thrusting of drift and bedrock. In Glacial Till. Roy. Soc. Can. Spec. Pub. 12, 121-132

Church, M. 1972. Baffin Island Sandust: a study of arctic fluvial processes. Geol. Surv. Can. Bull. 216

Church, M. and Ryder, J.M. 1972. Paraglacial sedimentation: a consideration of fluvial processes conditioned by glaciation. Bull. Geol. Soc. Am., 83, 3059-3072

Church, M., Stock, R.F. and Ryder, J.M. 1979. Contemporary sedimentary environments on Baffin Island, N.W.T., Canada. Debris slope accumulations. Arctic and Alp. Res., 11, 371-402

Clague, J. 1975. Sedimentology and paleohydrology of Late Wisconsinan outwash, Rocky Mountain Trench, southeastern British Columbia. In A.V. Jopling and B.C. McDonald. eds. Glaciofluvial and Glaciolacustrine Sedimentation. Soc. Ec. Pal. Miner. Spec. Pub. 23, 223-237

Clague, J.J. 1981. Landslides at the south end of Kluane Lake, Yukon Territory. Can. J. Earth Sci., 18, 959-971

Clapperton, C.M. 1975. The debris content of surging glaciers. In Svalbard and Iceland. J. Glac., 14, 395-406

Clarke, G.K.C. 1976. Thermal regulation of glacier surging. J. Glac., 16, 231-250

Clarke, A.R. and Johnson, D.K. 1975. Geotechnical mapping as an integral part of site investigation - Two case histories. Q. Jl. Engng. Geol., 8, 211-224

Clarke, C.L. and le Masurier, M. 1976. Inspection of old embankment dams. Paper 5.2. In BNCOLD/Univ. Symposium 9

Clark, D.L., Whitman, R.R., Morgan, K.A. and Mackey, S.D. 1980. Stratigraphy and glacial-marine sediments of the Amerasian Basin, Central Arctic Ocean. Geol. Soc. Am. Spec. Pub. 181

Clarke, A.R. and Johnson, D.K. 1975. Geotechnical mapping as an integral part of site investigation: Two case histories. Q. J. Eng. Geol., 8, 211-224

Clayton, K.M. 1963. A map of the drift geology of Great Britain and Northern Ireland. Geog. Jnl., 129, 75-81

Clayton, K.M. 1965. Glacial erosion in the Finger Lakes region. (New York State, U.S.A.). Zeit. Geomorph., 9, 50-62

Clayton, K.M. 1974. Zones of glacial erosion. Inst. Brit. Geogr. Spec. Pub. 7, 163-176

Clayton, K.M. 1977. River Terraces. In British Quaternary Studies. ed. F.W. Shotton, 163-168. Oxford

Clayton, L. 1964. Karst topography on stagnant glaciers. Jour. Glac., 5, 107-112

Clayton, L. 1967. Stagnant glacier features of the Missouri Coteau in North Dakota. In Clayton, L. and Freers, T.F. eds. Glacial geology of the Missouri Coteau and Adjacent Areas. North Dakota Geol. Survey. Misc. Series 30, 25-52

Clayton, L. and Moran, S.R. 1974. A glacial process-form model. In Glacial Geomorphology. D.R. Coates, ed. State Univ., N.Y. 89-120

Clevenger, W.A. 1958. Experiences with soil as a foundation material. Trans. Am. Soc. Civ. Eng., 123, 151-180

Coates, D.F. 1964. Some cases of residual stress effects on engineering works. State of Stress in the Earth's Crust. Elsevier, N.Y., 679-688

Coates, D.R. 1974. Reappraisal of the glaciated Appalachian Plateau. In Glacial Geomorphology, D.R. Coates, ed. 205-244

Coates, D.R. 1977. ed. Landslides. Reviews in Eng. Geol. III. Geol. Soc. Am. Boulder

Cocksedge, J.E. and Hight, D.W. 1975. Some geotechnical aspects of road design and construction in tills. In The Engineering Behaviour of Glacial Materials. 197-206. Geo-Abstracts

Colback, P.S.B., Carter, T.G. and Eastaff, D.J. 1975. Investigations of discontinuities within the glacial deposits forming an integral part of the Brenig Dam and its foundations in Foundations on Quaternary Deposits. Engng. Group Geol. Soc. London. Regional meeting. Norwich. 27-32

Coleman, J.M. and L.E. Garrison. 1977. Geological aspects of marine slope instability. Northwestern Gulf of Mexico. Marine Geotech., 2, 9-44

Collinson, J.D. 1970. Bedforms of the Tana River, Norway. Geog. Annaler, 52A, 31-55

Collinson, J.D. and Thompson, D.B. 1982. Sedimentary Structures. George Allen and Unwin

Conybeare, H. 1859. In Discussion of Jackson, M.B. On the Water Supply to the City of Melbourne, South Australia. Proc. Instn. Civ. Engrs., 18, 363-403

Cooke, R.U. and Doornkamp, J.C. 1977. Geomorphology in Environmental Management. Oxford

Cooper, A.J. 1979. Quaternary Geology of the Grand Bend-Parkhill Area, Southern Ontario. Ontario Geological Survey Report 188

Corner, R. 1975. The Tana Valley terraces: a study of the morphology and sedimentology of the lower Tana Valley terraces between Mansholman and Maskejokka, Finnimark, Norway. Univ. Uppsala Dept. Phys. Geogr. Ungi Rapport 38

Crampton, C.B. and Taylor, J.A. 1967. Solifluction terraces in South Wales. Biuletyn Periglacjalny, 16, 15-36

Crandell, D.R. and Waldron, H.H. 1956. A recent volcanic mudflow of exceptional dimensions from Mt. Rainier, Washington. Am. J. Sci., 254, 349-362

Crann, H.H. 1978. Llyn Brenig. Part 1: The concept and its promotion. J. Instn. Water Engnrs. and Scientists, 32, 279-287

Cratchley, C.R. and Denness, B. 1972. Engineering geology in urban planning with an example from the new city of Milton Keynes. International Geological Congress. 24th Session, Montreal. Section 13 Proceedings. 13-22

Crawford, C.B. and Burn, K.N. 1962. Settlement Studies on the Mt. Sinai Hospital, Toronto. Engng. Jour., 31-37

Cripps, J.C. and R.K. Taylor. 1981. The engineering properties of mudrocks. Q. J. Eng. Geol., 14, 325-346

Cronin, T.M. 1977. Late-Wisconsin marine environments of the Champlain Valley, New York, Quebec. Quat. Res., 7, 238-253

Cruden, D.M. 1976. Major rock slides in the Rockies. Can. Geotech. J., 13, 8-20

Curtis, L.F., Courtney, F.M. and Trudgill, S.T. 1976. Soils in the British Isles. Longmans

Dance, D.R. 1981. Resonant Vibrational Drill Systems. 3rd Annual Conf. Alaskan Placer Mining Proc., 1-12

Day, T.E. and Morris, W.A. 1982. Magnetic methods rapid in drift exploration. The Northern Miner. June 17

Deacon, G. 1896. The Vrynwy Works for the water supply of Liverpool. Proc. Inst. Civ. Engrs., 126, 24-67

Dearman, W.R. 1974. The characterization of rock for civil engineering practice in Britain. Colloque de Geol. de l'Ingenieur. Roy. Geol. Soc. Belg., 1-75

Dearman, W.R. and Fookes, P.F. 1974. Engineering geological mapping for civil engineering practice in the United Kingdom. Q.J. l Eng. Geol., 7, 223-256

Dearman, W.R. and Matula, M. 1976. Environmental aspects of engineering geological mapping. Bull. Int. Ass. Eng. Geol., 14, 141-146

Dearman, W.R., Money, M.S., Coffey, R.J., Scott, P. and Wheeler, M. 1977. Engineering geological mapping of the Tyne and Wear conurbation N.E. England. Q. Jl. Eng. Geol., 10, 145-168

Dearman, W.R., Baynes, F.J. and Pearson, R. 1977a. Geophysical detection of disused mineshafts in the Newcastle-Upon-Tyne area, north-east England. Q. J. Eng. Geol., 10, 257-269

Dearman, W.R., Money, M.S., Strachan, A.D., Coffey, J.R. and Marsden, A. 1979. A regional engineering geological map of the Tyne and Wear County, N.E. England. Bull. Int. Assoc. Eng. Geol., 19, 5-17

Dearman, W.R. and Eyles, N. 1982. An engineering geological map of the soils and rocks of the United Kingdom. Bull. Int. Ass. Eng. Geol., 25, 3-18

Deere, D.U. and Patton, F.D. 1971. Slope stability in residual soils. Proc. 4th. Pan Am. Conf. Soil. Mech. Found. Eng. State of the Art Report 1, 88-171

De Jong, J. and Harris, M.C. 1971. Settlement of two multistorey buildings in Edmonton. Can. Geotech. J., 8, 217-235

De Jong, J. and Morgenstern, N.R. 1973. Heave and settlement of two tall building foundations in Edmonton, Alberta. C. Geotech. J., 10, 261-281

Dell, C.I. 1976. Sediment distributions and bottom topography of Southeastern Lake Superior. Jour. Great Lakes Res., 2, 164-176

Denton, G. and Hughes, T.J. 1981. The Last Great Ice Sheets. Wiley Interscience

Department of Transport. 1976. Specification for Roads and Bridgeworks. HMSO Choice of an upper limit of moisture content for highway earthworks

Derbyshire, E. 1975. Distribution of glacial soils in Great Britain. In The Engineering Behaviour of Glacial Materials. GeoAbstracts. Norwich, 6-17

Desaulniers, D.E., Cherry, J.A. and Fritz, P. 1981. Origin, Age and Movement of Porewater in Argillaceous Quaternary Deposits at Four Sites in Southwestern Ontario. Jl. Hydrology, 50, 231-257

Dines, H.G. et al., 1940. The mapping of Head deposits. Geol. Mag., 77, 198-226

Domack, E.W. 1982. Facies of Late Pleistocene glacial marine sediments on Whidbey Island, Washington. (Ph.D. Thesis). Rice University Texas

380   References

Domack, E.W. 1982a. Sedimentology of glacial and glacialmarine deposits on the
    George V - Adelie continental shelf. East Antarctica. Boreas, 11, 79-97
Donovan, J.J. 1978. On the retrogression of landslides in sensitive muddy
    sediments. Can. Geot. J., 15, 441-446
Donovan, J.J. and Lajoie, G. 1979. Geotechnical implications of diagenetic
    iron sulphide formation in Champlain Sea sediments. Can. J. Earth Sci.,
    16, 575-584
Dreimanis, A. 1958. Tracing ore boulders as a propsecting method in Canada.
    Can. Inst. Min. Metall. Bull., 51, 73-80
Dreimanis, A. 1976. Tills: their origin and properties. In Glacial till.
    Edited by R.F. Legget. Royal Soc. Can. Spec. Pub. 12, 11-49
Dreimanis, A. 1977a. Correlation of Wisconsin glacial evnets between the eastern
    Great Lakes and the St. Lawrence Lowlands. Geographie, Physique et
    Quaternaire. XXXI, 37-51
Dreimanis, A. 1977b. Late Wisconsin glacial retreat in the Great Lakes region.
    North America. Annals of the New York Acad. Sci., 288, 70-89
Dreimanis, A. 1979. The problem of waterlaid tills. In Schluchter, C., ed.
    Moraines and Varves. Balkema, Rotterdam. 167-177
Dreimanis, A. and Vagners, U.J. 1971. Bimodal distribution of rock and mineral
    fragments in basal till. In R.P. Goldthwait. ed. Till: A Symposium, 237-
    50. Univ. Ohio Press
Drewry, D.J. and Cooper, A.P.R. 1981. Processes and models of Antactic glacio-
    marine sedimentation. Annals of Glaciology, 2, 117-122
Dowding, C.H. 1979. Site characterization and exploration. Am. Soc. Civ. Eng.
    New York
Early, K.R. and Skempton, W.A. 1972. Investigations of the landslide of Walton's
    Wood, Staffordshire. Q. J. Eng. Geol., 5, 19-42
Easterbrook, D.J. 1963. Late Pleistocene glacial events and related sea-level
    changes in the northern Puget Lowland. Geol. Soc. Amer. Bull., 74, 1465-1484
Easterbrook, D.J. 1964. Void ratios and bulk densities as a means of identifying
    Pleistocene tills. Bull. Geol. Soc. Am., 75, 745-750
Easterbrook, D.J. and Boellstorff, J. 1978. Paleomagnetism of tills and the age
    of the Brunhes-Matuyama boundaries. Geol. Soc. Am. Ab. Programs, 10, No. 7,
    394
Eckel, E.B. 1951. Interpreting geologic maps for engineers. Symposium on surface
    and subsurface reconnaissance. Spec. Tech. Pub. 122, 5-15. Am. Soc. Testing
    Materials.
Eden, R.A. 1975. North Sea environmental geology in relation to pipelines and
    structures. Oceanology International '75, 302-309
Eden, R.A., Holmes, R. and Fannin, N.G.T. 1977. Depositional Environment of
    Offshore Quaternary Deposits of the Continental Shelf Around Scotland.
    Inst. Geol. Sci. Rep. No. 77/15
Eden, W.J. 1976. Construction difficulties with loose glacial tills on Labrador
    Plateau. In Glacial Till. Roy. Soc. Can. Spec. Pub. 12, 391-400
Edwards, M.B. 1978. Glacial Environments. In Sedimentary Environment and Facies.
    H.G. Reading, ed. Oxford, Blackwells. 416-438
Eisbacher, G.H. 1979. First order regionalisation of landslide characteristics
    in the Canadian Cordillera. Geoscience Can., 6, 69-79
Eisbacher, G.H. 1979a. Cliff collapse and rock avalanches. (Sturzstroms). In
    the Mackenzie Mountains, northwestern Canada. Can. Geot. J., 16, 309-334
Eisbacher, G.H. 1982. Slope stability and land use in mountain valleys. Geoscience
    Can., 9, 14-27
Eisenstein, Z. and Morrison, N.A. 1973. Prediction of foundation deformations
    in Edmonton using an in situ pressure probe. Can. Geotech. J., 10, 193-210
Elson, J.A. 1961. The geology of tills. Proc. 14th Canadian Soil Mech. Conf.
    Nat. Res. Council. Ottawa, 5-17

Elson, J.A. 1967. Geology of glacial Lake Agassiz. In Life, land and water.
   University of Manitoba Press, Winnipeg. Man., 36-95
Embleton, C.E. and King, C.A.M. 1978. Glacial and Periglacial Geomorphology.
   Arnold
Engquist, P., Olsson, T. and Svensson, T. 1978. Pumping and recovery tests in
   wells sunk in till. Nordic Hydrol. Conf., Hanasaari, Finland. 134-42
Evenson, E.B. 1971. The relationship of macro- and microfabric of till and
   genesis of glacial landforms in Jefferson County, Wisconsin. In Till,
   A. Symposium. Ohio State University Press. R.P. Goldthwait. ed., 345-364
Evenson, E.B., Dreimanis, A. and Newsome, J.W. 1977. Subaquatic Flow Tills:
   A New Interpretation for the Genesis of Some Laminated Till Deposits.
   Boreas, 6, 115-133
Evenson, E.B., Pasquini, T.A., Stewart, R.A. and Stephens, G. 1979. Systematic
   provenance investigations in areas of alpine glaciation. Applications to
   glacial geology and mineral exploration. In Ch. Schluchter, ed. Moraines
   and Varves. Balkema Press, Rotterdam, 25-42
Eyles, C.H. and Eyles, N. 1983. Sedimentation in a large lake: A reinterpretation
   of the Late Pleistocene stratigraphy of Scarborough Bluffs, Ontario, Canada.
   Geology, 11, 146-152
Eyles, C.H. and Eyles, N. 1983a. Glaciomarine diamicts of the Isle of Man
   (Irish Sea Basin) and their paleoenvironmental significance. Geology,
   in press
Eyles, N. 1979. Facies of supraglacial sedimentation on Icelandic and Alpine
   temperate glaciers. Can. J. Earth Sci., 16, 1341-1361
Eyles, N. 1983. Modern Icelandic glaciers as depositional models for 'hummocky
   moraine' in the Scottish Highlands. In Tills and Related Sediments.
   E.B. Evenson and C.H. Schluchter. eds. A.A. Balkema, in press
Eyles, N. and Rogerson, R.J. 1977. Glacier movement, ice structures and medial
   moraine form at a glacier confluence. Berendon Glacier, British Columbia.
   Canada. Can. J. Earth Sci., 14, 2807-2816
Eyles, N. and Slatt, R.M. 1977. Ice marginal sedimentary, glacitectonic and
   morphologic features of Pleistocene drift: an example from Newfoundland.
   Quat. Res., 8, 267-281
Eyles, N. and Rogerson, R.J. 1977a. How to save your mine from a glacier.
   Can. Min. J., 98, 32
Eyles, N. and Rogerson, R.J. 1978. The sedimentology of medial moraines on
   Berendon Glacier, British Columbia: Implications for debris transport in
   a glacierized basin. Bull. Geol. Soc. Am., 89, 1688-1693
Eyles, N. and Dearman, W.R. 1981. A glacial terrain map of Britain for engineer-
   ing purposes. Bull. Int. Ass. Eng. Geol., 24, 173-184
Eyles, N. and Sladen, J.A. 1981. Stratigraphy and geotechnical properties of
   weathered lodgement tills in Northumberland, England. Q. Jour. Eng. Geol.
   London, 14, 129-141
Eyles, N., Sladen, J.A. and Gilroy, S. 1982. A depositional model for stratigraphic complexes and facies superimposition in lodgement tills. Boreas,
   11, 317-333
Evles, N., Sasseville, D.R., Slatt, R.M. and Rogerson, R.J. 1982. Geochemical
   denudation rates and solute transport mechanisms in a maritime temperate
   glacier basin. Can. J. Earth Sci., 19, 1570-1582
Eyles, N., Eyles, C.H. and Miall, A.D. 1983. Lithofacies types and vertical
   profile models: an alternative approach to the description and environmental
   interpretation of glacial diamict and diamictite sequences. Sedimentology,
   30, No. 3, in press
Eyles, N., Eyles, C.H. and Day, T.E. 1983a. Sedimentologic and paleomagnetic
   characteristics of glaciolacustrine diamict assemblages at Scarborough
   Bluffs, Ontario, Canada. In Tills and Related Sediments. E.B. Evenson
   and C.H. Schluchter. eds. A.A. Balkema, in press

Falconer, A. 1972. Use of Q-mode factor analysis in the interpretation of glacial deposits. In Research Methods in Pleistocene Geomorphology, GeoAbstracts, Norwich, U.K., 148-155

Fahnestock, R.K. 1963. Morphology and hydrology of a braided stream. U.S. Geol. Surv. Prof. Pap. 422A

Fahnestock, R.K. 1969. Morphology of the Slims River: Icefield Ranges Research Project. Sci. Results, 1, 161-172

Farrand, W.R. 1969. The Quaternary History of Lake Superior. Proc. 12th Conf. Great Lakes Res. Int. Ass., Great Lakes Res., 181-197

Farrar, D.M. 1971. A laboratory study of the use of wet fill in embankments. TRRL Report 406

Farrar, D.M. and Darley, P. 1975. The operation of earthmoving plant on wet fill. Trans. Road Res. Lab. Report LR 688

Feda, J. 1966. Structural stability of subsident loess from Praha-Dejvice. Eng. Geol., 1, 201-219

Ferguson, H.F. 1967. Valley stress release in the Allegheny Plateau. Bull. Assoc. Eng. Geol., 4, 63-68

Ferrians, O.J. 1963. Glaciolacustrine Diamicton Deposits in the Copper River Basin, Alaska. U.S. Geol. Surv. Prof. Pap., 476c, C120-C125

Fisher, D.A. and Jones, G.J. 1971. The possible future behaviour of Berendon Glacier, Canada: A further study. Jnl. Glac., 10, 85-92

Fisher, W.R. 1983. Modulus of elasticity of a very dense glacial till determined by plate load tests. Can. Geotech. J., 20, 183-186

Flemal, R.C. 1976. Pingos and pingo scars: Their characteristics, distribution and utility in reconstructing former periglacial environments. Quat. Res., 6, 37-54

Fleming, R.W., Spencer, G.S. and Banks, D.C. 1970. Empirical study of behaviour of clay shale slopes. U.S. Army. Engr. Nucl. Cratering Gp. Tech. Rep. 15, 1 and 2

Flint, R.F. 1971. Glacial and Quaternary Geology. John Wiley

Facht, J.A. and Kraft, L.M. 1977. Progress in marine geotechnical engineering. Jnl. Geot. Eng. Div. Am. Soc. Civ. Eng., 103, 1097-1118

Fookes, P.G. 1969. Geotechnical mapping of soils and sedimentary rock for engineering purposes with examples of practice from the Mangla Dam project. Geotechnique, 19, 52-74

Fookes, P.G. and Best, R. 1969. Consolidation characteristics of some late Pleistocene periglacial metastable soils of east Kent. Q. J. Eng. Geol., 2, 103-128

Fookes, P.G. and Higginbottom, I.E. 1970. Engineering aspects of periglacial features in Britain. Quart. J. Eng. Geol., 3

Fookes, P.G., Hinch, L.W. and Dixon, J.C. 1972. Geotechnical considerations of the site investigation for Stage IV of the Taff Vale Trunk Road to South Wales. Second British Regional Congress. Cardiff 1-25. British National Committee, Permanent International Assoc. Road Congresses

Fookes, P.G., Gordon, D.L. and Higginbottom, I.E. 1975. Glacial landforms, their deposits and engineering characteristics. In The Engineering Behaviour of Glacial Materials, 18-51. GeoAbstracts

Fookes, P.G., Hinch, L.W., Huxley, M.A. and Simons, N.E. 1975. Some soil properties in glacial terrain - the Taff Valley, South Wales. In The Engineering Behaviour of Glacial Materials, 93-116. GeoAbstracts

Fookes, P.G. and Sweeney, M. 1976. Stabilization and control of local rock falls and degrading rock slopes. Q. J. Eng. Geol., 9, 37-55

Ford, D.C., Schwarcz, H.P., Drake, J.J., Gascoyne, M., Harmon, R.S. and Latham, A.G. 1981. Estimates of the age of existing relief within the southern Rocky Mountains of Canada. Arc. Alp. Res., 13, 1-10

Ford, S.E.H., Arah, R.M., Reeves, J.W., Fleming, J.H. and Cockshaw, A. 1978. Llyn Brenig, Part II. Design and construction. J. Instn. Water Engineers and Scientists, 32, 288-302

Frakes, L.A. 1978. Diamictite. In The Encyclopedia of Sedimentology (R.W.
    Fairbridge and J. Bourgeois, eds.). Dowden, Hutchinson and Ross, 262-263
Frakes, L.A. and J.C. Crowell. 1975. Characteristics of modern glacial marine
    sediments: Application to Gondwana Glacials. In Gondwana Geology Austr.
    Nat. Univ. Press. 373-380
French, H.M. 1976. The Periglacial Environment. Longman
Fulton, R.J. and Halstead, E.C. 1972. Field Excursion 402: Quaternary geology
    of the Southern Canadian Cardillera. 24th Internat. Geol. Congr., Montreal,
    Quebec, guidebook
Gadd, N.R. 1971. Pleistocene Geology of the St. Lawrence Lowland. Memoir Geol.
    Surv. Can., 359
Gadd, N.R. 1975. Geology of Leda clay. Mass Wasting, 4th Guelph Symp. Geom.
    Geo. Abstracts. Norwich, U.K., 137-151
Galloway, W.E. 1976. Sediments and stratigraphic framework of the Copper River
    fan-delta, Alaska. J. Sed. Pet., 46, 726-737
Galloway, R.W. 1956. The structure of moraines in Lyngdalen, north Norway.
    J. Glac., 2, 730-733
Gamblin, D.G. and A.L. Little. 1970. Emergency measures at Lluest Wen Reservoir.
    Water and Water Engng., 74, 93-111
Geddes, W.G.N. and Pradoura, H.H.M. 1967. Backwater Dam in the Country of Angus,
    Scotland. Grouted cut-off. Proc. 9th Int. Congr. Large Dams 1, 273-274
Geddes, W.G.N., Rocke, G. and Scrimgeour, J. 1972. The Backwater Dam. Proc. Inst.
    Civil Engrs., 51, 433-464
Gerber, E. and Scheidegger, A.E. 1969. Stress-induced weathering of rock masses.
    Ecl. Geol. Helv., 62, 401-416
Gerber, E. and Scheidegger, A.E. 1973. Erosional and stress-induced features of
    steep slopes. Zeitschrift. Geomorph., 18, 38-49
Gibbard, P. 1980. The origin of stratified Catfish Creek till by basal melting.
    Boreas, 9, 71-85
Gibbs, H.J. and Holtz, W.G. 1957. Research on determining density of sands by
    spoon penetration testing. Proc. 4th Int. Conf. Soil Mech., Found. Eng.,
    London, 1, 35-39
Gilbert, G.K. 1906. Crescentic gouges on glaciated surfaces. Bull. Geol. Soc.
    Am., 17, 303-316
Gilbert, R. 1971. Observations on ice-dammed Summit Lake, British Columbia,
    Canada. J. Glac., 10, 351-356
Gilbert, R. 1975. Sedimentation in Lillooet Lake, British Columbia. Can.
    J. Earth Sci., 12, 1697-1711
Gilbert, R. and Shaw, J. 1981. Sedimentation in proglacial Sunwapta Lake,
    Alberta. Can. J. Earth Sci., 18, 81-93
Gillberg, G. 1977. Redeposition: a process in till formation. Geol. Forn.
    Stockh. Forh., 99, 246-253
Gillott, J.E. 1970. Fabric of Leda clay investigated by optical, electron-
    optical and X-ray diffraction methods. Eng. Geol., 4, 133-153
Gillott, J.E. 1971. Mineralogy of Leda clay, Canadian Mineralogist, 10,
    797-811
Glen, J.W. 1953. Experiments on the deformation of ice. J. Glac., 2, 111-114
Glen, J.W. 1955. The creep of polycrystalline ice. Proc. R. Soc. A, 228, 519-38
Glossop, R. 1968. Landslide on A40 Trunk road near Monmouth. Geotechnique,
    18, 107-150
Goldthwait, R.P. 1979. Grant grooves made by concentrated basal ice streams.
    J. Glac., 23, 297-306
Goodman, D.J., King, G.C.P., Millar, D.H.M. and Robin, G. de Q. 1979. Pressure-
    melting effects in basal ice of temperate glaciers. J. Glac., 23, 259-272
Gordon, J.E. 1981. Ice scoured topography and its relationships to bedrock
    structure and ice movement in parts of northern Scotland and West Greenland.
    Geografiska.Annaler A., 63, 55-65

Gowers, A.G. 1963. Some points of interest in the design and construction of
     Orrin Dam. Rossshire. Proc. Instn. Civil Engnrs., 24, 449
Gravenor, C.P. 1953. The origin of Drumlines. Am. J. Sci., 251, 674-681
Gravenor, C.P. 1975. Erosion by continental ice sheets. Am. J. Sci., 275, 594-604
Gravenor, C.P. and Bayrock, L.A. 1956. Stream-trench systems in east-central
     Alberta. Res. Counc. Alb. Prelim. Rep. 56-4
Gravenor, C.P. and Meneley, W.A. 1958. Glacial flutings in central and northern
     Alberta. Am. J. Sci., 256, 715-728
Gravenor, C.P. and Kupsch, W.O. 1959. Ice-disintegration features in Western
     Canada. J. Geol., 67, 48-64
Gravenor, C.P., Green, R. and Godfrey, J.D. 1960. Air photographs of Alberta.
     Res. Council. Alb. Bull. 5
Gravenor, C.P., Stupavsky, M. and Symons, D.T.A. 1973. Paleomagnetism and its
     relationship to till deposition. Can. Jour. Earth Sci., 10, 1068-1078
Gravenor, C.P. and Stupavsky, M. 1974. Magnetic susceptibility of the surface
     tills of southern Ontario. Can. J. Earth Sci., 11, 658-663
Gravenor, C.P. and Stupavksy, M. 1976. Magnetic, physical and lithologic properties
     and age of till exposed along the east coast of Lake Huron, Ontario, Canada.
     Can. Journ. Earth Sci., 13, 1655-1666
Grisak, G.E. 1975. The fracture porosity of glacial till. Can. J. Earth Sci.,
     12, 513-515
Grisak, G.E. and Cherry, J.A. 1975. Hydrological characteristics of response of
     fractured till and clay containing a shallow aquifer. Can. Geotech. J.,
     12, 23-43
Grisak, G.E., Cherry, J.A., Vonhof, J.A. and Bleumie, J.P. 1976. Hydrogeological
     and hydrochemical properties of fractured tills in the interior plains
     region. In Glacial Till. Spec. Pub., 12, Roy. Soc. Can., 304-335
Grove, J.M. 1972. The incidence of landslides, avalanches and floods in western
     Norway during the Little Ice Age. Arct.& Alp. Res., 4, 131-138
Gunn, G.B. 1967. The origin of diamonds in the drift of the north central
     United States. J. Geol., 75, 232-233
Gustavson, T.C. 1974. Sedimentation on gravel outwash fans, Malaspina Glacier.
     Alaska. J. Sed. Pet., 44, 374-389
Gustavson, I.C. 1975. Sedimentation and physical limnology in proglacial
     Malaspina Lake, southeastern Alaska. In Glaciofluvial and glaciolacustrine
     sedimentation. Edited by A.V. Jopling and B.C. McDonald. Soc. Econ. Paleon.
     Min. Spec. Pub. 23, 249-63
Gustavson, T.C., Ashley, G.M. and Boothroyd, J.C. 1975. Depositional sequences
     in glaciolacustrine deltas. In A.V. Jopling and B.C. McDonald. eds.
     Glaciofluvial and glaciolacustrine sedimentation. Soc. Econ. Paleont.
     Mineral. Spec. Pub. 23, 264-280
Haefeli, C.J. 1972. Groundwater inflow into Lake Ontario from the Canadian side:
     Environment Canada, Inland Waters Branch Sci. Series No. 9
Haldorsen, S. 1978. Glacial comminution of mineral grains. Norsk. Geol. Tidss.,
     58, 241-243
Haldorsen, S. 1982. The genesis of tills from Astadalen, southeast Norway.
     Norsk. Geol. Tidss., 62, 17-38
Haldorsen, S. and Shaw, J. 1982. The problem of recognizing melt-out till.
     Boreas, 11, 261-277
Hallberg, G.R. 1980. Status of Pre-Wisconsinan Pleistocene stratigraphy in Iowa.
     Geol. Soc. Am. Abstr. Programs., 12, 5, 228
Hallberg, G.R. and Boellstorff, J.D. 1978. Stratigraphic confusion in the region
     of the Type areas of Kansan and Nebraskan deposits. Geol. Soc. Am. Abs.
     Programs, 10, 255
Hallet, B. 1976. The effect of subglacial chemical processes on glacier sliding.
     J. Glaciol., 17, 209-221
Hallet, B. 1979. A theoretical model of glacial abrasion. J. Glac., 23, 51-56

Hallet, B. 1981. Glacial abrasion and sliding: their dependence on the debris
    concentration in basal ice. Annals, Glaciology, 2, 23-28
Hambrey, M.J. and Harland, W.B. 1981. Earth's Pre-Pleistocene Glacial Record.
    Cambridge University Press
Hamilton, T.D. and Obi, C.M. 1982. Pingos in the Brooks Range, Northern
    Alaska, U.S.A. Arctic and Alp. Res., 14, 13-20
Handy, R.L. 1973. Collapsible loess in Iowa. Soil Science Soc. Am. Proc., 37,
    281
Hansard, 1981.   House of Commons Official Report on Parliamentary Debates.
    Replies to questions; 20 May 1981, Secretary of State for the Environment.
    1 July 1981, Secretary of State for Scotland
Harms, J.C., Southard, J.B. and Walker, R.G. 1982. Structures and sequences in
    clastic rocks. Soc. Econ. Paleont. Mineral. Short Course, 9
Harris, C.C. 1968. The aplication of size distribution equations to multi-event
    comminution. Trans. Soc. Min. Engs., 241, 343-358
Harris, C. 1977. Engineering properties, groundwater conditions and the nature
    of soil movement on a solifluction slope in North Norway. Q. J. Eng. Geol.,
    10, 27-43
Harris, C. 1981. Microstructures in solifluction sediments from South Wales
    and North Norway. Biul. Peryglac., 28, 221-226
Harris, C. 1981a. Periglacial mass wasting: a review of research. British Geom.
    Res. Group Res. Mono. No. 4
Harrison, W. 1958. Marginal zones of vanished glaciers reconstructed from pre-
    consolidation values of overridden silt. J. Geol., 66, 72-95
Hartshorn, J.H. 1952. Supraglacial and proglacial geology of the Malaspina
    Glacier, Alaska and its bearing on glacial features of New England.
    Geol. Soc. Am. Bull., 63, 1259-1260
Hartshorn, J.H. 1958. Flow till in southeastern Massachusetts. Bull. Geol. Soc.
    Am., 69, 477-482
Hawkins, A.B. and Privett, K.D. 1981. A building site on cambered ground at
    Radstock, Avon. Q. J. Eng. Geol., 14, 151-168
Haynes, J.E. and Quigley, R.M. 1978. Framboids in Champlain Sea sediments.
    Can. J. Earth Sci., 15, 464-465
Hedberg, H.D. 1976. International Stratigraphic Guide. New York
Hedges, J. 1972. Expanded joints and other periglacial phenomena along the
    Niagara Escarpment. Biuletyn Peryglacjalny, 21, 87-126
Hein, F.J. and Walker, R.G. 1977. Bar evolution and development of stratification
    in the gravelly, braided, Kicking Horse River, British Columbia. Can. J.
    Earth Sci., 14, 562-570
Hendershot, W.H. and Carson, M.A. 1978. Changes in the Plasticity of a sample
    of Champlain clay after selective chemical dissolution to remove amorphous
    material. Can. Geot. Jour., 15, 609-616
Hicock, S.R., Dreimanis, A. and Broster, B.E. 1981. Submarine flow tills at
    Victoria, British Columbia. Can. J. Earth Sci., 18, 71-80
Higginbottom, I.E. and Fookes, P.G. 1971. Engineering aspects of periglacial
    features in Britain. Q. J. Eng. Geol., 3, 85-117
Hight, D.W. and Green, P.A. 1978. Earthworks. In Developments in Highway
    Pavement Engineering. Vol. 2. Applied Science Publishers Ltd.
Hight, D.W., El-Ghamrawg, M.K., and Gens, A. 1979. Some results from a laboratory
    study of a sandy clay and implications regarding its in-situ behaviour.
    2nd Int. Conf. on Behaviour of Off-shore Structures Paper 13, 133-150
Hillaire-Marcel, C. and Fairbridge, R.W. 1978. Isostasy and eustasy of Hudson
    Bay. Geology, 6, 117-122
Hillaire-Marcel, C., Occhietti, S. and Vincent, J.S. 1981. Sakami Moraine,
    Quebec: A 500 km-long Moraine without Climatic Control. Geology, 9, 210-214
Hill, G.H. 1896. The Thirlmere Works for the Water-Supply of Manchester. Min.
    Proc. Inst. Civ. Engrs., 126, 2-22

Hill, H.P. 1949. The Ladybower Reservoir. J. Inst. Water Engrs., 35, 414-425

Hillefors, A. 1973. The stratigraphy and genesis of stass- and lee-side moraines. Bull. Geol. Inst. Univ. Uppsala, 5, 139-154

Hjeldnes, E.I. and Lavania, B.V.K. 1980. Cracking, leakage and erosion of earth dam materials. Jnl. Geot. Eng. Div. Am. Soc. Civ. Eng., 106, 117-136

Hodgson, J.H., Rayner, J.H. and Catt, J.A. 1974. The geomorphological significance of clay-with-flints on the South Downs. Trans. Inst. Brit. Geogr., 61, 119-129

Hollingworth, S.E., Taylor, J.H. and Kellaway, G.A. 1944. Large scale superficial structures in the Northampton ironstone field. Q. J. Geol. Soc. Lond., 100, 1-44

Holmes, R. 1977. The Quaternary Geology of the U.K. Sector of the North Sea between 56° and 58°N. Inst. Geol. Sci. Rep. No. 77/14

Holtz, W.G. and Gibbs, H.J. 1951. Consolidation and related properties of loessial soils. Am. Soc. Testing Materials Spec. Tech. Pub. No. 126, 9-33

Holtz, W.G. and Ellis, E. 1961. Triaxial shear characteristics of clayey gravelly soils. 5th Int. Conf. Soil Mech. and Foundn. Enging. Paris, 1, 143-149

Hoppe, G. and Schytt, V. 1953. Some observations on fluted moraine surfaces. Geogr. Annlr., 35, 105-115

Horswill, P. and Horton, A. 1976. Cambering and valley bulging in the Gwash Valley at Empingham, Rutland. Phil. Trans. Roy. Soc. Lond., A283, 427-462

Horton, A. 1974. The sequences of Pleistocene deposits proved during the construction of the Birmingham motorways. Inst. Geol. Sci. Rep. 74/11

Humlum, O. 1981. Observations on debris in the basal transport zone of Myrdalsjokull, Iceland. Annals of Glaciology, 2, 71-77

Hutchinson, J.N. 1970. A coastal mudflow on the London Clay cliffs at Beltinge, North Kent. Geotechnique, 20, 412-438

Hutchinson, J.N. 1974. Periglacial solifluction: an approximate mechanism for clayey soils. Geotechnique, 24, 438-443

Hutchinson, J.N. 1975. The response of London Clay cliffs to differing rates of toe erosion. Build. Res. Estab. Curr. Paper CP27/75

Hutchinson, J.N. 1977. General report: theme 3. Symposium of the International Association of Engineering Geologists

Hutchinson, J.N. 1980. Possible late-Quaternary pingo remnants in central London. Nature, 284, 253-255

Hutchinson, J.N., Somerville, S.H. and Petley, D.J. 1973. A landslide in periglacially disturbed Etruria Marl at Bury Hill, Staffordshire. Q.J. Eng. Geol., 6, 377-404

Hutchinson, J.N., Bromhead, E.N. and Lupini, J.F. 1980. Additional observations on the Folkestone Warren landslides. Q. J. Eng. Geol., 13, 1-31

Hyvarinen, L., Kauranne, K. and Yletyinen, Y. 1973. Modern boulder tracing in prospecting. In Jones, M.J. ed. Prospecting in areas of glaciated terrain. Inst. Min. Metall.

Idel, K.H., Muhs, H. and Von Soos, P. 1969. Proposals for quality classes in soil sampling and the importance of boring methods and sampling equipment. Speciality Session No. 1, Proc. 7th Int. Conf. on Soil Mech., Foundn. Eng. Mexico

Imbrie, J. and Imbrie, K.P. 1979. Ice Ages: Solving the Mystery. MacMillan

Ingold, T.S. 1975. The stability of highways in landslipped areas. The Highway Engineer. XII, 14-22

International Commission on Large Dams. 1979. World Register of Dams.

Institute of Geological Sciences. 1971. Geology of Belfast and District. Special Engineering Geology Sheet, Solid and Drift. Geological Survey in Northern Ireland

Institute of Geological Sciences. 1977. Quaternary map of the United Kingdom. 2 Sheets. 1st Ed., Scale 1:625,000

Jackson, L.E. 1979. A catastrophic glacial outburst flood mechanism for debris flow generation at the Spiral Tunnels, Kicking Horse basin, B.C. Can. Geot. Jour., 4, 806-813

Jackson, L.E. Jr., MacDonald, G.M. and Wilson, M.C. 1982. Paraglacial origin of terraced river sediments in Bow Valley, Alberta. Can. J. Earth Sci., 19, 2219-2231

Jacob, S.S. 1894. The water supply of Jeypore Rajputna. Min. Proc. Instn. Civ. Engnrs., 115, 53-62

Jahn, A. 1976. Contemporary geomorphological processes in Longyeardalen, Vestspitsbergen, Biuletyn Peryglacjalny, 25, 253-267

Jaky, J. 1944. The coefficient of earth pressure at rest. Journ. Soc. Hungarian Architects and Engineers, 355-358

James, A.N. and Kirkpatrick, I.M. 1980. Design of foundations of dams containing soluble rocks and soils. Q. Jnl. Engng. Geol., 13, 189-198

Johnson, A.M. 1970. Physical Processes in Geology. Freeman Coop.

Johnson, G.H. 1981. Permafrost. Engineering Design and Construction. John Wiley

Johnson, R.H. and Wathall, S. 1979. The Longdendale Landslides. Geol. Journ., 14, 135-156

Jones, M.J. 1975. Prospecting in areas of glaciated terrain. Inst. Min. Metall., 154

Karrow, P.F., Anderson, T.W., Clarke, A.H., DeLorme, L.D. and Sreenivisa, M.R. 1975. Stratigraphy, paleontology and age of Lake Algonquin sediments in southwestern Ontario, Canada. Quat. Res., 5, 49-87

Kaszycki, C.A. and Shilts, W.W. 1979. Average depth of glacial erosion. Canadian Shield. Geol. Surv. Can. Paper 79-1B, 395-396

Kauranne, L.K. 1976. Conceptual models in exploration geochemistry. J. Geochem. Expl., 5, 173-420

Kay, G.F. 1916. Gumbotil, a new term in Pleistocene geology. Science. N.S. 44, 637-638

Kaye, C.A. 1960. Surficial geology of the Kingston Quadrangle, Rhode Island. U.S. Geol. Surv. Bull. 1071-1, 344-396

Kazi, A. and Knill, J.L. 1973. Fissuring in glacial lake clays and tills on the Norfolk coast, United Kingdom. Enging. Geol., 7, 35-48

Kemmis, T.J. 1981. Importance of the regelation process to certain properties of basal tills deposited by the Laurentide Ice Sheet in Iowa and Illinois. Annals, Glac. 2, 147-152

Kemmis, T.J., Hallberg, G.R. and Lutenegger, A.J. 1981. Depositional environment of glacial sediments and landforms on the Des Moines Lobe, Iowa. Iowa Geol. Surv. Guidebook 6

Kennard, M.F., Lovenbury, H.T., Chartres, F.R.D. and Hoskins, C.G. 1978. Shear strength specification for clay fills. In Clay Fills. Inst. Civ. Eng. London, 143-148

Kennard, M.F. 1972. Examples of the internal conditions of some old earth dams. J. Inst. Water Engrs., 26, 135-147

Kennard, M.F. 1976. Eighteenth Century Dams in England. Paper 5.1 in BNCOLD/ Univ. Symposium

Kennard, M.F. and Knill, J.L. 1969. Reservoirs on Limestone, with particular reference to the Cow Green scheme. J. Instn. Water Engrs., 23, 87-136

Kennard, M.F. and Reader, R.A. 1975. Cow Green dam and reservoir. Proc. Instn. Civ. Engrs., 58, 147-175

Kenney, T.C. 1964. Sea level movements and the geologic histories of the post-glacial marine soils at Boston, Nicolet, Ottawa and Oslo. Geotechnique, 14, 203-230

Kenney, T.C. 1967. The influence of mineral composition on the residual strength of ntural soils. Proc. Geotech. Conf. Oslo, 123-131

Kenney, T.C. 1976. Formation and geotechnical characteristics of glacial-lake
    varved soils. In Laurits Bjerrum Memorial Volume - Contribution to soil
    mechanics. Edited by N. Janbu, F. Jorstad and B. Kjaernsli, Norwegian
    Geotechnical Institute Publication, 15-39

Kenney, T.C. and Folkes, D.J. 1979. Mechanical properties of soft soils.
    State-of-the-Art Report to Session 2, Proceedings, 32nd Canadian Geotechnical
    Conference, Quebec City, P.Q., Sept. 1979, 5

Kerney, M.P., Brown, E.H. and Chandler, T.J. 1964. The late-glacial and post-
    glacial history of the chalk escarpment near Brook, Kent. Phil. Trans. Roy.
    Soc. Lond., B248, 135-204

Kirwan, L.D. 1978. The discovery of the mid-west Lake Uranium deposit. In
    Uranium Exploration Techniques. Sask. Geol. Soc. Spec. Pub. 4, 59-79

Klohn, E.J. 1965. The elastic properties of a dense glacial till deposit.
    Can. Geotech. J., 11, 396-408

Knill, J.L. 1968. Geotechnical significance of certain glacially-induced
    discontinuities in rock. Bull. Int. Engng. Geol., 5, 49-62

Knill, J.L. 1974. The application of engineering geology to the construction
    of dams in the United Kingdom. Centenaire de la Societe Geologique de
    Belgique, Colloque Geologie de l'ingenieur Liege, 113-147

Knill, J.L. and Jones, K.S. 1965. The recording and interpretation of geological
    conditions in the foundations of geological conditions in the foundation
    of the Reseires, Kariba and Latiyan dams. Geotechnique, 15, 94-124

Knox, G. 1927. Landslides in South Wales Valleys. Proc. 5th Wales. Inst. Eng.
    43, 161-247, 257-290

Knutsson, G. 1971. Studies of groundwater flow in till soils. Geologiska foren
    Stock. Forhand. 93, 533-573

Kohlbeck, F., Scheiddeger, A.E. and Sturgil, J.R. 1979. Geomechanical model of
    an alpine valley. Rock Mechanics, 12, 1-14

Kukla, G.J. 1977. Pleistocene Land-Sea correlations. Earth Sci. Rev., 13, 307-374

Kujansuu, R. 1976. Glaciogeological surveys for ore-prospecting purposes in
    Northern Finland. In Glacial Till. Roy. Soc. Can. Spec. Pub. 12, 225-239

Lamb, T.W. and Whitman, R.V. 1969. Soil Mechanics. John Wiley and Sons. New York

Lapworth, H. 1911. The geology of dam trenches. Trans. Instn. Wat. Engnrs., 16,
    25-66

La Rochelle, P., Chagnon, J.-Y. and Lefebvre, G. 1970. Regional geology and
    landslides in the marine clay deposits of eastern Canada. Can. Geot. J.,
    7, 145-156

Larson, G.J. and Stone, B.D. 1982. Late Wisconsinan Glaciation of New England.
    Kendall Hunt Publishing Co.

Latham, J.H. 1872. On the Soonkesala Canal of the Madras Irrigation and Canal
    Company. Proc. Instn. Civ. Engrs., 34, 67-110

Lawson, D.E. 1979. Sedimentological analysis of the western terminus region of
    the Matanuska Glacier, Alaska. Cold Regions Research and Engineering
    Laboratory Report 79-9

Lawson, D.E. 1981. Sedimentological characteristics and classification of
    depositional processes and deposits in the glacial environment. Cold Reg.
    Res. Eng. Lab. Rep. 81-27

Lawson, D.E. 1981a. Sedimentological characteristics and classification of
    depositional processes and deposits in the glacial environment. U.S. Army
    Cold Region. Res. Eng. Lab. Rep. 81-27

Lawson, D.E. 1982. Mobilization, movement and deposition of active subaerial
    sediment flows, Matanuska Glacier, Alaska. J. Geol., 90, 279-300

Lee, H.A. 1968. An Ontario Kimberlike occurrence discovered by the glaciofocus
    method to a study of the Munro Esker. Geol. Surv. Can. Paper 68-7

Lee, H.A. 1971. Mineral discovery in the Canadian Shield using the physical
    aspects of overburden. Can. Inst. Min. Metall. Bull. Nov. 32-37

Leeder, M.R. 1982. Sedimentology. George Allen and Unwin

Legget, R.F. 1973. Cities and Geology. McGraw-Hill

Lettet, R.F. 1974. Glacial landforms and civil engineering. In Coates, D.R., ed. Glacial Geomorphology, 351-374

Legget, R.F. 1979. Geology and Geotechnical Engineering. Jnl. Geot. Eng. Div. Am. Soc. Civ. Eng., 105, 339-392

Leighton, M.M. and MacClintock, P. 1930. Weathered zones of the drift sheets of Illinois. Journ. Geol., 38, 28-53

Leighton, M.M. and MacClintock, P. 1962. The weathered mantle of glacial tills beneath original surfaces in North Central United States. Journ. Geol., 70, 267-293

Levy, J.F. and Morton, K. 1974. Loading tests and settlement observations on granular soils. Proc. Conf. on Settlement of Structures. British Geotechnical Society, Cambridge, 43-52

Lill, G.O. and Smalley, I.J. 1978. Distribution of loess in Britain. Proc. Geol. Ass., 89, 57-66

Lin, Z. and Liang, W. 1982. Engineering properties and zoning of loess and loess-like soils in China. Can. Geotech. Jour., 19, 76-91

Lineback, J.A., Gross, D.L., Meyer, R.P. and Unger, W.L. 1971. High resolution seismic profiles and gravity cores of sediments in southern Lake Michigan. Illinois State Geol. Surv., Env. Geol. Note 47

Lineback, J.A., Gross, D.L. and Meyer, R.P. 1974. Glacial tills under Lake Michigan. Ill. State Geol. Surv. Env. Geol. Note No. 69

Linton, D.L. 1955. The problem of tors. Geogr. Journ., 121, 470-487

Linton, D.L. 1959. Morphological contrasts of eastern and western Scotland. In R. Miller and J.W. Watson, eds. Geographical essays in memory of A.G. Ogilvie. 16-45

Linton, D.L. 1963. The forms of glacial erosion. Trans. Inst. Brit. Geog., 33, 1-27

Linton, D.L. 1964. The origin of the Pennine tors. Zeit. Geomorph., 8, 5-24

Liboutry, L., Arnao, B.M., Pautre, A. and Schneider, B. 1977. Glaciological problems set by the control of dangerous lakes in Cordillera Blanca, Peru. J. Glac., 18, 239-254

Lloyd, J.W. 1980. The influence of Pleistocene deposits on the hydrogeology of major British aquifers. Jl. Inst. Water Engrs. and Sci., 34, 346-356

Lloyd, J.W., Harker, D. and Baxendale, R.A. 1981. Recharge mechanisms and groundwater flow in the Chalk and drift deposits of southern East Anglia. Q. Jl. Eng. Geol., 14, 87-96

Lo, K.Y. 1975. Regional distribution of in situ horizontal stresses in rocks of Southern Ontario. Can. Geot. Journ., 15, 371-381

Lovell, J.S., Hale, M. and Webb, J.S. 1980. Vapour geochemistry in mineral exploration. Mining Mag., Sept. 229-239

Lowe, D.R. 1979. Sediment Gravity Flows: their classification and some problems of application to natural flows and deposits. Soc. Econ. Paleont. Mineral. Spec. Pub. 27, 75-82

Lowe, J.J., Gray, J.M. and Robinson, J.E. (eds.). 1980. Studies in the Lateglacial of Northwest Europe. Pergamon

Lundquist, J. 1969. The problem of the so-called rogen moraine. Sverig. Geol. Unders. Ser. C., 648, 32

Lundquist, J. 1977. Till in Sweden. Boreas, 6, 73-85

Lupini, J.F., Skinner, A.E. and P.A. Vaughan. 1981. The drained residual strength of cohesive soils. Geotechnique, 31, 181-213

Lutenegger, A.J. 1983. Engineering properties and zoning of loess and loess-like soils in China: Discussion. Can. Geotech. Jour., 20, 192-193

Lutton, R.J. 1969. Fractures and failure mechanisms in loess and applications to rock mechanics. U.S. Army. Eng. Waterways Ex. Station. Res. Rep. S-69-1. Vicksburg. Miss. U.S.A.

Lutton, R.J. 1971. A mechanism for progressive mass-failure as revealed by loess slumps. Int. J. Rock Mech. Min. Sci., 8, 143-151

Maarleveld, G.S. 1960. Wind directions and cover sands in the Netherlands. Biuletyn Peryglacjalny, 8, 49-58

Maarleveld, G.C. 1981. The sequence of ice-pushing in the central Netherlands. Med. Rijks. Geol. Dienst., 34, 2-6

MacArthur, A. 1969. An investigation into the use of glacial ablation tills in highway construction, M.Sc. Thesis, Univ. Strataclyde

Mackay, J.R. 1972. Offshore permafrost and ground ice. South Beaufort Sea, Canada. Can. J. Earth Sci., 9, 1550-1561

Mackay, J.R. 1978. Contemporary Pingos-a discussion. Biul. Peryglac., 28, 133-154

Mackay, J.R. -979. Pingos of the Tuktoyaktuk Peninsula area, Northwest Territories. Geog. Phys. Quaternaire., 33, 3-61

Madgett, P.A. and Catt, J.A. 1978. Petrography, stratigraphy and weathering of Late Pleistocene tills in E. Yorkshire, Lincolnshire and N. Norfolk. Proc. Yorks. Geol. Soc., 42, 55-108

Mahaney, W.C. ed. 1976. Quaternary stratigraphy of North America. Dowden, Hutchinson and Ross

Marcussen, I. 1977. Deglaciation landscapes formed during the wasting of the late middle Weichselian ice sheet in Denmark. Geological Survey of Denmark. II Series. No. 110

Marsal, R.J. 1969. Mechanical properties of rockfill and gravel materials. Proc. 7th Int. Con. Sci. Mech. and Fnd. Eng. Mexico, 499-506

Marsland, A. 1977. The evaluation of the engineering design parameters for glacial clays. Q. Jour. Eng. Geol., 10, 1-26

Martin, G.R., Lam, I. and Tsai, C.F. 1980. Pore pressure dissipation during offshore cyclic loading. Jnl. Geot. Eng. Div. Am. Soc. Civ. Eng., 106, 981-996

Martini, A. 1967. Preliminary experimental studies on frost weathering of certain rock types from the west Sudetes. Biuletyn Peryglacjalny, 16, 147-194

Matheson, D.S. 1970. A tunnel roof failure in till. Can. Geotech. J., 7, 312-317

Matheson, D.S. and Thompson, S. 1973. Geological implications of valley rebound. Can. J. Earth Sci., 10, 961-978

Mathews, W.H. 1964. Water pressure under a glacier. Jour. Glac., 5, 235-40

Matile, G. and Nielsen, E. 1981. Proglacial Lake Sedimentation in Southeastern Manitoba. Geol. Ass. Can. Ann. Meeting Abstracts, 6, 1738

Matter, A. and Tucker, M. eds. 1978. Modern and Ancient Lake Sediments. Int. Ass. Sed. Spec. Pub. 2

Matula, M. 1969. Regional Engineering Geology of Czechoslovakia. Carpathian Pub. House. Slovak Acad. Sci. Bratislava

Matula, M. 1979. Regional engineering geologic evaluation for planning purposes. Bull. Int. Ass. Eng. Geol., 19, 18-24

May, R.W. 1977. Facies model for sedimentation in the glaciolacustrine environment. Boreas, 6, 175-180

May, R.W. and Thompson, S. 1978. The geology and geotechnical properties of till and related deposits in the Edmonton, Alberta area. Can. Geot. J., 15, 362-370

McCave, I.N., Caston, V.N.D. and Fannin, N.G.T. 1977. The Quaternary of the North Sea. In F.W. Shotton (ed.) British Quaternary Studies. Clarendon Press. Oxford, 187-204

McClaren, D. 1968. M6 Trial embankment at Killington. Trans. Road Res. Lab. Rep. LR 238

McDonald, B.C. and Shilts, W.W. 1975. Interpretation of faults in glacioflurival sediments. In Soc. Ec. Pal. Mineral. Spec. Pub. 23, 123-131

McGown, A. and Radwan, A.M. 1975. The presence and influence of fissures in the boulder clays at West Central Scotland. Can. Geot. J., 12, 84-97

McGown, A. and Derbyshire, E. 1977. Genetic influences on the properties of tills. Q. Jour. Eng. Geol., 10, 389-410

McGown, A., Radwan, A.M. and A.W.A. Gabr. 1977. Laboratory testing of
    fissured and laminated soils. Proc. 9th Int. Conf. Soil. Mech. 205-210
McGown, A., Saldivar-Sali, A. and A.M. Radwan. 1974. Fissure patterns and
    slope failures in boulder clay at Hurlford, Ayrshire. Q. Jour. Eng. Geol.,
    7, 1-26
McIldowie, G. 1934. The construction of the Silent Valley Reservoir. Belfast
    Water Supply. Trans. Instn. Civ. Engnrs., 239, 465-516
McKinley, D.G., Tomlinson, M.J. and W.F. Anderson. 1974. Observations on the
    undrained strength of a glacial till. Geotechnique, 24, 503-516
McKinley, D.G. and Anderson, W.F. 1975. Determination of the Modulus of
    Deformation of a Fill using a Pressuremeter. Ground Engineering, 8, 11
McKinlay, D.G., McGown, A., Radwan, A.M. and Hossain, D. 1975. Representative
    sampling and testing in fissured lodgement tills. In The Engineering
    Behaviour of Glacial Materials., 129-140. Geo-Abstracts
McRoberts, E.C. and Morgenstern, N.R. 1974. The stability of thawing slopes.
    Can. Geot. J., 12, 130-141
McRoberts, E.C. and Morgenstern, N.R. 1975. Porewater expulsion during freezing.
    Can. Geot. J., 12, 130-141
McQuillan, R. 1964. Geophysical investigation of seismic shot holes in the
    Cheshire Basin. Bull. Geol. Surv. Gt. Brit., 21, 197-203
Meier, M. 1960. Mode of flow of Saskatchewan Glacier. U.S. Geol. Surv. Prof.
    Pap. 351
Mellor, M. and Testa, R. 1969. Effect of temperature on the creep of ice.
    J. Glaciol., 8, 131-145
Menard, L. and Brosse, Y. 1975. Theoretical and practical aspects of dynamic
    consolidation. Geotechnique, 25
Menzies, J. 1979. The mechanics of drumlin formation with particular reference
    to the change in pore-water content of the till. J. Glac., 22, 373-384
Menzies, J. 1979a. A review of the literature on the formation and location
    of drumlins. Earth Sci. Reviews, 14, 315-359
Menzies, J. 1981. Freezing fronts and their possible influence upon processes
    of subglacial erosion and deposition. Annals. Glaciology, 2, 52-6
Menzies, J. 1982. A till hummock (proto-drumlin) at the ice/glacier bed interface.
    In Research in Glacial, Glaciofluvial and Glaciolacustrine Systems. Geo-
    Abstracts, 33-47
Merla, A., Merlo, C., and Oliveri, F. 1976. Detailed engineering geological
    mapping in selected Italian mountainous areas. Bull. Int. Ass. Eng. Geol.,
    14, 129-136
Metcalf, R.C. 1979. Energy dissipation during subglacial abrasion of Nisqually
    Glacier, Washington, U.S.A. J. Glac., 23, 233-246
Meyerhof, G.G. 1951. The ultimate bearing capacity of foundations. Geotechnique.
    2, 301-332
Meyerhof, G.G. 1956. Penetration tests and bearing capacity of cohesionless
    soils. J. Soil Mech. and Found. Div., Proc. Amer. Soc. Civil Engrs., 82,
    1-19
Meyerhof, G.G. 1976. Bearing capacity and settlement of pile foundations.
    A.S.C.E., 102, 197-228
Miall, A.D. 1977. A review of the braided river depositional environment.
    Earth Sci. Rev., 13, 1-62
Miall, A.D. 1978. Lithofacies types and vertical profile models in braided
    rivers: A summary. In Miall, A.D. (ed.). Fluvial Sedimentology. Can. Soc.
    Petr. Geol. Memoir., 5, 597-604
Miall, A.D. 1980. Cyclicity and the facies model concept in fluvial deposits.
    Bull. Can. Petrol. Geol., 28, 59-80
Miall, A.D. 1981. Alluvial sedimentary basins: tectonic setting and basin archi-
    tecture. In Sedimentation and Tectonics in Alluvial Basins. A.D. Miall (ed.).,
    Geol. Ass. Can. Spec. Pub. 23, 1-33

Miall, A.D. 1983. Principles of sedimentary basin analysis. Springer Verlag. N.Y.
Mickelson, D.M. 1973. Nature and rate of basal till deposition in a stagnating
    ice mass. Burroughs Glacier, Alaska. Arctic. Alp. Res., 5, 17-27
Middleton, G.V. and Hampton, M.A. 1976. Subaqueous sediment transport and
    deposition by sediment gravity flows. In Stanley, D.J. and Swift, D.J. eds.
    Marine Sediment Transport and Environmental Management. John Wiley.
Miller, A.A. 1953. The skin of the earth. London
Miller, R.D. 1973. Gastineau Channel Formation: a composite glaciomarine
    deposit near Janeau, Alaska. U.S.G.S. Bull., 1394-C
Milligan, V. 1976. Geotechnical aspects of glacial tills. In Legget, R.F. (ed.)
    Glacial Till. Roy. Soc. of Can. Spec. Pub. 12, 269-292
Milling, M.E. 1975. Geological appraisal of foundation conditions, northern
    North Sea. Oceanology International '75, 310-319
Mitchell, C. 1973. Terrain Evaluation. Longman
Mitchell, G.F., Penny, L.F. and West, R.G. 1973. A correlation of Quaternary
    deposits in the British Isles. Geol. Soc. Lond. Spec. Pap. No. 4
Mitchum, R.M., Vail, P.R. and Thompson, S. III. 1977. The depositional
    sequence as a basic unit for stratigraphic analysis. In Seismic Stratigraphy-
    Applications to Hydrocarbon Exploration. Am. Assoc. Petrol. Geol. Mem. 26,
    213-248
Moffat, A.I.B. 1976. Effective operation of reservoir legislation. Paper 1.4.
    BNCOLD/Univ. Symp. Newcastle-Upon-Tyne - and Discussion D1/14-15
Mollard, J.D. 1973. Landforms and surface materials of Canada. A stereoscopic
    Atlas and Glossary. Commercial Printers Ltd. Regina, Canada.
Mollard, J.D. 1977. Regional landslide types in Canada. Landslides.
    Geol. Soc. Am. Reviews in Eng. Geol. III, 29-56
Money, M.S. 1976. The investigation of old embankment dams in glaciated
    valleys. Paper 5.3. In BNCOLD/Univ. Symp. q.v.
Moon, C.F. 1972. The microstructure of clay sediments. Earth Sci. Rev., 8, 303-321
Moon, C.F. 1974. Quickclays as products of glacial action - a short comment.
    Eng. Geol., 7, 359-361
Moon, C.F. 1975. The failure mechanism of quickclays: a model approach. In
    Engineering Behaviour of Glacial Materials. Geo-Abstracts, 75-80
Moran, S.R. 1971. Glaciotectonic structures in drift. In Till. A Symposium.
    Ohio State Univ. Press. 127-148
Moran, S.R., Clayton, L., Hooke, R. Le B. Fenton, M.M. and Andriashek, L.D.
    1980. Glacier bed landforms of the Prairie region of North America.
    Jour. Glac., 25, 457-476
Morgan, A.V. 1971. Engineering problems caused by fossil permafrost features
    in the English Midlands. Q.J. Eng. Geol., 4, 111-114
Morgan, A.V. 1972. Late Wisconsinan ice wedge polygons near Kitchener, Ontario,
    Canada. Can. J. Earth Sci., 9, 607-617
Morgenstern, N.R. 1967. Submarine slumping and the initiation of turbidity
    currents. In Marine geotechnique. Univ. Illinois Press, 189-220
Morgenstern, N.R. and J.F. Nixon. 1971. One dimensional consolidation of thawing
    soils. Can. Geotech. J., 8, 558-565
Morner, N.A. 1971. Eustatic changes during the last 20,000 years and a method
    of separating the isostatic and eustatic factors in an uplifted area.
    Palaeogeography, Palaeoclimatology, Palaeoecology, 9, 153-181
Morner, N.A. 1979. Eustasy, palaeogeodesy and glacial volume changes. In
    Sea level, ice and climatic change. I. Allison (ed.). I.A.H.S. Publ.
    131, 277-280
Morton, E. 1973. A review of the influence of geology on the design and
    construction of impounding dams. J. Inst. Water Engrs., 27, 243-262
Mottershead, D.N. 1971. Coastal head deposits between Start Point and Hope Core,
    Devan. Field Studies, 3, 433-453

Moum, J., Loken, T. and Torrance, J.K. 1971. A geochemical investigation of
    the sensitivity of a normally consolidated clay from Drammen, Norway.
    Geotechnique, 21, 329-340
Moum, J. and Zimmie, T.F. 1972. Geochemical tests on a Canadian quick clay from
    St. Jean Vianney, Quebec. Norwegian Geotechnical Institute. Internal.
    Report, 7
Nardin, T.R., Hein, F.J., Gorsline, D.S. and Edwards, B.D. 1979. A review of
    mass movement processes, sediment and acoustic characterstics and contrasts
    in slope and base of slope systems versus canyon-fan-basin floor basins.
    Geology of Continental Slopes. Soc. Econ. Pal. and Mineral. Spec. Pub. 27,
    61-73
Nelson, C.H., Hopkins, D.M. and Scholl, D.W. 1974. Tectonic and cenozoic
    sedimentary history of the Bering Sea. In Marine Geology and Oceanography
    of the Arctic Seas. Springer Verlag. 119-140
Nelson, A.R. 1981. Quaternary glacial and marine stratigraphy of the Quivitu
    Peninsula, Northern Cumberland Peninsula. Baffin Island, Canada. Geol. Soc.
    Am. Bull., 92, 512-518. Pt. I.
Newbery, J. and Subramaniam, A.S. 1977. Geotechnical aspects of route location
    studies for M4 north of Cardiff. Q. Jl. Eng. Geol., 10, 423-441
Newman, W.S., Marcus, L.F., Pardi, R.R. 1979. Palaeogeodesy: Late Quaternary
    geoidal configurations as determined by ancient sea levels. In Sea Level,
    Ice and Climatic Change. I. Allison (ed.). I.A.H.S. Publ. 131, 263-276
Nichols, T.C. 1980. Rebound, its nature and effect on engineering works.
    Q. J. Eng. Geol., 13, 133-152
Nixon, J.F. and Morgenstern, N.R. 1973. The residual stress in thawing soils.
    Can. Geot. J., 10, 572-580
Norris, S.E. and White, G.W. 1961. Hydrologic significance of buried valleys
    in glacial drift. U.S. Geol. Surv. Prof. Pap. 424-B, 34-35
Nye, J.F. 1952. The mechanics of glacier flow. J. Glac., 2, 82-93
Nye, J.F. 1976. Water flow in glaciers: Jokulhlaups, tunnels and veins.
    J. Glaciol., 17, 181-207
Olsson, T. 1974. Groundwater in till soils. Striae, 4, 13-16
Orheim, O. and Elverhoi, A. 1981. Model for submarine glacial deposition.
    Annals of Glaciol., 2, 123-128
Ostrem, G. 1975. Sediment transport in glacial meltwater streams in glaciofluvial
    and glaciolacustrine sedimentation. A.V. Jopling and B.C. McDonald (eds.)
    Soc. Ec. Paleontologists and Mineralogists Spec. Pub. 23, 101-122
Ostry, R.C. 1979. The hydrogeology of the IFYGL Duffins Creek Study area.
    Ministry of Environment. Ontario Water Res. Rep. 5c
Palmquist, R.C. and Bible, G. 1974. Bedrock topography beneath the Des Moines
    Lobe Drift Sheet, North-Central Iowa. Proc. Iowa Acad. Sci., 81, 164-170
Parsons, A.W. 1978. Moisture condition test for assessing the engineering
    behaviour of earthworks materials. In Clay Fills. Inst. Civ. Eng. London.
    169-176
Parsons, A.W. 1981. The assessment of soils and soft rocks for embankment
    construction. Q.J. Eng. Geol., 14, 219-230
Parsons, J.D. 1976. New York's Glacial Lake Formation of varved silt and clay.
    Jnl. Geot. Eng. Div. Am. Soc. Civ. Eng., 102, 605-640
Pasek, J. 1968. The development of engineering geological maps in Czechoslovakia.
    Zentr. Geol. Inst. Abh., Berlin, 14, 75-85
Paterson, W.B. 1981. The physics of glaciers. Pergamon Press
Paton, J. 1956. The Glen Shira Hydro-electric Project. Proc. Inst. Civ. Engrs.,
    5, 593-618
Paul, M.A. 1981. (ed.) Soil Mechanics in Quaternary Geology. Quaternary Res.
    Assoc. U.K. ISBN 0 907780 00 8

Paul, M.A. and Evans, H. 1974. Observations on the internal structure and
    origin of some flutes in glacio-fluvial sediments. Blomstrandbreen,
    Northwest Spitsbergen. J. Glac., 13, 393-400
Peacock, J.D. and McL. Michie, U. 1975. Superficial deposits of the Scottish
    Highlands and their influence on geochemical exploration. In Prospecting
    in areas of glaciated terrain 1975. Inst. Min. Metall., 41-53
Peake, D.S. 1981. The Devensian glaciation on the north Welsh border. In J.
    Neale and J. Flenley (eds.). The Quaternary in Britain. Pergamon Press,
    49-59
Peck, R.B. and Reed, W.C. 1954. Engineering properties of Chicago subsoils.
    Univ. Illinois Bull., 51, 62
Peckover, F.L. and Kerr, J.W.G. 1976. Treatment and maintenance of rock slopes
    on transportation routes. Can. Geot. Journ., 14, 487-507
Peltier, W.R. 1980. Models of glacial isostasy and the relative sea level.
    In Dynamics of Plate Interiors. Am. Geophys. Union. Geol. Soc. American
    Boulder. 111-128
Penck, A. and Bruckner, E. 1909. Die Alpen im Eiszeitalter. Leipzig
Penman, H.L. 1950. The water balance of the Stour catchment area. Jl. Instn.
    Water Engrs., 4, 457-469
Penman, A.D.M. 1971. Rockfill. Building Res. Sta. U.K. Current Paper 15/71
Penner, E. 1965. A study of sensitivity in Leda Clay. Can. J. Earth Sci., 2,
    425-441
Penner, E. and Burn, K.N. 1977. Review of engineering behaviour of marine clays
    in Eastern Canada. Can. Geot. J., 15, 269-282
Pepler, S.W.E. and Mackenzie, I.D. 1976. Glacial till in winter dam construction.
    In Glacial till. Roy. Soc. Can. Spec. Pub., 12, 381-390
Perrin, R.M.S., Rose, J. and Davies, H. 1979. The distribution, variation and
    origins of pre-Devensian tills in eastern England. Phil. Trans. Roy. Soc.
    Lond., B 287, 535-570
Peters, J. and McKeown, J. 1976. Glacial till and the development of the Nelson
    River. In Legget, R.F., Glacial Till. Roy. Soc. of Can. Spec. Pub. 12,
    364-380
Peterson, R. 1958. Rebound in the Bearpaw Shale, Western Canada. Bull. Geol.
    Soc. Amer., 69, 113-124
Phemister, T.C. and Simpson, S. 1949. Pleistocene deep weathering in northeast
    Scotland. Nature, 164, 318-319
Piteau, D.R. 1976. Significance of river morphology on slope stability in the
    Fraser Canyon. In Proc. 29th Can. Geotech. Conf., 2-13
Plafker, G. 1981. Late Cenozoic glaciomarine deposits of the Yakatagn Formation,
    Alaska. In Earth's Pre-Pleistocene Glacial Record. M.J. Hambrey and W.B.
    Harland (eds.). C.U.P. 694-699
Poole, E.G. and Whiteman, A.J. 1966. Geology of the Country around Nantwich
    and Whitchurch. Memoir Geol. Surv. Great Britain. HMSO
Porter, S.E. and Orombelli, G. 1981. Alpine rockfall hazards. Science, 69,
    67-75
Powell, R.D. 1981. A model for sedimentation by tidewater glaciers. Annals
    of Glaciology, 2, 129-134
Powell, R.D. 1983. Submarine flow tills at Victoria. Can. J. Earth Sci., 20,
    509-510
Prest, V.K. 1968. Glacial map of Canada. Geol. Surv. Can. Map 1253A
Prest, V.K. 1970. Quaternary geology of Canada. In Geology and Economic
    Minerals of Canada. Geol. Surv. Can. Ec. Geol. Rep. 1
Prest, V.K. 1973. Glacier retreat. The National Atlas of Canada. Canada
    Department of Energy, Mines and Resources. 31-32
Pretorius, D.A. 1979. Depositional environment of Witwatersrand goldfields.
    Geol. Soc. S. Afr. Spec. Pub. 37-55

Price, D.G. 1971. Engineering geology in the urban environment. Q. Jl. Eng. Geol., 4, 191-208

Price, D.G. and Knill, J.L. 1967. The engineering geology of Edinburgh Castle Rock. Geotechnique, 17, 411-422

Price, R.J. 1973. Glacial and fluvioglacial landforms. Oliver and Boyd. Edinburgh

Puranen, R. 1977. Magnetic suscpetibility and its anisotropy in the study of glacial transport in northern Finland. In Prospecting in areas of glaciated terrain. Inst. Min. Metall., 111-119

Quigley, R.M. 1975. Weathering and changes in strength of glacial till. In Yatsu, E., Ward, A.J. and Adams, F. (eds.) Mass Wasting Geo. Abstracts. Norwich, 117-131

Quigley, R.M. 1980. Geology, mineralogy and geochemistry of Canadian soft soils: A geotechnical perspective. Can. Geotech. Jour., 17, 261-285

Quigley, R.M., Matich, M.A.J., Harvath, R.G. and Hawson, H.H. 1971. Swelling clays in two slope failures at Toronto. Can. Geot. J., 8, 417-424

Quigley, R.M. and Ogunbadejo, T.A. 1972. Clay layer fabric and oedometer consolidation of a soft varved clay. Can. Geot. J., 9, 165-175

Quigley, R.M. and Ogunbadejo, T.A. 1976. Till geology, mineralogy and geotechnical behaviour. Sarnia, Ontario. In Legget, R.F. (ed.). Glacial Till, Royal Soc. Can. Spec. Pub. 12, 336-345

Quigley, R.M., Gelinas, P.J., Bou, W.T. and Packer, R.W. 1977. Cyclic erosion-instability relationships. Lake Erie North Shore Bluffs. Can. Geotech. Jour., 14, 310-323

Radbruch-Hall, D., Edwards, K. and Batson, R.M. 1979. Experimental engineering geologic maps of the coterminous United States prepared by computer. Bull. Int. Ass. Eng. Geol., 19, 358-363

Radforth, N.W. and Brawner, C.O. 1972. Muskeg and the Northern environment in Canada. University of Toronto Press

Radhakrishna, H.S. and Klym, T.W. 1974. Geotechnical properties of a very dense till. Can. Geotech. J., 11, 396-408

Rahman, M.S., Seed, H.B. and Booker, J.R. 1977. Pave pressure development under offshore gravity structures. Jnl. Geot. Div. Am. Soc. Civ. Eng., 103, 1419-1436

Rampton, V. 1974. The influence of ground ice and thermokarst upon the geomorphology of the MacKenzie-Beaufort region. In Research in Polar and Alpine Geomorphology. B.D. Fahey and R.D. Thompson (eds.). 43-59

Rapp, A. 1960. Recent development of mountain slopes in Karkevagge and surroundings, Northern Scandinavia. Geogr. Annaler, 42, 65-200

Reekie, C.J., Coffey, J.R. and Marsden, Ann E. 1979. Computer aided techniques in urban engineering geological mapping. Bull. Int. Assoc. Engng. Geol., 19, 322-330

Rees, I. 1965. The use of anisotropy of magnetic susceptibility in the estimation of sedimentary fabric. Sedimentology, 4, 257-283

Reger, R.D. and Pewe, T.L. 1976. Cryoplanation terraces: indicators of a permafrost environment. Quaternary Research, 6, 99-110

Rich, J.L. 1943. Buried stagnant ice as a normal product of a progressively retreating glacier in a hilly region. Am. J. Sci., 241, 95-100

Rose, J. 1974. Small scale variability of some sedimentary properties of lodgement and slumped till. Proc. Geol. Ass., 85, 223-237

Rosenqvist, I.T. 1966. Norwegian research into the properties of quick clay - a review. Eng. Geol., 1, 445-450

Rothlisberger, F. and Schneebeli, W. 1979. Genesis of lateral moraine complexes, demonstrated by fossil soils and trunks. In Moraines and Varves. A.A. Balkema, Rotterdam, 387-420

Rowe, P.W. 1970. Derwent Dam - embankment stability and displacements. Proc. Instn. Civ. Engrs., 45, 423-452

Rowe, P.W. 1972. The relevance of soil fabric to site investigation practice. Geotechnique, 22, 195-300

Ruffle, N. 1965. Derwent Reservoir. J. Instn. Water Engrs., 19, 361-408

Ruffle, N.J. 1970. Derwent Dam - design considerations. Proc. Instn. Civ. Engnrs., 45, 381-400

Rutie, R.H. 1969. Quaternary Landscapes in Iowa. Iowa St. Univ. Press

Rushton, K.R. and Booth, S.J. 1976. Pumping test analysis using a discrete-time discrete-space numerical method. Jnl. Hydrol., 28, 13-27

Russell, D.J. and Parker, A. 1979. Geotecnical, mineralogical and chemical interrelationships in weathering profiles of an over-consolidated clay. Q. J. Eng. Geol., 12, 107-116

Rust, B.R. 1972. Structure and process in a braided river. Sedimentology, 18, 221-246

Rust, B.R. 1978. Depositional models for braided alluvium. In A.D. Miall, ed. Fluvial sedimentology. Can. Soc. Petrol. Geol. Mem., 5, 605-625

Rust, B.R. and Romanelli, R. 1975. Late Quaternary subaqueous outwash deposits near Ottawa, Canada. In Soc. Ec. Pal. Mineral. Spec. Pub., 23, 177-192

Rust, B.R. 1977. Mass flow deposits in a Quaternary succession near Ottawa, Canada. Diagnostic criteria for subaqueous outwash. Can. J. Earth Sci., 14, 175-184

Ryckborst, H. and Leusink, A. 1980. Optimum well sampling distance of ground-water levels in till and coversands, Leerinkbeck, the Netherlands. Geol. en Mijnb., 59, 43-48

Ryder, J.M. 1971. The stratigraphy and morphology of para-glacial alluvial fans in south-central British Columbia. Can. J. Earth Sci., 8, 279-298

Sage, R.C. and Lloyd, J.W. 1978. Drift deposits influences on the Triassic Sandstone aquifer of northwest Lancashire as inferred by hydrochemistry. Q. Jl. Engng. Geol., 11, 209-218

Sanglerat, G. 1972. The penetrometer and soil exploration. Elsevier Publishing Company

Sarma, S.K. 1973. Stability analysis of embankment slopes. Geotechnique, 23

Saver, E.K. 1975. Urban fringe development and slope instability of southern Saskatchewan. Can. Geot. J., 12, 106-118

Saver, E.K. 1977. A valley crossing in Pleistocene deposits. Eng. Geol., 7, 1-21

Saver, E.K. 1978. The engineering significance of glacier ice-thrusting. Can. Geotech. J., 15, 457-472

Saver, E.K. 1979. A slope failure in till at Lebret, Saskatchewan, Canada. Can. Geot. J., 7, 116-126

Saunderson, H.C. 1975. Sedimentology of the Brampton esker and its associated deposits: an empirical test of theory. In A.V. Jopling and B.C. McDonald, eds. Glaciofluvial and glaciolacustrine sedimentation. Soc. Econ. Paleont. Mineral. Spec. Pub. 23, 155-176

Schluchter, C.H. 1979. (ed.). Moraines and Varves. A.A. Balkema, Rotterdam

Schmertmann, J.H. 1953. Estimating the true consolidation behaviour of clay from laboratory test results. Proc. A.S.C.E. 79

Scholz, C.H. and Engelder, J.T. 1970. The role of asperity indentation and ploughing in rock friction. Int. Jnl. Rock Mech. Min. Sci., 13, 149-154

Schultz, C.B. and Frye, J.C. 1965. Loess and related eolian deposits of the world. Univ. Nebraska Press

Schwan, J., Loon, A.J., Steenbeck, R. and Van der Gaauw, P. 1980a. The sedimentary sequence of a Weichselian intraglacial lake at Ormehoj (Funen, Denmark). Geol. En Mijn., 59, 129-138

Scott, C.R. 1975. An introduction to soil mechanics and foundations. Applied Science Publishers

Scott, J.S. 1976. Geology of Canadian tills. In Legget, R.F. (ed.). Glacial till. Roy. Soc. of Can. Spec. Pub., 12, 50-66

Scott, J.S. and Brooker, E.W. 1968. Geological and engineering aspects of
    Upper Cretaceous Shales in western Canada. Geol. Surv. Can. Paper 66-37
Shackleton, N.J. and Opdyke, N.D. 1973. Oxygen isotope and palaeomagnetic
    stratigraphy of equatorial Pacific core V28-238: oxygen isotope temperatures
    and ice volumes on a 10^5 and 10^6 year scale. Quat. Res., 3, 39-55
Shakesby, R.A. 1979. The pattern of glacial dispersal and comminution of rock
    fragments and mineral grains from two point sources. Zeit. Gletscherk.
    Glazialgeol., 15, 31-45
Shaw, J. and Freschauf, R.C. 1973. A kinematic discussion of the formation of
    glacial flutings. Can. Geog., 17, 19-35
Shaw, J. 1977. Till body morphology and structure related to glacier flow.
    Boreas, 6, 189-201
Shaw, J. 1977a. Tills deposited in arid polar environments. Can. J. Earth Sci.,
    14, 1239-1245
Shaw, J. 1977b. Sedimentation in a alpine lake during deglaciation. Okanagan
    Valley, British Columbia, Canada. Geografiska Annaler 59A, 221-240
Shaw, J. 1979. Genesis of the Sveg tills and Rogen moraines of central Sweden:
    a model of basal melt-out. Boreas, 8, 409-426
Shaw, J. 1980. Drumlins and large-scale flutings related to glacier folds.
    Arctic and Alp. Res., 12, 287-298
Shaw, J. 1982. Glacigenic deposits of the Edmonton area. Int. Ass. Sediment.
    11th Int. Congr., Hamilton, Ontario, Excurs. Guidebook 20B, 7-32
Shaw, J. 1982a. Melt-out till in the Edmonton area, Alberta, Canada. Can. J.
    Earth Sci., 19, 1548-1569
Shaw, J. 1975. Sedimentary successions in Pleistocene ice-marginal lakes,
    In A.V. Jopling and B.C. McDonald, eds. Glaciofluvial and glaciolacustrine
    sedimentation. Soc. Econ. Paleont. Mineral. Spec. Pub. 23, 281-303
Shaw, J. and Archer, J. 1979. Deglaciation and glaciolacustrine sedimentation
    conditions. Okanagan Valley, British Columbia, Canada. In Moraines and
    Varves, 347-355, Balkema, Rotterdam
Sheeler, J.B. 1968. Summarization and comparison of engineering properties of
    loess in the U.S. Highway Res. Record No. 212, Highway R. Board, 1-9
Shephard-Thorn, E.R. 1975. The Quaternary of the Weald: A review. Proc. Geol.
    Ass., 86, 537-647
Sherwood, P.T. 1970. The reproducibility of the results of soil classification
    and compaction tests. Trans. Road. Res. Lab. Report LR 339
Shilts, W.W. 1973. Glacial dispersal of rocks, minerals and trace elements
    in Wisconsinan till, southeastern Quebec, Canada. Geol. Soc. Am. Mem.,
    136, 189-219
Shilts, W.W. 1975. Common glacial sediments of the Shield, their properties,
    distribution and possible uses as a geochemical sampling media. J. Geochem.
    Expl., 4, 189-199
Shilts, W.W. 1975a. Principles of geochemical exploration for sulphide deposits
    using shallow samples of glacial drift. Can. Inst. Min. Bull., 68, 73-80
Shilts, W.W. 1976. Glacial till and mineral exploration. In Glacial Till. Roy.
    Soc. Can. Spec. Pub. 12, 205-224
Shilts, W.W. 1978. Detailed sedimentological study of till sheets in a
    stratigraphic section, Samson River, Quebec. Geol. Surv. Can. Bull. 285
Shilts, W.W. 1980. Geochemical profile of till from Longlac, Ontario to
    Somerset Island. Can. Inst. Min. October, 85-94
Shilts, W.W. and McDonald, B.C. 1975a. Dispersion of clasts and trace elements
    in the Windsor Esker, southern Quebec. In Geol. Surv. Can. Paper 75-1A,
    495-499
Shotton, F.W. 1960. Large-scale patterned ground in the valley of the
    Worcestershire Avon. Geol. Mag., 97, 404-408
Shotton, F.W. 1965. Normal Faulting in British Pleistocene Deposits. Q. J.
    Geol. Soc. Lond., 121, 419-434

Shroder, J.F. and Sewell, R.E. 1981. Diamicton differentiation. La Sal
    Mountains and high plateaus. Utah and Idaho. Geol. Soc. Am. Abs. Prog.
    13, No. 7, 553
Simmons, M.D. 1982. Remedial treatment exploration. Wolf Creek Dam, Ky.
    Jnl. Geot. Eng. Div. Am. Soc. Civ. Eng., 108, 966-984
Sissons, J.B. 1976. Scotland. London, Methuen
Sissons, J.B. 1979. The Loch Lomond Stadial in the British Isles. Nature,
    280, 199-203
Sissons, J.B. 1981. The last Scottish ice-sheet: facts and speculative
    discussion. Boreas, 10, 1-17
Sissons, J.B. 1981a. Ice-dammed lakes in Glen Roy and vicinity: A summary.
    In The Quaternary of Britain (J. Neale and J. Flenley, eds.). Pergamon
    Press, 174-183
Skempton, A.W. and Northey, R.D. 1952. The sensitivity of clays. Geotechnique,
    3, 30-53
Skempton, A.W. 1953. The colloidal activity of clays. Proc. 3rd Int. Conf.
    Soil. Mech. and Foundn. Enging. Switzerland, 1, 57-61
Skempton, A.W. 1964. Long-term stability of clay slopes. Geotechnique, 14,
    77-101
Skempton, A.W. and Bjerrum, L. 1957. A contribution to the settlement analysis
    of foundations on clay. Geotechnique, 2, 168-178
Skempton, A.W. and Brown, J.D. 1961. A landslide in Boulder Clay at Selset,
    Yorkshire. Geotechnique, 11, 280-293
Skempton, A.W. and Weeks, A.G. 1976. The Quaternary history of the Lower
    Greensand escarpment and Weald Clay vale near Sevenoaks, Kent. Phil.
    Trans. Roy. Soc. Lond.,   A283, 493-526
Skempton, A.W. and Hutchinson, J.N. 1976. A discussion on valley slopes and
    cliffs in southern England. Phil. Trans. Roy. Soc. London. A, 283-421-631
Skinner, R.G. 1972. Drift prospecting in the Abitibi clay belt. Geol. Surv.
    Can. Open File, 116, 27
Slatt, R.M. 1971. Texture of ice-cored deposits from ten Alaskan valley
    glaciers. J. Sed. Petr., 41, 828-843
Slatt, R.M. 1972. Texture and composition of till from parent rocks of
    contrasting textures: southeastern Newfoundland. Sed. Geol., 7, 283-290
Slatt, R.M. and N. Eyles. 1981. Petrology of glacial sands: implications for
    the origin and mechanical durability of lithic fragments. Sedimentology,
    28, 171-183
Smalley, I.J. 1966. The properties of glacial loess and the formation of
    loess deposits. J. Sed. Pet., 36, 669-676
Smalley, I.J. 1971. The nature of quickclays. Nature, 231, 310
Smalley, I.J. and D. Unwin. 1968. The formation and shape of drumlins and
    their distribution and orientation. J. Glac., 7, 377-390
Smith, D.T. 1975. Geophysical assessemnt of sea-floor sediment properties.
    Oceanology International '75. 320-328, Brighton, U.K.
Smith, H.T.U. 1948. Giant glacial grooves in northwestern Canada. Am. J. Sci.,
    246, 503-514
Smith, H.T.V. 1953. The Hickory Run boulder field, Carson County, Pennsylvania.
    Am. J. Sci., 251, 625-642
Smith, H.T.V. 1962. Periglacial frost features and related phenomena in the
    United States. Biul. Peryglac., 11, 325-342
Smith, H.T.V. 1965. Giant glacial grooves in northwest Canada. Am. J. Sci.,
    246, 503-514
Smith, N.D. 1970. The braided stream depositional environment: comparison
    of the Platte River with some Silurian clastic rocks, north-central
    Appalachians. GSA Bull., 81, 2993-3014
Smith, N.D. 1971. Transverse bars and braiding in the lower Platte River,
    Nebraska. GSA Bull., 82, 3407-3420

Smith, N.D. 1972. Some sedimentological aspects of planar cross-stratification
in a sandy braided river. J. Sed. Pet., 42, 624-634
Smith, N.D. 1978. Sedimentation processes and patterns in a glacier-fed lake
with low sediment input. Can. J. Earth Sci., 15, 741-756
Smith, N.D. and Minter, W.E.L. 1980. Sedimentological controls of gold and
uranium in two Witwatersrand paleoplacers. Econ. Geol., 75, 1-14
Snow, D.T. 1969. Anisotropic permeability of fractured media. Water Resources
Research, 5, 1273-1289
Soderblom, R. 1966. Chemical aspects of quick-clay formation. Eng. Geol., 1,
415-431
Soderman, L.G., Kenney, T.C. and Loh, A.K. 1961. Geotechnical properties of
glacial clays in Lake St. Clair region of Ontario. Nat. Sci. and Eng.
Res. Council. Tech. Memo. No. 69, 55-90.
Soderman, L.G. and Quigley, R.M. 1965. Geotechnical properties of three
Ontario clays. Can. Geot. J., 2, 167-189
Soderman, L.G., Kim, Y.D. and Milligan, V. 1968. Field and laboratory studies
of Modulus of Elasticity of a clay till. Symp. on Soil Properties from
In-situ Measurements. Highway Res. Bd. Publ. No. 243
Soderman, L.G. and Kim, Y.D. 1970. Effect of Groundwater Levels on Stress
History of the St. Clair Clay Till Deposit. Can. Geotech. J., 7, 173-187
Sparks, B.W., Williams, R.G.B. and Bell, F.G. 1972. Presumed ground ice
depressions in East Anglia. Proc. Roy. Soc. London. A, 327, 329-343
Spears, D.A. and Reeves, M.J. 1975. The influence of superficial deposits on
groundwater quality in the Vale of York. Q. Jl. Engng. Geol., 8, 255-270
Stalker, A.M. 1976. Megablocks or the enormous erratics of the Albertan
Prairies. Geol. Surv. Can. Paper 76-1c, 185-188
Steer, E.C. and Binnie, G.M. 1951. Design and construction of earth dams, core
walls and diaphragms of earth and rock fill dams. Proc. 4th Int. Cong.
Large Dams, New Delhi, 1, 47-66
Stephenson, D.A. 1967. Hydrogeology of glacial deposits of the Mahomet
Bedrock Valley in east-central Illinois. Ill. Geol. Surv. Circular 409
Stevenson, J. 1952. The construction of Loch Sloy Dam. Proc. Inst. Civil Engrs.
1, III, 169-205
Stigzelius, H. 1977. Recognition of mineralized areas by a regional geochemical
survey of the till-blanket in northern Finland. J. Geochem. Expl., 8, 473-481
St. Onge, D.A. 1972. Sequence of glacial lakes in north-central Alberta. Geol.
Surv. Can. Bull., 213
Straw, A. 1966. Periglacial mass-movement on the Niagara Escarpment near Meaford,
Grey County. Geog. Bull. VIII, 367-376
Straw, A. 1968. Late Pleistocene glacial erosion along the Niagara Escarpment
of southern Ontario. Bull. Geol. Soc. Am., 79, 889-910
Stroud, M.A. 1974. The standard penetration test in insensitive clays and
soft rocks. Proc. European Symp. on Penetration Testing, Stockholm.
Stroud, M.A. and Butler, F.G. 1975. Standard penetration test. Engineering
properties of glacial materials. Geo-Abstracts, 124-135
Stuart, J.G. and Graham, J. 1974. Settlement performance of a raft foundation
on sand. Proc. Conf. on Settlement of Structures. British Geotech. Soc.,
Cambridge, 62-67
Stupavsky, M. and Gravenor, C.P. 1974. Water release from the base of active
glaciers. Geol. Soc. Am. Bull., 85, 433-436
Stupavsky, M., Symons, D.T.A. and Gravenor, C.P. 1974. Paleomagnetism of the
Port Stanley Till, Ontario. Geol. Soc. Am. Bull., 85, 141-144
Stupavsky, M. and Gravenor, C.P. 1983. Paleomagnetic dating of Quaternary
sediments: a review. In Quaternary Dating Methods. W.C. Mahaney (ed.).
in press

Sugden, D.E. and B.S. John. 1976. Glaciers and Landscape. Edward Arnold London

Sugden, D.E. 1977. Reconstruction of the morphology, dynamics and thermal characteristics of the Laurentide ice sheet at its maximum. Arct. Alp. Res., 9, 21-47

Sutherland, H.B. 1963. The use of in-situ tests to estimate the allowable bearing pressure on cohesionless soils. Structural Engineer. 41, 85-92

Swiger, W.F., Wild, P.A. and Lamb, T.J. 1980. Northfield Mountain pumped storage project. Jnl. Gt. Div. ASCE, 106, 673-689

Swiss Association for Water Economy. 1970. Swiss Dam Technique. 10th Cong. Int. Comm. Large Dams. Montreal. Pub. No. 42 of the Association 162 p

Symons, I.F. and Booth, A.I. 1971. Investigation of the stability of earthworks construction on the original line of Sevenoaks Bypass. Trans. Road Res. Lab. Report LR 393

Szalay, K. 1980. Performance monitoring for dam safety. Water Power and Dam Construction (GB). 32, 21-26

Tarbet, M.A. 1973. Geotechnical properties and sedimentation characteristics of tills in south-east Northumberland. Thesis. Ph.D. (Unpubl.). Univ. of Newcastle-Upon-Tyne. Dept. of Geol., 2 vols.

Tavenas, F., Chagnon, J.Y. and La Rochelle, P. 1971. The St. Jean Vianney Landslide: observations and eyewitness accounts. Can. Geotech., 8, 463-478

Taylor, G.E. 1951. The Haweswater Reservoir. J. Inst. Water Engrs., 4.4, 355-380

Teller, J.T. 1976. Lake Agassiz deposits in the main offshore basin of southern Manitoba. Can. J. Earth Sci., 13, 27-43

Terzaghi, K. 1943. Theoretical soil mechanics. John Wiley and Sons Ltd.

Terzaghi, K. and Peck, R.B. 1948. Soil mechanics in engineering practice. Wiley

Terzaghi, K. and Peck, R.B. 1967. Soil mechanics in engineering practice. 2nd ed. John Wiley and Sons, New York

Thompson, I. 1979. Till prospecting for sulphide ore in the Abitibi clay belt of Ontario. Can. Inst. Min. Bull., 72, 65-72

Thompson, I. and Wadge, D.R. 1981. Experimental work with the Wink Sonic Drill. Ontario Geol. Surv. Misc. Paper 100, 177-178

Thompson, S., Martin, R.L. and Eisenstein, Z. 1982. Soft zones in the glacial till in downtown Edmonton. Can. Geot. J., 19, 175-179

Thompson, M.E. and Eden, R.A. 1977. The Quaternary sequence in the West-Central North Sea. Inst. Geol. Sci. Rep. No. 77/12

Thompson, S. and Mekechuk, J. 1982. A landslide in glacial lake clays in central British Columbia. Can. Geot. J., 19, 296-306

Thorburn, S. 1963. Tentative correction chart for the standard penetration test in non-cohesive soils. Civil Eng. and Public Works Review, 58, 752-753

Thorburn, S. and Reid, W.W. 1973. Stability of slopes in lodgement till within the Glasgow District. Civ. Enging. and Pub. Works. Rev. April 321-325

Threadgold, L. and Weeks, R.C. 1975. Deep laminated clay deposits in the Skipton area. Proc. Symp. on the Engineering Behaviour of Glacial Materials. Geo-Abstracts, Norwich. 203-208

Tomlinson, M.J. 1980. Foundation, Design and Construction. 4th Edition, Pitman

Tooley, M.J. 1977. The Isle of Man, Lancashire coast and Lake District. INQUA X Congress. Guidebook A4, Geo. Abstracts Ltd. Norwich, U.K.

Torrance, J.K. 1975. On the role of chemistry in the development and behaviour of the sensitive marine clays of Canada and Scandinavia. Can. Geot. J., 12, 326-335

Torrance, J.K. 1975a. Leaching, weathering and origin of Leda clays in the Ottawa area. Mass Wasting, 4th Guelph Symp. on Geomorphology. Geo. Abstracts. 105-116

Townsend, D.L. 1965. Geotechnical properties of three Ontario clays:
     discussion. Can. Geot. J., 2, 190-193
Trow, W.A. 1965. Disc. to the Elastic properties of the dense glacial till
     deposit. by J. Klohn. Can. Geot. J., 2, 116-140
Twort, A.C. 1977. The repair of Lluest Wen dam. J. Inst. Water Engnrs.
     and Scientists, 31, 269-279
Utgard, R.O., McKenzie, G.D. and Foley, D. 1978. Geology in the Urban
     Environment. Burgess Pub. Co., Minnesota
UNESCO. 1967. International Quaternary Map of Europe. 1:2500,000. 15 sheets
Unifield Soil Classification. 1968. System for Roads, airfields, embankments
     and foundations. MIL-STD-619B. U.S. Department of Defense, Washington
Varnes, D.J. 1974. The logic of geological maps with reference to their
     interpretation and use for engineering purposes. U.S. Geol. Surv. Prof.
     Paper 837
Varnes, D.J. 1975. Slope movements in the western United States. In Mass
     Wasting. Geo Abstracts. 1-17
Vaughan, P.R. 1976. The deformations of the Empingham Valley slope. Phil. Trans.
     Roy. Soc. London.   A283, 451-462
Vaughan, P.R. and Walbancke, H.J. 1975. The stability of cut and fill slopes
     in boulder clay. Engineering Behaviour of Glacial Materials. Geo-Abstracts.
     209-219
Vaughan, P.R., Lovenbury, H.T. and Horswill, P. 1975. The design, construction
     and performance of Cow Green embankment dam. Geotechnique, 25, 555-580
Vaughan, P.R., Hight, D.W., Sodha, V.G. and Walbancke, H.J. 1978. Factors
     controlling the stability of clay tills in Britain. In Clay Fills.
     Inst. Civ. Eng. London, 205-218
Vaughan, P.R. and Soares, H.F. 1982. Design of filters for clay cores of dams.
     Jnl. Geot. Eng. Div. Am. Soc. Civ. Eng., 108, 17-32
Veklich, M.F. 1979. Pleistocene loesses and fossil soils of the Ukraine.
     Acta. Geol. Acad. Scient. Hungaricae,. 22, 35-62
Vermeiden, J. 1948. Improved sounding apparatus as developed in Holland since
     1936. Proc. 2nd Int. Conf. on S M and F E, 1, 280-287
Verosub, K.L. 1977. Depositional and post-depositional processes in the
     magnetization of sediments. Rev. Geophys. Space Phys., 15, 129-145
Vigdorchik, M.E. 1980. Submarine permafrost on the Alaskan continental shelf.
     Westview Press, Boulder.
Vigdorchik, M.E. 1980a. Arctic Pleistocene history and the development of
     submarine permafrost. Westview Press, Boulder
Vincent, J.S. and Hardy, L. 1977. L'evolution et l'extension des lacs
     glaciaires Barlow et Ojibway en territoire Quebecois. Geogr. Phys.
     Quat. XXXI, 357-372
Visser, J.N.J. 1982. Upper Carboniferous glacial sedimentation in the Karoo
     Basin near Prieska, South Africa. Palaeogeog. Palaeoclim., Palaeoec.,
     38, 63-92
Vitorello, I. and Van der Voo, R. 1977. Magnetic stratigraphy of Lake Michigan
     sediments obtained from cores of lacustrine clay. Quat. Res., 7, 398-412
Vivian, R. 1970. Hydrologie et erosion sous-glaciaires. Rev. Geogr. Alp., 58,
     241-264
Vivian, R. 1980. The nature of the ice-rock interface: the results of invest-
     igation on 20,000 m^2 of the rock bed of temperate glaciers. J. Glac.,
     25, 267-278
Voight, B. 1973. Correlation between Atterberg plasticity limits and residual
     shear strength of natural soils. Geotechnique, 23, 265-267
Voight, B. 1978. Rockslides and avalanches. I. Developments in geotechnical
     engineering. 14 A, Elsevier. Sci. Pub. Co. Oxford
Walker, R.G. 1975. Generalized facies models for resedimented conglomerates
     of turbidite association. Geol. Soc. Am. Bull., 86, 737-748

Walsh, P. and Evans, J. 1973. The Dolgarrog Dam disaster of 1925 in retrospect.
     Quest., 25 (The City University), 14-19
Walters, J.C. 1978. Polygonal patterned ground in central New Jersey.
     Quat. Res., 10, 42-54
Walters, R.C.S. 1971. Dam Geology. Butterworths, London
Walton, M. 1970. Groundwater resource evaluation. McGraw-Hill
Walton, M. 1978. Engineering Geology of the Twin Cities Central Urban
     Area, Minnesota. Geol. Soc. Am. Abstracts. with Programs, 10, No. 7, 512
Ward, W.H., Burland, J.B. and Gallois, R.W. 1968. Geotechnical assessment
     of a site at Mundford, Norfolk, for a large proton accelerator. Geotechnique,
     18, 399-431
Washburn, A.L. 1980. Geocryology - A survey of Periglacial Processes and Environ-
     ments. New York. John Wiley
Watson, E. 1971. Remains of pingos in Wales and the Isle of Man. Geol. Journ.
     7, 381-392
Watson, E. and Watson, S. 1967. The periglacial origin of the drifts at Morfa
     Bychan, near Aberystwyth. Geol. Journ., 5, 419-440
Watts, W. 1897. Notes on sinking, timbering and refilling concrete and puddle
     trenches for reservoir embankments. Trans. Brit. Assn. Water Engrs., 1,
     82-100
Watts, W. 1904. Concrete and puddle for reservoir embankments. Trans. Brit.
     Ass. Water Engrs., 9, 57-74
Watts, W. 1906. Geological notes on sinking Langsett and Underbank concrete
     trenches in the Little Don Valley. Trans. Inst. Min. Engnrs., 31, 668-678
Weeks, A.G. 1969. The stability of natural slopes in south-east England as
     affected by periglacial activity. Q. J. Eng. Geol., 2, 49-61
Weertman, J. 1957. On the sliding of glaciers. J. Glac., 3, 33-38
Weltman, A.J. and Healy, P.R. 1978. Piling in "Boulder Clay" and other glacial
     tills. D.O.E. and C.I.R.I.A. Piling Development Group Report PG5 78 pp
Whalley, W.B. 1974. The mechanics of high magnitude, low frequency rock failure.
     Geogr. Paper 27. Dept. Geog. Univ. Reading, U.K.
Whalley, W.B. 1974a. Rock glaciers and the formation of part of a glacier debris-
     transport system. Geogr Paper . 24, Dept. Geogr. Univ. Reading, U.K.
Whalley, W.B. 1975. Abnormally steep slopes on moraines constructed by valley
     glaciers. In The Engineering Behaviour of Glacial Materials. 60-66
     Geo-Abstracts
White, G.W. 1972. Engineering Implications of the Stratigraphy of Glacial Deposits.
     Proc. 24th Int. Geol. Congress Montreal. Section 13 (Enging. Geol.).
     76-82
White, G.W. 1974. Buried glacial geomorphology. In Glacial Geomorphology,
     D.R. Coates (ed.). State Univ. N.Y. 331-350
Whittecar, G.R. and Mickelson, D.M. 1979. Composition of internal structures,
     and an hypothesis for formation for drumlins. Wankesha County, Wisconsin.
     U.S.A. J. Glaciol., 22, 357-372
Wickham, J.T., Gross, D.L., Lineback, J.A. and Thomas, R.L. 1978. Late
     quaternary sediments of Lake Michigan. Ill. State Geol. Surv. Env.
     Geol. Note 84, 26
Williams, G.J. 1968. The buried channel and superficial deposits of the
     lower Usk and their correlation with similar features in the Lower Severn.
     Proc. Geol. Ass. 79, 325-348
Williams, L. 1979. An energy balance model of potential glacization of
     northern Canada. Arct. Alp. Res., 11, 443-456
Williams, P.F. and Rust, B.R. 1969. The sedimentology of a braided river,
     J. Sed. Pet., 39, 649-679
Williams, R.B.G. 1968. Periglacial erosion in southern and eastern England.
     Biuletyn Periglacjalny, 17, 311-335

Williman, H.B. and Frye, J.C. 1970. Pleistocene stratigraphy of Illinois. Ill. State Geol. Surv. Bull., 94

Wintle, A.G. 1981. Thermoluminescence dating of Late Devensian loesses in southern England. Nature, 289, 479-480

Wold, B. and Ostrem, G. 1979. Subglacial constructions and investigations at Bondhusbreen, Norway. J. Glac., 23, 363-380

Woodland, A.W. 1970. The buried tunnel-valleys of East Anglia. Proc. Yorks. Geol. Soc., 37, 521-578

Woods, K.B., Dryer, R.W.J. and Eden, W.J. 1959. Soil engineering problems on the Quebec North Shore and Labrador Railway. Bull. Am. Rlwy. Eng. Ass., 60, 669-688

Worsley, P. 1969. The Cheshire-Shropshire lowlands. In Lewis (ed.). The Glaciations of Wales and adjoining regions. 83-106, Longmans

Worsley, P. 1967. Problems in naming the Pleistocene deposits of the north-east Cheshire Plain. Mercian Geologist, 2, 51-55

Worsley, P. 1977. The Cheshire-Shropshire Plain. In D.Q. Bowen (ed.). INQUA Congress Excursion Guides. Wales and the Cheshire-Shropshire Lowland.

Worssam, B.C. 1981. Pleistocene deposits and superficial structures. Allington Quarry, Maidstone, Kent. In The Quaternary in Britain. (ed. by J. Neale and J. Flenley) 20-31, Pergamon

Wright, H.E. 1973. Tunnel valleys, glacial surges and subglacial hydrology of the Superior Lobe, Minnesota. Geol. Soc. Am. Mem., 136, 251-276

Wright, H.E. 1980. Surge moraines of the Klutlan Glacier, Yukon Territory, Canada and application to the Late Glacial of Minnesota. Quat. Res., 14, 2-17

Wright, S.G. 1976. Analyses for wave induced sea floor movements. 8th Annual Offshore Tech. Conf., 1, 41-52

Wroth, C.P. and Wood, D.M. 1978. The correlation of index properties of soils. Can. Geot. J., 15, 137-145

Wroth, C.P. 1979. Correlations of some engineering properties of soils. 2nd Int. Conf. on behaviour of off-shore structures, 121-132

Wu, T.H. 1958. Geotechnical properties of glacial lake clays. J. Soil Mech. Fnd. Div. ASCE, 84, 1-34

Yong, R.N., Sethi, A.J. and La Rochelle, P. 1979. Significance of amorphous material relative to sensitivity in some Champlain clays. Can. Geot. J., 16, 511-520

Yong-yan, W. and Qong-hu, Z. 1980. Loess in China. Shaanxi People's Art Publishing House.

Young, W. and Falkiner, R.H. 1966. Some design and construction features of the Cruachan pumped storage project. Proc. Inst. Civ. Engnrs., 35, 407-450

Youssef, M.S., El Ramli, A.M., El Demery, M. 1965. Relationships between shear strength, consolidation, liquid limit and plastic limit for remoulded clays. Proc. 6th Int. Conf. Soil Mech. and Foundn. Enging. Montreal, 1, 126-129

Zeller, J. and Zeindler, H. 1957. Test fills with coarse shell materials for Goschenealp Dam. Proc. 4th Int. Conf. Soil Mech. Foundn. Eng. London, 2, 405-409

Zuidberg, H.M. and Windle, D. 1979. High capacity sampling used in a Drillstring Anchor. Conf. on offshore site investigation. Society for Underwater Technology.

Index

CHESTER COLLEGE LIBR